DISCRETE WAVELET TRANSFORM

DISCRETE WAVELET TRANSFORM

A SIGNAL PROCESSING APPROACH

D. Sundararajan

Adhiyamaan College of Engineering, India

This edition first published 2015

Registered office
John Wiley & Sons Singapore Pte. Ltd., I Fusionopolis Walk, #07-01 Solaris South Tower, Singapore 138628.

For details of our global editorial offices, for customer services and for information about how to apply for permission to reuse the copyright material in this book please see our website at www.wiley.com.

Library of Congress Cataloging-in-Publication Data applied for.

ISBN: 9781119046066

Typeset in 10/12pt TimesLTStd by SPi Global, Chennai, India

1 2015

Contents

Preface

The discrete wavelet transform, a generalization of the Fourier analysis, is widely used in many applications of science and engineering. The primary objective of writing this book is to present the essentials of the discrete wavelet transform – theory, implementation, and applications – from a practical viewpoint. The discrete wavelet transform is presented from a digital signal processing point of view. Physical explanations, numerous examples, plenty of figures, tables, and programs enable the reader to understand the theory and algorithms of this relatively difficult transform with minimal effort.

This book is intended to be a textbook for senior-undergraduate-level and graduate-level discrete wavelet transform courses or a supplementary textbook for digital signal/image processing courses in engineering disciplines. For signal and image processing professionals, this book will be useful for self-study. In addition, this book will be a reference for anyone, student or professional, specializing in signal and image processing. The prerequisite for reading this book is a good knowledge of calculus, linear algebra, signals and systems, and digital signal processing at the undergraduate level. The last two of these topics are adequately covered in the first few chapters of this book.

MATLAB® programs are available at the website of the book, www.wiley.com/go/sundararajan/wavelet. Programming is an important component in learning this subject. Answers to selected exercises marked with * are given at the end of the book. A Solutions Manual and slides are available for instructors at the website of the book.

I assume the responsibility for all the errors in this book and would very much appreciate receiving readers' suggestions at d_sundararajan@yahoo.com. I am grateful to my Editor and his team at Wiley for their help and encouragement in completing this project. I thank my family for their support during this endeavor.

D. Sundararajan

List of Abbreviations

bpp	bits per pixel
DFT	discrete Fourier transform
DTDWT	dual-tree discrete wavelet transform
DTFT	discrete-time Fourier transform
DWPT	discrete wavelet packet transform
DWT	discrete wavelet transform
FIR	finite impulse response
FS	Fourier series
FT	Fourier transform
IDFT	inverse discrete Fourier transform
IDTDWT	inverse dual-tree discrete wavelet transform
IDWPT	inverse discrete wavelet packet transform
IDWT	inverse discrete wavelet transform
ISWT	inverse discrete stationary wavelet transform
PR	perfect reconstruction
SWT	discrete stationary wavelet transform
1-D	one-dimensional
2-D	two-dimensional

1

Introduction

A signal conveys some information. Most of the naturally occurring signals are continuous in nature. More often than not, they are converted to digital form and processed. In digital signal processing, the information is extracted using digital devices. It is the availability of fast digital devices and numerical algorithms that has made digital signal processing important for many applications of science and engineering. Often, the information in a signal is obscured by the presence of noise. Some type of filtering or processing of signals is usually required. To transmit and store a signal, we would like to compress it as much as possible with the required fidelity. These tasks have to be carried out in an efficient manner.

In general, a straightforward solution from the definition of a problem is not the most efficient way of solving it. Therefore, we look for more efficient methods. Typically, the problem is redefined in a new setting by some transformation. We often use u-substitutions or integration by parts to simplify the problem in finding the integral of a function. By using logarithms, the more difficult multiplication operation is reduced to addition.

Most of the practical signals have arbitrary amplitude profile. Therefore, the first requirement in signal processing is the appropriate representation or modeling of the signals. The form of the signal is changed. Processing of a signal is mostly carried out in a transformed representation. The most commonly used representation is in terms of a set of sinusoidal signals. Fourier analysis is the representation of a signal using constant-amplitude sinusoids. It has four versions to suit discrete or continuous and periodic or aperiodic signals. Fourier analysis enables us to do spectral analysis of a signal. The spectral characterization of a signal is very important in many applications. Even for a certain class of signals, the amplitude profile will vary arbitrarily and system design can be based only on the classification in terms of the spectral characterization. Another important advantage is that complex operations become simpler, when signals are represented in terms of their spectra. For example, convolution in the time domain becomes multiplication in the frequency domain. In most cases, it is easier to analyze, interpret, process, compress, and transmit a signal in a transformed representation. The use of varying-amplitude sinusoids as the basis signals results in the Laplace transform for continuous signals and the z-transform for discrete signals.

The representation of a signal by a set of basis functions, of transient nature, composed of a set of continuous group of frequency components of the signal spectrum is called the wavelet

---

Discrete Wavelet Transform: A Signal Processing Approach, First Edition. D. Sundararajan.

Companion Website: www.wiley.com/go/sundararajan/wavelet

transform, the topic of this book. Obviously, the main task in representing a signal in this form is filtering. Therefore, all the essentials of digital signal processing (particularly, Fourier analysis and convolution) are required and described in the first few chapters of this book. The wavelet transform is a new representation of a signal. The discrete wavelet transform (DWT) is widely used in signal and image processing applications, such as analysis, compression, and denoising. This transform is inherently suitable in the analysis of nonstationary signals.

There are many transforms used in signal processing. Most of the frequently used transforms, including the DWT, are a generalization of the Fourier analysis. Each transform representation is more suitable for some applications. The criteria of selection include ease of transformation and appropriateness for the required application. The signal representation by a small number of basis functions should be adequate for the purpose. For example, the basic principle of logarithms is to represent a number in exponential form. The difference between $\log_e(16)$, $\log_2(16)$, and $\log_{10}(16)$ is the base. Each of these is more suitable to solve some problems.

1.1 The Organization of This Book

An overview of the topics covered is as follows. The key operation in the implementation of the DWT is filtering, the convolution of the input and the filter impulse response. The concept is not difficult to understand. We thoroughly study convolution in linear systems, signals and systems, and signal processing courses. However, there are several aspects to be taken care of. The basic step in the computation of the DWT is convolution followed by downsampling. As only half of the convolution output is required, how to compute the convolution efficiently? How to compute the convolution involving upsampled signals efficiently? How to compute the convolution at either end of the finite input signal? How to design the filter coefficients? This aspect is formidable and is different from the filter design in signal processing. How the time-domain signal is represented after the transformation? How the inverse transform is computed? How this transformation is useful in applications? In studying all these aspects, we use Fourier analysis extensively.

The contents of this book can be divided into three parts. In the first part, in order to make the book self-contained, the essentials of digital signal processing are presented. The second part contains the theory and implementation of the DWT and three of its versions. In the third part, two of the major applications of the DWT are described.

Except for the filter design, the basic concepts required to understand the theory of the DWT are essentially the same as those of multirate digital signal processing. Therefore, in the first part of the book, we present the following topics briefly, which are tailored to the requirements of the study of the DWT: (i) Signals, (ii) Convolution and Correlation, (iii) Fourier Analysis of Discrete Signals, (iv) The z-Transform, (v) Finite Impulse Response Filters, and (vi) Multirate Digital Signal Processing.

In Chapter 2, the classification, sampling, and operations of signals are covered. Basic signals used for signal decomposition and testing of systems are described. Sampling theorem is discussed next. Finally, time shifting, time reversal, compression, and expansion operations of signals are presented. Convolution and correlation operations are fundamental tools in the analysis of signals and systems, and they are reviewed from different points of view in Chapter 3. Not only the DWT, but also many aspects of signal processing are difficult to understand without using Fourier analysis. Chapter 4 includes an adequate review of the discrete Fourier transform (DFT) and the discrete-time Fourier transform (DTFT) versions of the Fourier analysis. The z-transform, which is a generalization of the Fourier analysis for discrete signals, is introduced in Chapter 5. DWT is implemented using digital filters. In Chapter 6, characterization of digital filters is presented, and the frequency responses of linear-phase filters are derived. In Chapter 7, multirate digital signal processing is introduced. Fundamental operations, such as decimation and interpolation, are presented, and two-channel filter banks and their polyphase implementation are described.

As the Haar DWT filter is the shortest and simplest of all the DWT filters used, most of the concepts of the DWT are easy to understand by studying the Haar DWT first. Accordingly, the Haar DWT is dealt in Chapter 8. In Chapter 9, some orthogonal DWT filters are designed using the orthogonality and lowpass filter constraints. Biorthogonal DWT filters can have symmetric filter coefficients, which provides the advantages of linear-phase response and effectively solving the border problem. The design of commonly used biorthogonal DWT filters is presented in Chapter 10. Chapter 11 is devoted to the implementation aspects of the DWT.

One extension of the DWT is the discrete wavelet packet transform (DWPT), which provides more efficient signal analysis in some applications. The DWPT is described in Chapter 12. The DWT is a shift-variant transform. One version of the DWT that is shift-invariant, called the discrete stationary wavelet transform (SWT), is presented in Chapter 13. In Chapter 14, the dual-tree discrete wavelet transform (DTDWT) is described. This version of the DWT provides good directional selectivity, in addition to being nearly shift-invariant.

Image compression and denoising are two of the major applications of the DWT. Image compression makes the storage and transmission of digital images practical, which, in their naturally occurring form, are very redundant and require huge amount of storage space. In Chapter 15, compression of digital images using the DWT is examined. One way or the other, a signal gets corrupted with noise at the time of creation, transmission, or analysis. Denoising estimates the true signal from its corrupted version. Denoising of signals using the DWT is explored in Chapter 16.

Basically, the application of any transform to a signal results in its transformed form, which is usually more suitable for carrying out the required analysis and processing. Now, a suitable transform matrix has to be designed. After the application of the transform, the transformed signal is subjected to necessary further processing to yield the processed signal. The effectiveness of the processing is evaluated using suitable measures such as energy, entropy, or signal-to-noise ratio. In this book, the selected transform is the DWT. In essence, we go through the cycle of operations given previously with respect to the DWT.

2

Signals

A continuous signal $x(t)$ is defined at all instants of time. The value of a discrete signal is defined only at discrete intervals of the independent variable (usually taken as time, even if it is not). If the interval is uniform (which is often the case), the signal is called a uniformly sampled signal. In this book, we deal only with uniformly sampled signals. In most cases, a discrete signal is derived by sampling a continuous signal. Therefore, even if the source is not a continuous signal, the term sampling interval is used. A uniformly sampled discrete signal $x(nT_s)$, where $-\infty < n < \infty$ is an integer and T_s is the sampling interval, is obtained by sampling a continuous signal $x(t)$. That is, the independent variable t in $x(t)$ is replaced by nT_s to get $x(nT_s)$. We are familiar with the discrete Fourier spectrum $X(n\omega_0)$ of a continuous periodic signal $x(t)$. The n in the discrete independent variable $n\omega_0$ is an integer, and ω_0 is the fundamental frequency. Usually, the sampling interval T_s is suppressed and the discrete signal is designated as $x(n)$. In actual processing of a discrete signal, its digital version, called the digital signal, obtained by quantizing its amplitude, is used. For most analytical purposes, the discrete signal is used first, and then the effect of quantization is taken into account. A two-dimensional (2-D) signal, typically an image, $x(n_1, n_2)$ is a function of two independent variables in contrast to a one-dimensional (1-D) signal $x(n)$ with a single independent variable. Figure 2.1(a) shows an arbitrary discrete signal. The discrete sinusoidal signal $\sin\left(\frac{2\pi}{16}n + \frac{\pi}{3}\right)$ is shown in Figure 2.1(b). The essence of signal processing is to approximate practical signals, which have arbitrary amplitude profiles, as shown in Figure 2.1(a), and are difficult to process, by a combination of simple and well-defined signals (such as the sinusoid shown in Figure 2.1(b)) so that the design and analysis of signals and systems become simpler.

2.1 Signal Classifications

2.1.1 *Periodic and Aperiodic Signals*

A signal $x(n)$ is periodic, if $x(n + N) = x(n)$ for all values of n. The smallest integer N satisfying the equality is called the period of $x(n)$. A periodic signal repeats its values over a period indefinitely at intervals of its period. Typical examples are sinusoidal signals. A signal that is not periodic is an aperiodic signal. While most of the practical signals are aperiodic, their analysis is carried out using periodic signals.

Discrete Wavelet Transform: A Signal Processing Approach, First Edition. D. Sundararajan.

Companion Website: www.wiley.com/go/sundararajan/wavelet

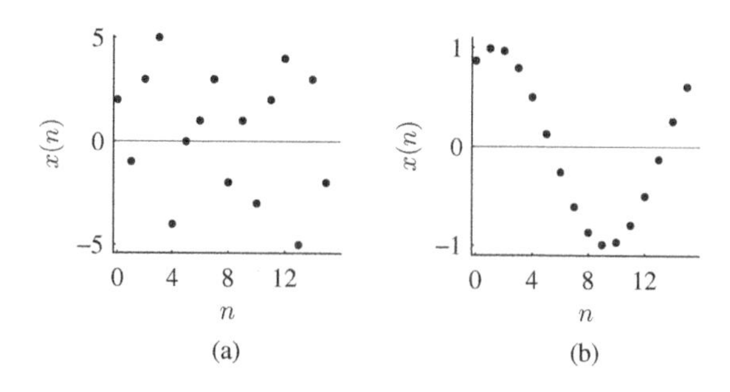

Figure 2.1 (a) An arbitrary discrete signal; (b) the discrete sinusoidal signal $\sin\left(\frac{2\pi}{16}n + \frac{\pi}{3}\right)$

2.1.2 Even and Odd Signals

Decomposing a signal into its components, with respect to some basis signals or some property, is the fundamental method in signal and system analysis. The decomposition of signals with respect to wavelet basis functions is the topic of this book. A basic decomposition, which is often used, is to decompose a signal into its even and odd components. A signal $x(n)$ is odd, if

$$x(-n) = -x(n)$$

for all values of n. When plotted, an odd signal is antisymmetrical about the vertical axis at the origin $(n = 0)$. A signal $x(n)$ is even, if

$$x(-n) = x(n)$$

for all values of n. When plotted, an even signal is symmetrical about the vertical axis at the origin. An arbitrary signal $x(n)$ can always be decomposed into its even and odd components, $x_e(n)$ and $x_o(n)$, uniquely. That is,

$$x(n) = x_e(n) + x_o(n)$$

Replacing n by $-n$, we get

$$x(-n) = x_e(-n) + x_o(-n) = x_e(n) - x_o(n)$$

Adding and subtracting the last two equations, we get

$$x_e(n) = \frac{x(n) + x(-n)}{2} \quad \text{and} \quad x_0(n) = \frac{x(n) - x(-n)}{2}$$

Example 2.1 Find the even and odd components of the sinusoid, $\sin\left(\frac{2\pi}{16}n + \frac{\pi}{3}\right)$, shown in Figure 2.1(b).

Solution

Expressing the sinusoid in terms of its cosine and sine components, we get

$$\sin\left(\frac{2\pi}{16}n + \frac{\pi}{3}\right) = \frac{\sqrt{3}}{2}\cos\left(\frac{2\pi}{16}n\right) + \frac{1}{2}\sin\left(\frac{2\pi}{16}n\right)$$

Note that the even component of a sinusoid is the cosine waveform and the odd component is the sine waveform. The even and odd components are shown, respectively, in Figures 2.2(a) and (b). The decomposition can also be obtained using the defining equation. The even component is obtained as

$$\sin\left(\frac{2\pi}{16}n+\frac{\pi}{3}\right)$$

$$=\frac{\sin\left(\frac{2\pi}{16}n+\frac{\pi}{3}\right)+\sin\left(\frac{2\pi}{16}(-n)+\frac{\pi}{3}\right)}{2}=\frac{\sin\left(\frac{2\pi}{16}n+\frac{\pi}{3}\right)-\sin\left(\frac{2\pi}{16}(n)-\frac{\pi}{3}\right)}{2}$$

$$=\frac{\left(\sin\left(\frac{2\pi}{16}n\right)\cos\left(\frac{\pi}{3}\right)+\cos\left(\frac{2\pi}{16}n\right)\sin\left(\frac{\pi}{3}\right)\right)-\left(\sin\left(\frac{2\pi}{16}n\right)\cos\left(\frac{\pi}{3}\right)-\cos\left(\frac{2\pi}{16}n\right)\sin\left(\frac{\pi}{3}\right)\right)}{2}$$

$$=\frac{\sqrt{3}}{2}\cos\left(\frac{2\pi}{16}n\right)$$

Similarly, the odd component can also be obtained.

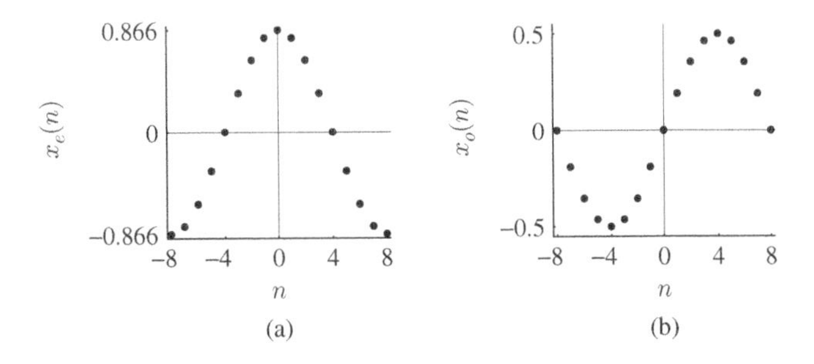

Figure 2.2 (a) The even component $\frac{\sqrt{3}}{2}\cos\left(\frac{2\pi}{16}n\right)$ of the signal $\sin\left(\frac{2\pi}{16}n+\frac{\pi}{3}\right)$; (b) the odd component $\frac{1}{2}\sin\left(\frac{2\pi}{16}n\right)$ ■

2.1.3 *Energy Signals*

The energy of a real discrete signal $x(n)$ is defined as

$$E=\sum_{n=-\infty}^{\infty}x^2(n)$$

An energy signal is a signal with finite energy, $E<\infty$. The energy of the signal $x(0)=2$ and $x(1)=-3$ is

$$E=2^2+(-3)^2=13$$

Cumulative Energy

This is a signal measure indicting the way the energy is stored in the input signal. Let $x(n)$, $n=0,1,\ldots,N-1$ be the given signal of length N. Form the new signal $y(n)$ by taking the absolute values of $x(n)$ and sorting them in descending order. Then, the cumulative energy of

$x(n)$, $C(n)$, is defined as

$$C(n) = \frac{\sum_{k=0}^{n} y^2(k)}{\sum_{l=0}^{N-1} y^2(l)}, \quad n = 0, 1, \ldots, N-1$$

Note that $0 \leq C(n) \leq 1$.

Example 2.2 Find the cumulative energy of $x(n)$.

$$x(n) = \{1, 6, 11, 16, 18, 14, 17, 20\}$$

Solution

Sorting the magnitude of the values of $x(n)$, we get $y(n)$ as

$$\{20, 18, 17, 16, 14, 11, 6, 1\}$$

The values of the cumulative sum of $y^2(n)$ are

$$\{400, 724, 1013, 1269, 1465, 1586, 1622, 1623\}$$

The cumulative energy is given by

$$\{0.2465, 0.4461, 0.6242, 0.7819, 0.9026, 0.9772, 0.9994, 1\}$$

Let the transformed representation of $x(n)$ be

$$\{7, 27, 32, 37, -5, -5, 4, -3\}/\sqrt{2}$$

The first four of these values are obtained by taking the sum of the pairs of $x(n)$, and second four are obtained by taking the difference. All the values are divided by $\sqrt{2}$. Sorting the magnitude of the values, we get

$$\{37, 32, 27, 7, 5, 5, 4, 3\}/\sqrt{2}$$

The values of the cumulative sum of the squared values are

$$\{1369, 2393, 3122, 3171, 3196, 3221, 3237, 3246\}/2$$

The cumulative energy is given by

$$\{0.4217, 0.7372, 0.9618, 0.9769, 0.9846, 0.9923, 0.9972, 1\}$$

In the case of the transformed values, the slope of the graph shown in Figure 2.3(b) is steeper than that shown in Figure 2.3(a). That is, most of the energy of the signal can be represented by fewer values. ■

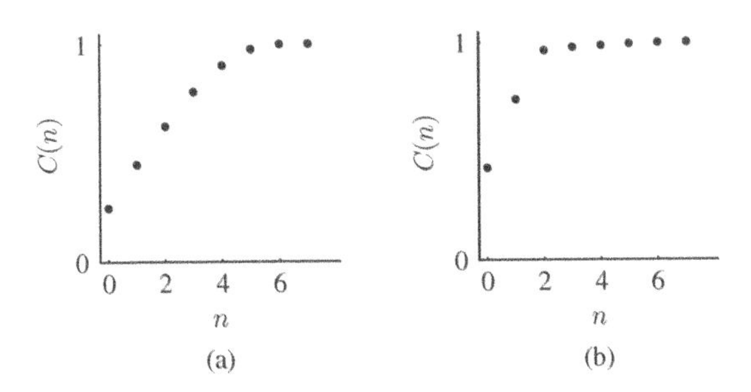

Figure 2.3 (a) Cumulative energy of an arbitrary discrete signal; (b) cumulative energy of its transformed version

2.1.4 Causal and Noncausal Signals

Practical signals are switched on at some finite time instant, usually chosen as $n = 0$. Signals with $x(n) = 0$ for $n < 0$ are called causal signals. Signals with $x(n) \neq 0$ for $n < 0$ are called noncausal signals. The sinusoidal signal, shown in Figure 2.1(b), is a noncausal signal. Typical causal signals are the impulse $\delta(n)$ and the unit-step $u(n)$, shown in Figure 2.4.

2.2 Basic Signals

Some simple and well-defined signals are used for decomposing arbitrary signals to make their representation and analysis simpler. These signals are also used to characterize the response of systems.

2.2.1 Unit-Impulse Signal

A discrete unit-impulse signal, shown in Figure 2.4(a), is defined as

$$\delta(n) = \begin{cases} 1, & \text{for } n = 0 \\ 0, & \text{for } n \neq 0 \end{cases}$$

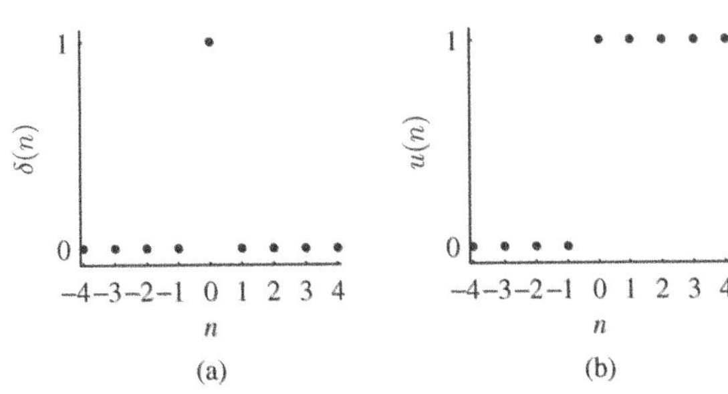

Figure 2.4 (a) The unit-impulse signal, $\delta(n)$; (b) the unit-step signal $u(n)$

It is an all-zero sequence, except that its value is one when its argument n is equal to zero. In the time domain, an arbitrary signal is decomposed in terms of impulses. This is the basis of the convolution operation, which is vital in signal and system analysis and design.

Consider the product of a signal $x(n)$ with a shifted impulse $\delta(n-m)$. As the impulse is nonzero only at $n = m$, we get

$$x(n)\delta(n-m) = x(m)\delta(n-m)$$

Summing both sides with respect to m, we get

$$\sum_{m=-\infty}^{\infty} x(n)\delta(n-m) = x(n) \sum_{m=-\infty}^{\infty} \delta(n-m) = x(n) = \sum_{m=-\infty}^{\infty} x(m)\delta(n-m)$$

The general term $x(m)\delta(n-m)$ of the last sum, which is one of the constituent impulses of $x(n)$, is a shifted impulse $\delta(n-m)$ located at $n = m$ with value $x(m)$. The summation operation sums all these impulses to form $x(n)$. Therefore, an arbitrary signal $x(n)$ can be represented by the sum of scaled and shifted impulses with the value of the impulse at any n being $x(n)$. The unit-impulse is the basis function, and $x(n)$ is its coefficient. As the value of the sum is nonzero only at $n = m$, the sum is effective only at that point. By varying the value of n, we can sift out all the values of $x(n)$. For example, consider the signal

$$\{x(-3) = 4, x(0) = -1, x(2) = 3, x(4) = 7\} \text{ and } x(n) = 0 \text{ otherwise}$$

This signal can be expressed, in terms of impulses, as

$$x(n) = 4\delta(n+3) - \delta(n) + 3\delta(n-2) + 7\delta(n-4)$$

With $n = 4$, for instance,

$$x(4) = 4\delta(7) - 1\delta(4) + 3\delta(2) + 7\delta(0) = 7$$

2.2.2 *Unit-Step Signal*

A discrete unit-step signal, shown in Figure 2.4(b), is defined as

$$u(n) = \begin{cases} 1, & \text{for } n \geq 0 \\ 0, & \text{for } n < 0 \end{cases}$$

It is an all-one sequence for positive values of its argument n and is zero otherwise.

2.2.3 *The Sinusoid*

A continuous system is typically modeled using a differential equation, which is a linear combination of derivative terms. A sinusoid differentiated any number of times is also a sinusoid of the same frequency. The sum of two sinusoids of the same frequency but differing in amplitude and phase is also a sinusoid of the same frequency. Due to these reasons, the steady-state output of a linear time-invariant system for a sinusoidal input is also a sinusoid of the same frequency, differing only in amplitude and phase. An arbitrary signal can be decomposed into a sum of sinusoidal waveforms (Fourier analysis). Therefore, the sinusoid is the most important waveform in signal and system analysis.

There are two forms of expressions describing a sinusoid. The polar form specifies a sinusoid, in terms of its amplitude and phase, as

$$x(n) = A\cos(\omega n + \theta), \quad n = -\infty, \ldots, -1, 0, 1, \ldots, \infty$$

where A, ω, and θ are, respectively, the amplitude, the angular frequency, and the phase. The amplitude A is the distance of either peak of the waveform from the horizontal axis ($A = 1$ for the wave shown in Figure 2.1(b)). The phase angle θ is with respect to the reference $A\cos(\omega n)$. The peak of the cosine waveform occurs at $n = 0$, and its phase is zero radian. The phase of $\sin(\omega n) = \cos(\omega n - (\pi/2))$ is $(-\pi/2)$ radians. A sinusoid expressed as the sum of its cosine and sine components is its rectangular form.

$$\begin{aligned} A\cos(\omega n + \theta) &= (A\cos(\theta))\cos(\omega n) + (-A\sin(\theta))\sin(\omega n) \\ &= C\cos(\omega n) + D\sin(\omega n) \end{aligned}$$

The inverse relations are

$$A = \sqrt{C^2 + D^2} \quad \text{and} \quad \theta = \tan^{-1}\left(\frac{-D}{C}\right)$$

A discrete sinusoid has to complete an integral number of cycles (say k, where $k > 0$ is an integer) over an integral number of sample points, called its period (denoted by N, where $N > 0$ is an integer), if it is periodic. Then, as

$$\cos(\omega(n + N) + \theta) = \cos(\omega n + \omega N + \theta) = \cos(\omega n + \theta) = \cos(\omega n + \theta + 2k\pi)$$

$N = \frac{2k\pi}{\omega}$. Note that k is the smallest integer that will make $\frac{2k\pi}{\omega}$ an integer. The cyclic frequency, denoted by f, of a sinusoid is the number of cycles per sample and is equal to the number of cycles the sinusoid makes in a period divided by the period, $f = \frac{k}{N} = \frac{\omega}{2\pi}$ cycles per sample. Therefore, the cyclic frequency of a discrete periodic sinusoid is a rational number. The angular frequency, the number of radians per sample, of a sinusoid is 2π times its cyclic frequency, that is, $\omega = 2\pi f$ radians per sample.

The angular frequency of the sinusoid, shown in Figure 2.1(b), is $\omega = \frac{\pi}{8}$ radians per sample. The period of the discrete sinusoid is $N = \frac{2k\pi}{\omega} = 16$ samples, with $k = 1$. The cyclic frequency of the sinusoid in Figure 2.1(b) is $f = \frac{k}{N} = \frac{1}{16}$ cycles per sample. The cyclic frequency of the sinusoid $\sin\left(\frac{2\sqrt{2}\pi}{16}n + \frac{\pi}{3}\right)$ is $\frac{\sqrt{2}}{16}$. As it is an irrational number, the sinusoid is not periodic. The phase of the sinusoid $\cos\left(\frac{2\pi}{16}n + \frac{\pi}{3}\right)$ is $\theta = \frac{\pi}{3}$ radians. As it repeats a pattern over its period, the sinusoid remains the same by a shift of an integral number of its period. A phase-shifted sine wave can be expressed in terms of a phase-shifted cosine wave as $A\sin(\omega n + \theta) = A\cos\left(\omega n + \left(\theta - \frac{\pi}{2}\right)\right)$. The phase of the sinusoid

$$\sin\left(\frac{2\pi}{16}n + \frac{\pi}{3}\right) = \cos\left(\frac{2\pi}{16}n + \left(\frac{\pi}{3} - \frac{\pi}{2}\right)\right) = \cos\left(\frac{2\pi}{16}n - \frac{\pi}{6}\right)$$

in Figure 2.1(b) is $-\frac{\pi}{6}$ radians. A phase-shifted cosine wave can be expressed in terms of a phase-shifted sine wave as $A\cos(\omega n + \theta) = A\sin\left(\omega n + \left(\theta + \frac{\pi}{2}\right)\right)$.

For analysis purposes, it is found that the complex sinusoid

$$Ae^{j\theta}e^{j\omega n} = Ae^{j(\omega n+\theta)}$$

is advantageous. The complex sinusoid is a mathematically equivalent representation of the real sinusoid. The advantage is that the single complex constant $Ae^{j\theta}$ is the coefficient of the single complex waveform $e^{j\omega n}$ in contrast to the two real constants A and θ for the real sinusoid. Furthermore, manipulation of exponential signals is much easier. The sinusoids are related by the Euler's formula

$$A\cos(\omega n+\theta) = \frac{A}{2}(e^{j(\omega n+\theta)} + e^{-j(\omega n+\theta)})$$

2.3 The Sampling Theorem and the Aliasing Effect

Most of the practical signals are of continuous type. As digital signal processing is efficient, signals have to be sampled and quantized. When a continuous signal is sampled, because samples represent the signal only at discrete intervals, an infinity of intermediate values are lost. However, if the rate of rapid variations of a signal is limited (it is for practical signals), then, for all practical purposes, the signal can be represented, with negligible error, by its samples taken with a sufficiently small sampling interval, T_s. The problem of determining the appropriate sampling interval is solved by the sampling theorem.

Let f_m be the cyclic frequency of the highest frequency component of a continuous signal $x(t)$. The sampling theorem states that $x(t)$ can be reconstructed exactly from its sampled version $x(n)$, if the sampling frequency $f_s = 1/T_s$ is greater than $2f_m$. That is, a sinusoid with frequency f Hz has a distinct set of $2f+1$ sample values. In practice, due to the nonideal characteristics of the physical devices such as filters, a somewhat higher sampling frequency than the theoretical minimum is used. For example, the bandwidth of the telephone signals is 3.4 kHz, and it is sampled at 8 kHz. For compact disc musical recordings, the required bandwidth is 20 kHz, and it is sampled at 44.1 kHz.

Consider the two discrete sinusoids $x_1(n) = \cos(\frac{2\pi}{8}n - \frac{\pi}{3})$ and $x_2(n) = \cos(\frac{6\pi}{8}n + \frac{\pi}{3})$, shown in Figure 2.5. The sample values of $x_1(n)$ over one cycle, starting from $n = 0$, are

$$\{0.5000, 0.9659, 0.8660, 0.2588, -0.5000, -0.9659, -0.8660, -0.2588\}$$

The sample values of $x_2(n)$ are

$$\{0.5000, -0.9659, 0.8660, -0.2588, -0.5000, 0.9659, -0.8660, 0.2588\}$$

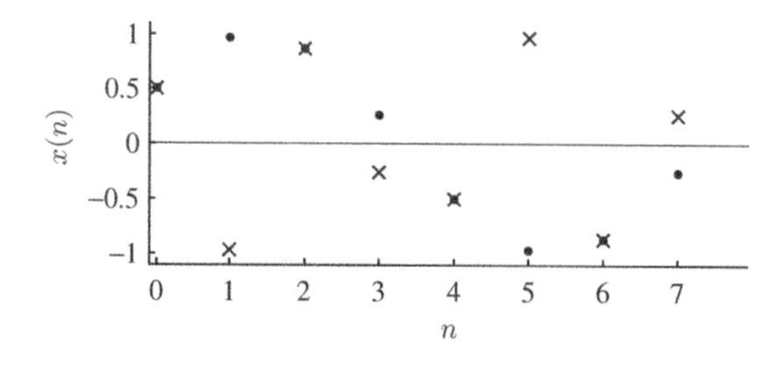

Figure 2.5 Discrete sinusoids $\cos\left(\frac{2\pi}{8}n - \frac{\pi}{3}\right)$ (dots) and $\cos\left(\frac{6\pi}{8}n + \frac{\pi}{3}\right)$ (crosses)

The second sinusoid can be expressed as

$$\cos\left(\frac{6\pi}{8}n + \frac{\pi}{3}\right)$$
$$= \cos\left(3\frac{2\pi}{8}n + \frac{\pi}{3}\right) = \cos\left((4-1)\frac{2\pi}{8}n + \frac{\pi}{3}\right) = \cos\left(\left(\frac{2\pi}{8} - \pi\right)n - \frac{\pi}{3}\right)$$
$$= \cos\left(\frac{2\pi}{8}n - \pi n - \frac{\pi}{3}\right) = \cos(\pi n)\cos\left(\frac{2\pi}{8}n - \frac{\pi}{3}\right) = (-1)^n \cos\left(\frac{2\pi}{8}n - \frac{\pi}{3}\right)$$

The even-indexed samples of the two sinusoids are the same, while the odd-indexed samples are the negatives of the other. If we are constrained to use only four samples, sinusoids $x_1(n)$ and $x_2(n)$ have the same representation

$$\{0.5000, 0.8660, -0.5000, -0.8660\}$$

$$x_1(2n) = \cos\left(\frac{2\pi}{8}(2n) - \frac{\pi}{3}\right) = x_2(2n) = (-1)^{(2n)}\cos\left(\frac{2\pi}{8}(2n) - \frac{\pi}{3}\right)$$

This is aliasing. Two frequency components cannot be distinguished, as the number of samples is inadequate. The alias of the high-frequency, $\frac{6\pi}{8}$, component is the low-frequency, $\frac{2\pi}{8}$, component. If it is specified that the frequency is in the range

$$0 \le \omega < \frac{\pi}{2} \quad \text{or} \quad \frac{\pi}{2} \le \omega < \pi$$

then we can identify the signal as $x_1(n)$ or $x_2(n)$ from the even-indexed samples using interpolation.

The aliasing problem is similar to representing binary numbers using bits. With more number of bits, we can represent numbers over a larger range. Given a certain number of bits, the number range is fixed. For example, with two bits, we can represent four binary numbers $\{00, 01, 10, 11\}$. With one bit, we can represent two binary numbers $\{0, 1\}$. Four numbers can be uniquely identified from one-bit representation, if it is known that the number is in the lower or upper range. Let the representation of a two-bit number be 0 using the least significant bit. Then, the number is 00 if it is less than 2, or it is 10 if it is greater than or equal to 2. Similarly, with a larger number of samples, we can represent a larger number of sinusoids uniquely. With the frequency range of the sinusoid known, it is possible to reconstruct a sinusoid with a reduced set of samples than that is required by the sampling theorem. Additional information (frequency range) acts as a substitute for the missing samples.

2.4 Signal Operations

Other than the arithmetic operations, there are operations of the signal involving the independent variable, n. Time shifting, time reversal, and time scaling operations are often used in the analysis of signals.

2.4.1 Time Shifting

If the independent variable n in $x(n)$ is replaced by $(n - n_0)$, the sample values of $x(n)$ are delayed by n_0 samples for positive values of n_0. For example, the value of $x(n)$ at $n = 0$ occurs

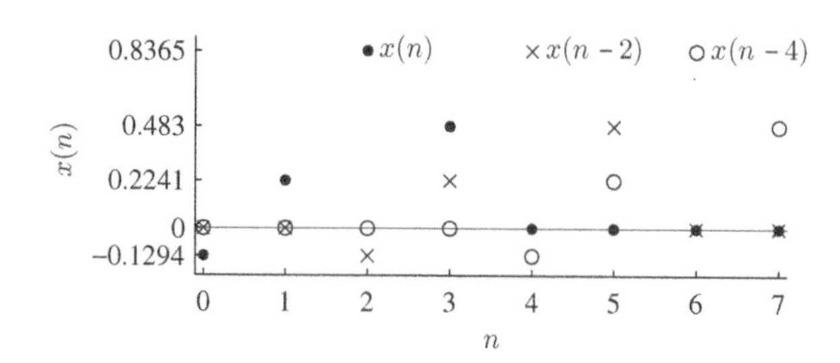

Figure 2.6 Signal $x(n)$ (dots) and its shifted versions, $x(n-2)$ (crosses) and $x(n-4)$ (unfilled circles)

at $n = n_0$ in $x(n - n_0)$. Graphically, the plot of $x(n)$ is shifted forward by n_0 sample intervals. For negative values of n_0, the sample values of $x(n)$ are advanced by n_0 samples. Graphically, the plot of $x(n)$ is shifted backward by n_0 sample intervals. The nonzero samples of the signal, shown in Figure 2.6 by dots, are specified as

$$x(n) = \{x(0) = -0.1294, x(1) = 0.2241, x(2) = 0.8365, x(3) = 0.4830\}$$

The signal $x(n-2)$, shown by crosses, is the signal $x(n)$ shifted by two sample intervals to the right (delayed by two sample intervals, as the sample values of $x(n)$ occur two sample intervals later). For example, the first nonzero sample value occurs at $n = 2$ as $x(2-2) = x(0) = -0.1294$. The signal $x(n-4)$, shown by unfilled circles, is the signal $x(n)$ shifted by four sample intervals to the right. For example, the second nonzero sample value occurs at $n = 5$ as $x(5-4) = x(1) = 0.2241$.

2.4.2 *Time Reversal*

If the independent variable n in $x(n)$ is replaced by $(-n)$, a mirror image of $x(n)$ about the vertical axis at the origin $(n = 0)$ is obtained. This operation resulting in $x(-n)$ is called time reversal or folding operation. For example, the value of $x(-n)$ at $n = 2$ occurs at $n = -2$ in $x(-n)$. Graphically, the plot of $x(n)$ is rotated by 180 degrees about the vertical axis at the origin. The nonzero samples of a signal are specified as

$$x(n) = \{x(0) = -0.4830, x(1) = 0.8365, x(2) = -0.2241, x(3) = -0.1294\}$$

The signal is shown in Figure 2.7 by dots. The folded signal $x(-n)$ is shown by crosses. The value of $x(n)$ at n_0 occurs in the time-reversed signal $x(-n)$ at $(-n_0)$.

2.4.3 *Time Scaling*

It is assumed that the scaling factor is 2 (which is of primary interest in the DWT) for both time compression and time expansion.

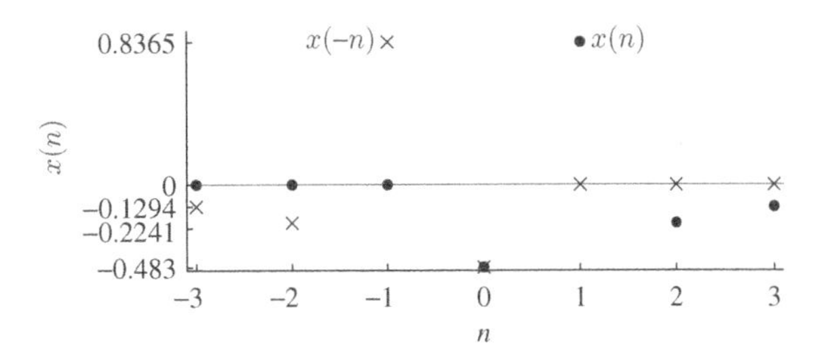

Figure 2.7 Signal $x(n)$ (dots) and its time-reversed version, $x(-n)$ (crosses)

Time Compression

In time compressing (also called downsampling) a signal $x(n)$

$$\dots, x(-2), x(-1), x(0), x(1), x(2), \dots$$

by a factor of 2, we replace the independent variable n by $2n$. The samples of the compressed signal $x(2n)$ are

$$\dots, x(-4), x(-2), x(0), x(2), x(4), \dots$$

The compressed signal is composed of the even-indexed samples of $x(n)$. The odd-indexed samples of $x(n)$ are lost in compression. Figure 2.8(a) shows the symbolic representation of compressing a signal $x(n)$ by a factor of 2, (b) the signal $x(n)$, and (c) the compressed signal $x_d(n) = x(2n)$.

Time Expansion

In time expanding (also called upsampling) a signal $x(n)$ by a factor of 2, we replace the independent variable n by $n/2$. The samples of the expanded signal $x(n/2)$ are

$$\dots, 0, x(-2), 0, x(-1), 0, x(0), 0, x(1), 0, x(2), 0, \dots$$

The expanded signal is composed of all the samples of $x(n)$ padded with zero-valued samples in between every pair of $x(n)$. The even-indexed samples of the expanded signal are the samples of $x(n)$. While it is possible to assign other values for the odd-indexed samples in the expanded signal, for DWT purposes, zero is the required value. Figure 2.9(a) shows the symbolic representation of expanding a signal $x(n)$ by a factor of 2, (b) the signal $x(n)$, and (c) the expanded signal $x_u(n) = x(n/2)$. Figure 2.10 shows two cycles of the cosine signal $x(t) = \cos\left(\frac{2\pi}{16}t\right)$ and its expanded version

$$x\left(\frac{t}{2}\right) = \cos\left(\frac{2\pi}{16}\frac{t}{2}\right) = \cos\left(\frac{2\pi}{32}t\right)$$

While $x(t)$ completes two cycles, its expanded version $x\left(\frac{t}{2}\right)$ completes one cycle during the same interval. The value of $x(t)$ at a specific instant t is the value of $x\left(\frac{t}{2}\right)$ at $2t$.

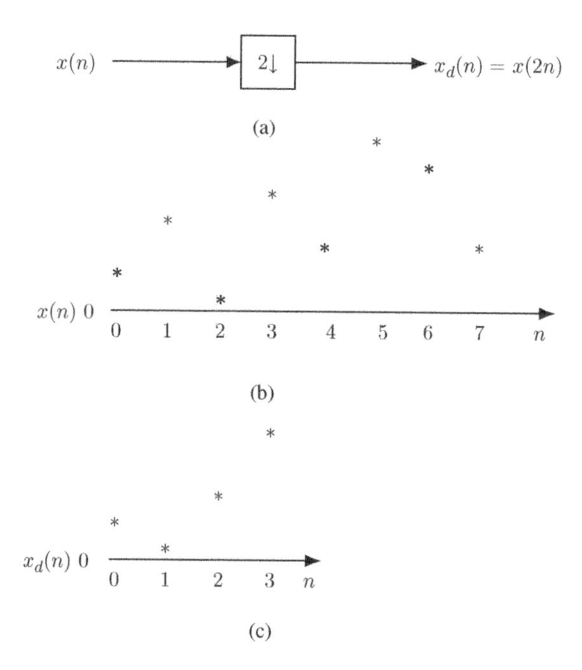

Figure 2.8 (a) Symbolic representation of compressing a signal by a factor of 2; (b) signal $x(n)$; (c) compressed signal, $x_d(n) = x(2n)$

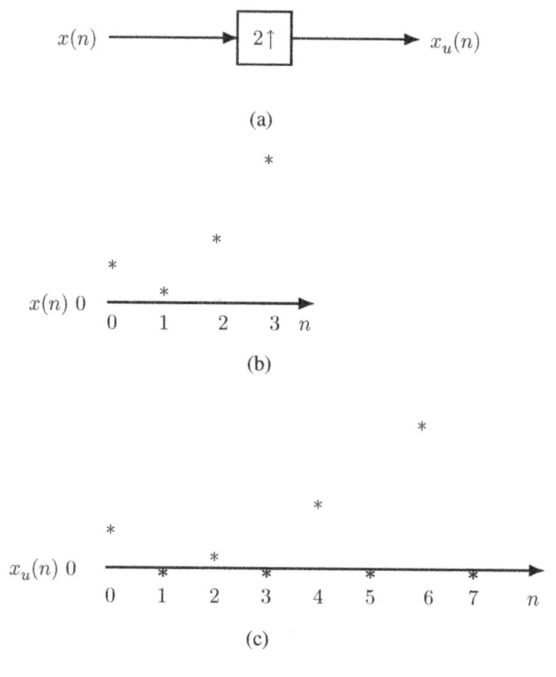

Figure 2.9 (a) Symbolic representation of expanding a signal by a factor of 2; (b) signal $x(n)$; (c) expanded signal, $x_u(n) = x(n/2)$

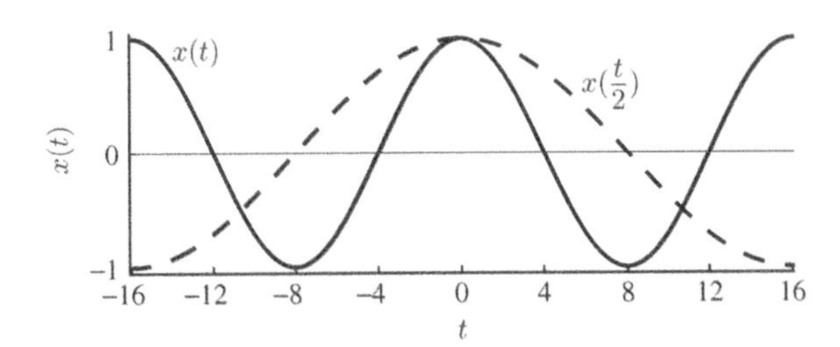

Figure 2.10 Cosine signal $x(t) = \cos\left(\frac{2\pi}{16}t\right)$ and its expanded version $x\left(\frac{t}{2}\right) = \cos\left(\frac{2\pi}{32}t\right)$

2.5 Summary

- Most naturally occurring signals are of continuous type.
- The discrete version of a signal is used in most of the analysis in digital signal processing.
- In actual processing, the digital version of a signal is used.
- Signals are classified depending on whether the independent variable is continuous or discrete and their values are continuous or discrete.
- Signals are also classified as energy or power signals, even or odd signals, causal or noncausal signals, and periodic or aperiodic signals.
- Simple and well-defined signals, such as impulse, sinusoid, and unit-step, are often used as basis signals in the analysis of signals and systems.
- Sinusoidal signal is of great significance, as it is the only signal that, when input to a linear time-invariant system, results in an output that has the same form as the input with changes in amplitude and phase.
- For mathematical convenience, the sinusoid is usually expressed in its equivalent complex exponential form.
- Sampling theorem states the condition under which a continuous signal can be sampled and reconstructed exactly from its samples.
- Signal operations, such as time shifting, folding, time expansion, and compression, are often used. These operations are carried out by changing the independent variable of a signal.

Exercises

2.1 Find the even and odd components of the signal $x(n)$.

2.1.1

$$x(n) = \{x(0) = (1+\sqrt{3}), x(1) = (3+\sqrt{3}), x(2) = (3-\sqrt{3}), x(3) = (1-\sqrt{3})\} \quad \text{and}$$

$$x(n) = 0 \quad \text{otherwise}$$

2.1.2

$$x(n) = \{x(0) = -1, x(1) = 3, x(2) = -3, x(3) = 1\} \quad \text{and}$$

$$x(n) = 0 \quad \text{otherwise}$$

2.1.3

$$x(n) = \{x(-1) = 4, x(0) = 3, x(1) = 6\} \quad \text{and} \quad x(n) = 0 \quad \text{otherwise}$$

***2.1.4**

$$x(n) = 2\cos\left(2\frac{2\pi}{8}n + \frac{\pi}{6}\right)$$

2.1.5

$$x(n) = \sin\left(\frac{2\pi}{8}n + \frac{\pi}{4}\right)$$

2.2 The nonzero samples of the sequence $x(n)$ are given. Find the energy of the signal.

2.2.1

$$x(n) = \frac{\sqrt{2}}{8}\{-1, 2, 6, 2, -1\}$$

2.2.2

$$x(n) = \frac{\sqrt{2}}{4}\{1, -2, 1\}$$

2.2.3

$$x(n) = \frac{1}{4\sqrt{2}}\{(1+\sqrt{3}), (3+\sqrt{3}), (3-\sqrt{3}), (1-\sqrt{3})\}$$

2.2.4

$$x(n) = \frac{\sqrt{2}}{4}\{-1, 3, 3, -1\}$$

2.2.5

$$x(n) = \frac{1}{\sqrt{2}}\{-1, 1\}$$

2.3 The nonzero samples of the sequence $x(n)$ are given. Find the cumulative energy. Find the cumulative energy of the transform of $x(n)$. The first four of the transformed values are obtained by taking the sum of the pairs of $x(n)$, and second four are obtained by taking the difference. All the values are divided by $\sqrt{2}$.

2.3.1

$$\{1, 35, 71, 57, 41, 51, 55, 56\}$$

2.3.2

$$\{48, 45, 42, 37, 29, 22, 13, 1\}$$

***2.3.3**

$$\{1, 10, 24, 36, 38, 44, 45, 42\}$$

2.3.4

$$\{4, 1, 7, 10, 8, 9, 8, 4\}$$

2.3.5

$$\{45, 44, 41, 41, 37, 33, 21, 1\}$$

2.4 Express the signal in terms of scaled and shifted impulses.

2.4.1 $x(-3) = 4, x(0) = -1, x(1) = 3, x(2) = -3, x(3) = -2$ and $x(n) = 0$ otherwise

2.4.2 $x(-1) = 5, x(0) = 3, x(1) = 3, x(3) = -4$ and $x(n) = 0$ otherwise

2.4.3 $x(-3) = 3, x(0) = -2, x(1) = 4, x(3) = -2$ and $x(n) = 0$ otherwise

2.4.4 $x(-3) = -2, x(0) = -2, x(3) = 4$ and $x(n) = 0$ otherwise
***2.4.5** $x(-3) = 3, x(1) = 2, x(2) = 4, x(3) = 1$ and $x(n) = 0$ otherwise

2.5 Find the rectangular form of the sinusoid. List the sample values of one cycle, starting from $n = 0$, of the sinusoid.

***2.5.1**

$$x(n) = 2\cos\left(2\frac{2\pi}{8}n - \frac{\pi}{6}\right)$$

2.5.2

$$x(n) = -3\cos\left(\frac{2\pi}{8}n + \frac{\pi}{3}\right)$$

2.5.3

$$x(n) = 2\sin\left(2\frac{2\pi}{8}n + \frac{\pi}{6}\right)$$

2.5.4

$$x(n) = -3\sin\left(\frac{2\pi}{8}n + \frac{\pi}{3}\right)$$

2.6 Find the polar form of the sinusoid. List the sample values of one cycle, starting from $n = 0$, of the sinusoid.

2.6.1

$$x(n) = 3\cos\left(2\frac{2\pi}{8}n\right) + 4\sin\left(2\frac{2\pi}{8}n\right)$$

***2.6.2**

$$x(n) = 3\cos\left(\frac{2\pi}{8}n\right) + 3\sin\left(\frac{2\pi}{8}n\right)$$

2.6.3

$$x(n) = \cos\left(\frac{2\pi}{8}n\right) - \sqrt{3}\sin\left(\frac{2\pi}{8}n\right)$$

2.6.4

$$x(n) = \sqrt{3}\cos\left(\frac{2\pi}{8}n\right) - \sin\left(\frac{2\pi}{8}n\right)$$

2.7 Find the period of the following sinusoids, if periodic.
2.7.1 $x(n) = \sin(6\pi n/8)$
2.7.2 $x(n) = \sin(n/5)$
2.7.3 $x(n) = \sin(\sqrt{2}\pi n/8)$
***2.7.4** $x(n) = 1 + \sin(2\pi n/8)$
2.7.5 $x(n) = \sin(4\pi n/7)$

2.8 Express the sinusoid in terms of complex exponentials.
2.8.1 $x(n) = 2\sin\left(\frac{\pi}{4}n\right)$

2.8.2 $x(n) = 3\cos\left(\frac{\pi}{4}n\right)$

***2.8.3** $x(n) = 4\cos\left(\frac{\pi}{4}n + \frac{\pi}{3}\right)$

2.8.4 $x(n) = \cos\left(\frac{\pi}{4}n - \frac{\pi}{6}\right)$

2.8.5 $x(n) = 4\sin\left(\frac{\pi}{4}n + \frac{\pi}{4}\right)$

2.9 After downsampling by a factor of 2, does the sinusoid get aliased? If aliased, what is the frequency component it impersonates? Give the samples of the sinusoid and its downsampled version in one cycle, starting with $n = 0$.

2.9.1

$$x(n) = 2\cos\left(2\frac{2\pi}{8}n + \frac{\pi}{6}\right)$$

2.9.2

$$x(n) = 3\sin\left(\frac{2\pi}{8}n - \frac{\pi}{4}\right)$$

***2.9.3**

$$x(n) = 3\sin\left(3\frac{2\pi}{8}n - \frac{\pi}{2}\right)$$

2.9.4

$$x(n) = -7$$

2.9.5

$$x(n) = 4\cos(\pi n)$$

2.10 The nonzero samples of the sequence $x(n)$ are given. Find the samples of the following sequences.

$$x(n) = \{x(-2) = -1, x(-1) = 2, x(0) = 6, x(1) = 2, x(2) = -1\}$$

2.10.1 $y(n) = x(n-2)$
2.10.2 $y(n) = x(-n)$
2.10.3 $x_d(n) = x(2n)$
2.10.4 $x_u(n) = x(n/2)$
2.10.5 $y(n) = x(-n+2)$
***2.10.6** $y(n) = x(2n+3)$
2.10.7 $y(n) = x((n/2)-2)$
2.10.8 $y(n) = x(-2n-1)$
2.10.9 $y(n) = x((-n/2)+3)$

2.11 The waveform $x(t)$ is given. Draw the following waveforms with $t = 0$ to $t =$ 16 seconds.

$$x(t) = \sin\left(\frac{2\pi}{8}t\right)$$

2.11.1 $x(t)$
2.11.2 $x\left(\frac{t}{2}\right)$
2.11.3 $x\left(-\frac{t}{2}\right)$
2.11.4 $x(2t)$
2.11.5 $x(-2t)$
2.11.6 $x(3t)$

3

Convolution and Correlation

The three most often used operations in digital signal processing are convolution, correlation, and the DFT. Convolution is one of the linear system models. In this model, the output of a system is expressed exclusively in terms of input values. An understanding of the convolution operation is essential in the study of signals and systems. For this reason, we present the convolution operation from different points of view. Convolution operation is based on decomposing an arbitrary input signal to a system into a set of scaled and delayed impulses. The responses to all the impulses are found, and superposition summation of the responses yields the output. The system is characterized by its impulse response. The DWT is implemented using the convolution operation. Convolution of a signal with the unit-impulse signal relocates the signal at the location of the impulse. Correlation, which is a similarity measure, of a signal with the unit-impulse signal relocates the time-reversed signal at the location of the impulse. Both the operations involve the computation of sum of products. In the case of convolution, one of the two sequences is time reversed, whereas no time reversal is required in the computation of correlation.

3.1 Convolution

3.1.1 The Linear Convolution

The linear convolution of two aperiodic sequences $x(n)$ and $h(n)$ is defined as

$$y(n) = \sum_{k=-\infty}^{\infty} x(k)h(n-k) = \sum_{k=-\infty}^{\infty} h(k)x(n-k) = x(n) * h(n) = h(n) * x(n)$$

The limits of the summation are $-\infty$ and ∞. The sequences involved may be infinite in extent. In that case, it is assumed that $y(n)$ is finite. Otherwise, $y(n)$ does not exist. The infinite limits are appropriate in any case, as there is no contribution to the summation during the interval where one or both the sequences are zero. This relationship is of fundamental importance in linear systems theory both in the time domain (where time is the independent variable of a function) and in the frequency domain (where frequency is the independent variable of a function). One interpretation of the convolution is that it expresses the relationship between

Discrete Wavelet Transform: A Signal Processing Approach, First Edition. D. Sundararajan.

Companion Website: www.wiley.com/go/sundararajan/wavelet

sequences $x(n)$ and $y(n)$ that is governed by the sequence $h(n)$. In this case, $x(n)$, $h(n)$ and $y(n)$ can be considered, respectively, as the input, the impulse response, and the output of a linear time-invariant system. In another interpretation, convolution can be considered as an interaction (moving weighted average of $x(n)$ with the weighting function $h(n)$ or vice versa) of the sequences $x(n)$ and $h(n)$ that results in the sequence $y(n)$.

Essentially, convolution operation is finding the sum of products of two sequences, each other's index running in opposite directions. Four operations (fold, shift, multiply, and add) are repeatedly used in implementing the convolution.

1. One of the two sequences to be convolved (say $h(k)$) is time reversed, that is, folded about the vertical axis at the origin ($k = 0$) to get $h(-k)$.
2. The time-reversed sequence, $h(-k)$, is shifted by n_0 sample intervals (right-shift for positive n_0 and left-shift for negative n_0), yielding $h(n_0 - k)$, to find the output at $n = n_0$.
3. The term-by-term products of the overlapping samples of the two sequences, $x(k)$ and $h(n_0 - k)$, are computed.
4. The sum of all the products is the convolution output at $n = n_0$.

The output of the convolution of two finite sequences of lengths N and M consists of $N + M - 1$ samples, as the overlap of nonzero portions can occur only over that length.

The convolution of the impulse response $\{h(k), k = 0, 1, 2\} = \{1, -2, 1\}$ with the input $\{x(k), k = 0, 1, 2, 3\} = \{1, 2, 4, 3\}$ is shown in Figure 3.1. From the figure, it is obvious that convolution of two sequences involves the time reversal of one of the sequences, shifting that sequence by one interval each time and forming the sum of products of the values over overlapping portions of the sequences. For example, consider the output $y(3) = -3$.

$$y(3) = x(1)h(2) + x(2)h(1) + x(3)h(0) = (2)(1) + (4)(-2) + (3)(1) = -3$$

The sum of the indices of the values making each of the product terms is equal to the output index. In computing $y(3)$,

$$1 + 2 = 2 + 1 = 3 + 0 = 3$$

There are two easy checks of the convolution output. The sum of the elements of the sequence $y(n)$ must be equal to the product of the sum of the elements of the sequences $x(n)$ and $h(n)$.

$$(1 - 2 + 1) \times (1 + 2 + 4 + 3) = 0 = (1 + 0 + 1 - 3 - 2 + 3)$$

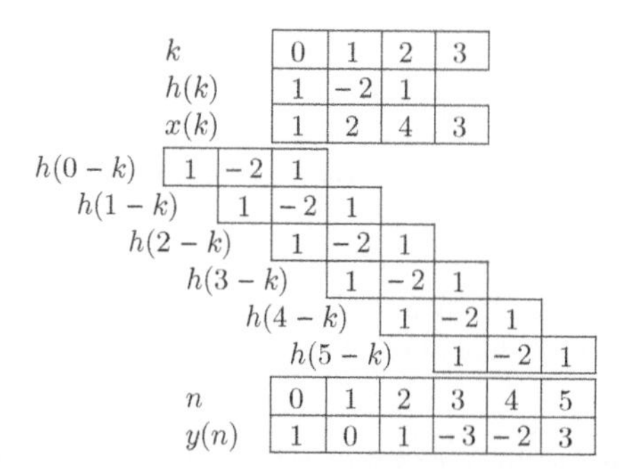

Figure 3.1 The linear convolution operation

Another check uses alternating sums. That is, the sign of the odd-indexed values of all the three sequences is changed first, and then the check is applied.

$$(1+2+1) \times (1-2+4-3) = 0 = (1-0+1+3-2-3)$$

A physical interpretation of the convolution, in terms of the impulse response of a system, is as follows. The impulse response of a system is its response to the unit-impulse signal $\delta(n)$ with the system initially relaxed (zero initial conditions). The impulse response of a linear time-invariant system for a delayed impulse $\delta(n-k)$ is $h(n-k)$. At the instant $n = k$, when the only nonzero value of the impulse occurs, the input $x(n)$ has amplitude $x(k)$. Therefore, the system response at the instant $n = k$ is $x(k)h(n-k)$. However, there are system responses at the instant $n = k$ due to impulses occurred earlier (for a causal system). The sum of all the responses is the system output at the instant $n = k$, assuming that the system is linear. We have to find the system responses at all values of n. The input sequence $\{1, 2, 4, 3\}$ is expressed in terms of scaled and delayed impulses as

$$x(n) = \delta(n) + 2\delta(n-1) + 4\delta(n-2) + 3\delta(n-3)$$

The response of the system to the impulse $\delta(n)$ is

$$\{y_0(0) = 1, y_0(1) = -2, y_0(2) = 1\}$$

The response of the system to the impulse $2\delta(n-1)$ is

$$\{y_1(1) = 2, y_1(2) = -4, y_1(3) = 2\}$$

At the instant $n = 0$, the output is $y(0) = y_0(0) = 1$. At the instant $n = 1$, the output is $y(1) = y_0(1) + y_1(1) = -2 + 2 = 0$, and so on. It is clear from Figure 3.1 that, at any instant, the sum of the contributions due to various impulse components of the input constitutes the output. The decomposing of a signal into a sum of simple and well-defined signals, such as the impulse and the sinusoid, is called taking the transform (change in form) of the signal. In most of the signal and system analysis, it is simpler to use the appropriate transform.

The linear convolution operation is an ordinary multiplication, but without carrying the digits from one column to the next as shown in Figure 3.2.

An another point of view is that convolution corresponds to multiplication of two polynomials. The product of the two polynomials

$$1 - 2x + x^2 \quad \text{and} \quad 1 + 2x + 4x^2 + 3x^3$$

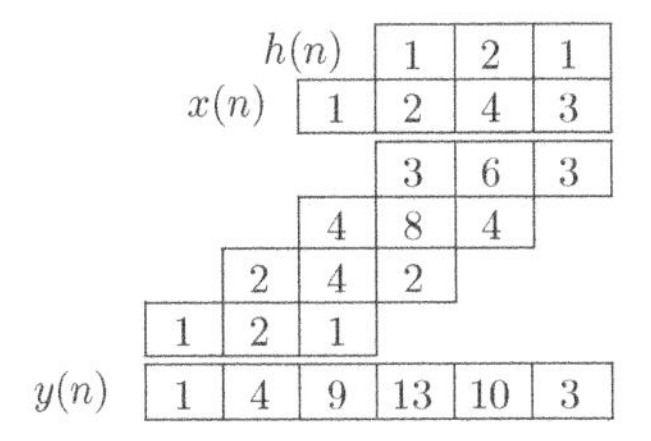

Figure 3.2 The linear convolution operation as an ordinary multiplication

is

$$1 + 0x + 1x^2 - 3x^3 - 2x^4 + 3x^5$$

The coefficients of the product polynomial are the output of the convolution of the coefficients of the corresponding polynomials.

A yet another interpretation of the convolution is that it is a weighted average. Let $\{h(k), k = 0, 1\} = \{1/2, 1/2\}$. Then,

$$y(n) = \sum_{k=-\infty}^{\infty} h(k)x(n-k) \quad \text{or} \quad y(n) = \frac{1}{2}x(n) + \frac{1}{2}x(n-1)$$

The second equation defining $y(n)$ is the difference equation, another system model. With the same weighting coefficients, the system computes the moving average of the input $x(n)$. The output is the average of the present and past samples. The bumps in the signal are smoothed out. As part of the spectrum of the signal is allowed and the rest suppressed, this system is called a filter. This is a lowpass filter, as it attenuates the high-frequency components of a signal more than the low-frequency components.

Let $\{h(k), k = 0, 1\} = \{1/2, -1/2\}$. Then,

$$y(n) = \frac{1}{2}x(n) - \frac{1}{2}x(n-1)$$

With the same but alternating coefficients, the system computes the moving difference of the input $x(n)$. The output is the scaled difference of the present and past samples. The bumps in the signal are picked up. This is a highpass filter, as it attenuates the low-frequency components of a signal more than the high-frequency components. As it turns out, the major part of the study of the DWT is the design and implementation of appropriate lowpass and highpass filters.

3.1.2 Properties of Convolution

The convolution is commutative.

$$h(n) * x(n) = x(n) * h(n)$$

Either of the sequences $h(n)$ and $x(n)$ can be time reversed in carrying out the convolution operation.

The convolution is distributive.

$$x(n) * h_1(n) + x(n) * h_2(n) = x(n) * (h_1(n) + h_2(n))$$

The sum of the convolution of a sequence with two sequences is the same as that of the convolution of the sequence with the sum of the two sequences.

The convolution is associative.

$$(x(n) * h_1(n)) * h_2(n) = x(n) * (h_1(n) * h_2(n))$$

The convolution of sequences is independent of their grouping.

The shift property of convolution is that

$$x(n) * h(n) = y(n) \quad \text{and} \quad x(n-k) * h(n-m) = y(n-k-m)$$

Convolution of $x(n)$ with the unit-impulse $\delta(n)$ relocates $x(n)$ at the location of the impulse.

$$x(n) * \delta(n) = x(n) \qquad \text{and} \qquad x(n) * \delta(n-k) = x(n-k)$$

3.1.3 The Periodic Convolution

Let $x(n)$ and $h(n)$ be two periodic sequences of the same period N. Then, the periodic convolution of the sequences is defined as

$$y(n) = \sum_{k=0}^{N-1} x(k)h(n-k) = \sum_{k=0}^{N-1} h(k)x(n-k), \quad n = 0, 1, \ldots, N-1$$

The principal difference of this type of convolution from that of the linear convolution is that the range of the summation is restricted to a single period.

3.1.4 The Border Problem

In practice, we deal with finite-length signals. When we convolve two sequences of length 4, the convolution output is of length 7. Only one output is computed with complete overlap of the two sequences. The other outputs are produced with the assumption that, at either end of the signal, the values of the signal are zero outside the defined set. This assumption may not be suitable in most cases. Usually, the data length is very long compared to that of the filter coefficients. Some assumption is required to carry out the convolution at the borders. We can assume that the data is periodic. Remember that the well-known DFT computation is based on assumed periodicity of a finite-length data. The problem is that it creates discontinuities at the borders, resulting in slow convergence of the transform coefficients. Because of this problem, symmetrical extension is often used in other transforms, including the DWT.

Let $x(n) = \{3, 2, 1, 4\}$. The three types of extensions, periodic, whole-point symmetric, and half-point symmetric, are shown in Table 3.1. The two symmetric extensions differ in whether to repeat the first and last data values or not. Let us convolve the extended sequences with $h(n) = \{3, 1\}$. The outputs are shown in Table 3.2. The three values, shown in boldface, are the same in each case, as no data extension is required. As the data length is usually very long, only a few output values are affected, but it is still significant. The extension suitable for the

Table 3.1 Three types of data extensions at the borders

Periodic	1	4	**3**	**2**	**1**	**4**	3	2
Whole-point symmetric	1	2	**3**	**2**	**1**	**4**	1	2
Half-point symmetric	2	3	**3**	**2**	**1**	**4**	4	1

Table 3.2 Outputs of convolving the data in Table 3.1 with $\{3, 1\}$

Periodic	13	13	**9**	**5**	**13**	13	9
Whole-point symmetric	7	11	**9**	**5**	**13**	7	7
Half-point symmetric	11	12	**9**	**5**	**13**	16	7

specific case should be used to get the required number of border output values. There are also other methods to solve the border problem.

3.1.5 Convolution in the DWT

Convolution operation is essential in the implementation of the DWT, as is the case in almost all aspects of signal and system analysis. In computing the DWT, a slightly modified form of the convolution is used. The reason is that only half of the convolution output is required. Therefore, either the input or the impulse response is shifted by two sample intervals, rather than one in regular convolution, after each output value is computed. The double-shifted convolution equation is

$$y(n) = \sum_{k=-\infty}^{\infty} x(k)h(2n-k) = \sum_{k=-\infty}^{\infty} h(k)x(2n-k) \tag{3.1}$$

Example 3.1 Find the double-shifted convolution (Equation 3.1) of the sequences

$$\{h(n), n = 0, 1, 2\} = \{1, -2, 1\} \quad \text{and} \quad \{x(n), n = 0, 1, 2, 3\} = \{1, 2, 4, 3\}$$

Solution
Using Equation (3.1), we get

$$y(0) = (1)(1) = 1$$
$$y(1) = (4)(1) + (2)(-2) + (1)(1) = 1$$
$$y(2) = (3)(-2) + (4)(1) = -2$$

The regular convolution output is

$$\{1, 0, 1, -3, -2, 3\}$$

■

This convolution output can be obtained by separately convolving even-indexed input values $x_e(n)$ with the even-indexed values $h_e(n)$ and the odd-indexed input values $x_o(n)$ with the odd-indexed values $h_o(n)$ and appropriately summing the two convolution outputs.

$$x_e(n) * h_e(n) = \{1, 4\}*\{1, 1\} = \{1, 5, 4\}$$
$$x_o(n) * h_o(n) = \{2, 3\}*\{-2\} = \{-4, -6\}$$
$$\{1, 5, 4\} + \{0, -4, -6\} = \{1, 1, -2\}$$

The advantage is that the smaller convolutions can be computed in parallel.

In computing the IDWT, upsampled sequence is involved in the convolution.

Example 3.2 Find the linear convolution of the sequences

$$\{h(n), n = 0, 1, 2\} = \{1, -2, 1\} \quad \text{and}$$
$$\{x(n), n = 0, 1, 2, 3, 4, 5, 6, 7\} = \{1, 0, 2, 0, 4, 0, 3, 0\}$$

Solution
Convolving $x(n)$ and $h(n)$, we get

$$
\begin{aligned}
y(0) &= (1)(1) = 1 \\
y(1) &= (1)(-2) + (0)(1) = -2 \\
y(2) &= (1)(1) + (0)(-2) + (2)(1) = 3 \\
y(3) &= (0)(1) + (2)(-2) + (0)(1) = -4 \\
y(4) &= (2)(1) + (0)(-2) + (4)(1) = 6 \\
y(5) &= (0)(1) + (4)(-2) + (0)(1) = -8 \\
y(6) &= (4)(1) + (0)(-2) + (3)(1) = 7 \\
y(7) &= (0)(1) + (3)(-2) + (0)(1) = -6 \\
y(8) &= (3)(1) + (0)(-2) = 3
\end{aligned}
$$

■

If we look at the pattern of computation, it is evident that the even-indexed output values can be obtained by convolving the downsampled data $\{x_d(n) = 1, 2, 4, 3\}$ with the even-indexed values of $h(n)$, $\{1, 1\}$. Similarly, the odd-indexed output values can be obtained by convolving the downsampled data $\{1, 2, 4, 3\}$ with the odd-indexed values of $h(n)$, $\{-2\}$.

$$
\begin{aligned}
x_d(n) * h_e(n) &= \{1, 2, 4, 3\} * \{1, 1\} = \{1, 3, 6, 7, 3\} \\
x_d(n) * h_o(n) &= \{1, 2, 4, 3\} * \{-2\} = \{-2, -4, -8, -6\}
\end{aligned}
$$

Merging the two outputs, we get

$$\{1, 0, 3, 0, 6, 0, 7, 0, 3\} + \{0, -2, 0, -4, 0, -8, 0, -6, 0\} = \{1, -2, 3, -4, 6, -8, 7, -6, 3\}$$

The following equation gets the same output without explicitly showing the zeros due to upsampling.

$$y(n) = \sum_{k=-\infty}^{\infty} x_d(k)h(n - 2k) \tag{3.2}$$

For the specific example,

$$
\begin{aligned}
y(0) &= x_d(0)h(0) = (1)(1) = 1 \\
y(1) &= x_d(0)h(1) = (1)(-2) = -2 \\
y(2) &= x_d(0)h(2) + x_d(1)h(0) = (1)(1) + (2)(1) = 3 \\
y(3) &= x_d(1)h(1) = (2)(-2) = -4 \\
y(4) &= x_d(1)h(2) + x_d(2)h(0) = (2)(1) + (4)(1) = 6 \\
y(5) &= x_d(2)h(1) = (4)(-2) = -8 \\
y(6) &= x_d(2)h(2) + x_d(3)h(0) = (4)(1) + (3)(1) = 7 \\
y(7) &= x_d(3)h(1) = (3)(-2) = -6 \\
y(8) &= x_d(3)h(2) = (3)(1) = 3
\end{aligned}
$$

3.2 Correlation

3.2.1 *The Linear Correlation*

Correlation operation measures the similarity between two signals. Correlation between two signals is positive if their values tend to move in the same direction, and it is negative if their values tend to move in the opposite direction. There is little or no correlation between two signals if their values move in the same direction half the time and in the opposite direction during the other half. The cross-correlation of two signals $x(n)$ and $y(n)$ is defined as

$$r_{xy}(m) = \sum_{n=-\infty}^{\infty} x(n)y(n-m), \quad m = 0, \pm 1, \pm 2, \ldots$$

(Sometimes, cross-correlation is defined as

$$r_{xy}(m) = \sum_{n=-\infty}^{\infty} x(n)y(n+m), \quad m = 0, \pm 1, \pm 2, \ldots$$

The output will be the same as that obtained from the previous definition, but is time reversed.) In the cross-correlation expression, the second subscript indicates the signal that is shifted by m samples to the right for positive m. The correlation or similarity or statistical relationship between the two functions $x(n)$ and $y(n)$ is a function of the delay time. For example, our hungriness depends on the delay time after our last meal. Therefore, the independent variable in the correlation function is the time-lag or time-delay variable m, which has the dimension of time (delay time). The independent time variable n indicates the running time. The correlation of $\{y(n), n = 0, 1\} = \{1, -1\}$ and $\{x(n), n = 0, 1, 2, 3\} = \{3, 1, 2, 4\}$ is shown in Figure 3.3. The convolution operation without time reversal is the correlation operation. The same correlation is computed by the convolution of the impulse response $\{h(k), k = 0, 1\} = \{-1, 1\}$ and the input $\{x(k), k = 0, 1, 2, 3\} = \{3, 1, 2, 4\}$, as shown in Figure 3.4. Convolution of the time-reversed version of the sequence $y(n)$ with $x(n)$ is the same as correlation of $x(n)$ and $y(n)$. Two real signals $x(n)$ and $y(n)$ are said to be orthogonal over the entire time interval if

$$\sum_{n=-\infty}^{\infty} x(n)y(n) = 0$$

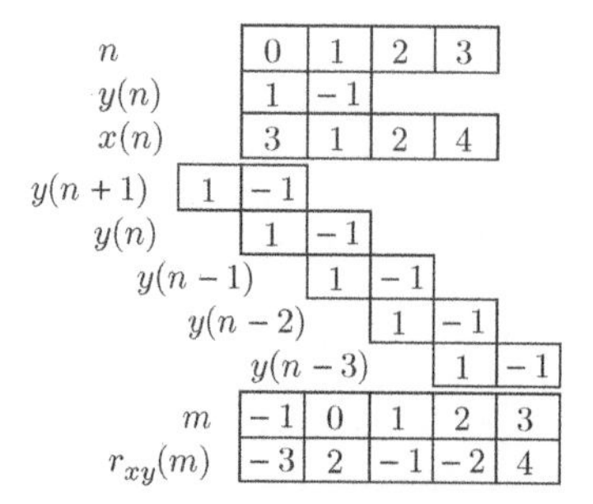

Figure 3.3 The linear correlation operation

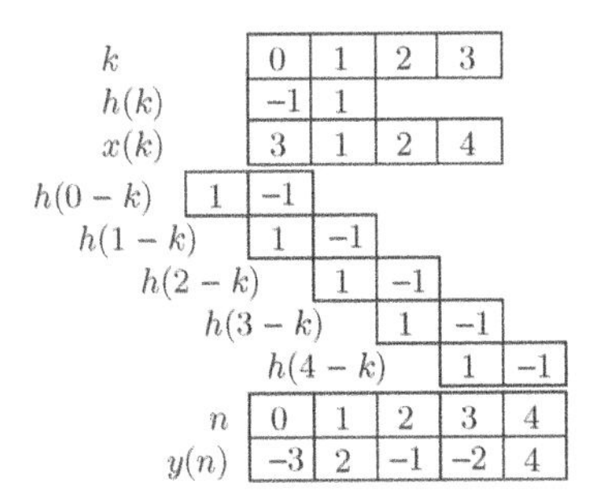

k	0	1	2	3
$h(k)$	-1	1		
$x(k)$	3	1	2	4

$h(0-k)$	1	-1				
$h(1-k)$		1	-1			
$h(2-k)$			1	-1		
$h(3-k)$				1	-1	
$h(4-k)$					1	-1

n	0	1	2	3	4
$y(n)$	-3	2	-1	-2	4

Figure 3.4 The linear correlation operation as a convolution

When two signals to be correlated are the same, the operation is called autocorrelation. The autocorrelation function is even-symmetric. Unlike convolution, correlation operation, in general, is not commutative. The correlation of a function with an impulse shifts the time-reversed version of the function to the location of the impulse.

3.2.2 Correlation and Fourier Analysis

Definitions of transforms are often in the form of correlation. Correlation operation is used in finding the coefficients of the frequency components of a signal from its amplitude profile in Fourier analysis.

Let

$$x(n) = a(0) + b(1)\sin\left(\frac{2\pi}{4}n\right) + a(1)\cos\left(2\frac{2\pi}{4}n\right)$$

The coefficient of the sine component $b(1)$ is obtained by weighting the samples of $x(n)$ over one period with those of $d(n) = \sin\left(\frac{2\pi}{4}n\right)$, finding the sum, and dividing it by the autocorrelation of $d(n)$ at lag zero. Multiplying both sides by $\sin\left(\frac{2\pi}{4}n\right)$ and summing, we get

$$\sum_{n=0}^{3} x(n)\sin\left(\frac{2\pi}{4}n\right) = \sum_{n=0}^{3}\left(a\,(0)\sin\left(\frac{2\pi}{4}n\right) + b(1)\sin^2\left(\frac{2\pi}{4}n\right)\right.$$
$$\left. + a\,(1)\cos\left(2\frac{2\pi}{4}n\right)\sin\left(\frac{2\pi}{4}n\right)\right)$$

$$r_{xd}(0) = b(1)r_{dd}(0), \quad b(1) = \frac{r_{xd}(0)}{r_{dd}(0)}$$

Note that there is no correlation of the sine function with the other two functions. Let

$$x(n) = 1 - 2\sin\left(\frac{2\pi}{4}n\right) + \cos\left(2\frac{2\pi}{4}n\right)$$

The samples of $x(n)$ over one cycle are

$$\{x(0) = 2, x(1) = -2, x(2) = 2, x(3) = 2\}$$

The corresponding samples of $d(n)$ are

$$\{d(0) = 0, d(1) = 1, d(2) = 0, d(3) = -1\}$$

$$r_{xd}(0) = \sum_{n=0}^{3} x(n)d(n) = (2)(0) + (-2)(1) + (2)(0) + (2)(-1) = -4$$

$$r_{dd}(0) = \sum_{n=0}^{3} d^2(n) = (0)(0) + (1)(1) + (0)(0) + (-1)(-1) = 2$$

$$\frac{r_{xd}(0)}{r_{dd}(0)} = \frac{-4}{2} = -2 = b(1)$$

The numerator is the cross-correlation of $x(n)$ and $d(n)$ at lag zero, and the denominator is the autocorrelation of $d(n)$ at lag zero. The other two coefficients can be found in a similar manner. The cross-spectrum corresponding to a cross-correlation function is a measure of the spectral coincidence of the signals. Therefore, in forming the cross-correlation function $r_{xd}(m)$, it can be found whether a function of the form $d(n)$ is part of $x(n)$.

3.2.3 *Correlation in the DWT*

Similar to the use of the convolution in the DWT, correlation operation is also used with double shifts of the data after the computation of each output value. Consider the equation, the type of which often occurs in the DWT,

$$\sum_{n} x(n)x(n-2m) = \delta(m), \quad m = 0, \pm 1, \pm 2, \ldots \tag{3.3}$$

This equation finds the double-shifted autocorrelation of $x(n)$ for all possible m. The autocorrelation of

$$x(n) = \frac{1}{4\sqrt{2}}\left\{\left(1-\sqrt{3}\right), (3-\sqrt{3}), (3+\sqrt{3}), (1+\sqrt{3})\right\}$$

is given by

$$\begin{aligned} r_{xx}(0) &= x^2(0) + x^2(1) + x^2(2) + x^2(3) \\ &= \frac{1}{32}((1-\sqrt{3})^2 + (3-\sqrt{3})^2 + (3+\sqrt{3})^2 + (1+\sqrt{3})^2) = 1 \\ r_{xx}(-2) = r_{xx}(2) &= x(2)x(0) + x(3)x(1) \\ &= \frac{1}{32}((3+\sqrt{3})(1-\sqrt{3}) + (1+\sqrt{3})(3-\sqrt{3})) = 0 \end{aligned}$$

as autocorrelation function is even-symmetric. After one double shift (for $m > 1$), there are no overlapping terms. Equation (3.3) is satisfied.

An example of double-shifted cross-correlation of $x(n)$ and $y(n)$, the type of which often occurs in the DWT, is given by

$$\sum_{n} x(n)y(n-2m) = 0, \quad m = 0, \pm 1, \pm 2, \ldots \tag{3.4}$$

Let

$$y(n) = \frac{1}{4\sqrt{2}} \left\{ -\left(1+\sqrt{3}\right), (3+\sqrt{3}), -(3-\sqrt{3}), (1-\sqrt{3}) \right\}$$

Then,

$$r_{xy}(-2) = x(0)y(2) + x(1)y(3) = \frac{1}{32}((1-\sqrt{3})(-(3-\sqrt{3})) + (3-\sqrt{3})(1-\sqrt{3})) = 0$$

$$r_{xy}(0) = x(0)y(0) + x(1)y(1) + x(2)y(2) + x(3)y(3) = \frac{1}{32}((1-\sqrt{3})(-(1+\sqrt{3}))$$
$$+ (3-\sqrt{3})(3+\sqrt{3}) + (3+\sqrt{3})(-(3-\sqrt{3})) + (1+\sqrt{3})(1+\sqrt{3})) = 0$$

$$r_{xy}(2) = x(2)y(0) + x(3)y(1) = \frac{1}{32}((3+\sqrt{3})(-(1+\sqrt{3})) + (1+\sqrt{3})(3+\sqrt{3})) = 0$$

Equation (3.4) is satisfied.

3.3 Summary

- Convolution and correlation are of fundamental importance in both time-domain and frequency-domain analysis of signals and systems.
- Convolution is the sum of products of two functions with each other's index running in opposite directions.
- Convolution can be interpreted as a linear system model. It produces the output of a system from the input and the impulse response.
- Convolution can also be interpreted as the interaction of two functions resulting in another function.
- Correlation is a similarity measure between signals.
- Correlation can be implemented by convolution with one of the two signals time reversed.
- Transform definitions are often based on correlation.
- The basic definitions of convolution and correlation are slightly modified in DWT usage.
- A double-shifted convolution can be carried out by convolving the even- and odd-indexed samples of the input, respectively, with the even- and odd-indexed samples of the impulse response separately and adding the two partial results.
- The convolution of an upsampled sequence can be obtained by downsampling the sequence and convolving it with the even- and odd-indexed samples of the impulse response separately and merging the partial results.

Exercises

3.1 Find the coefficients of the product polynomial of the two given polynomials using the convolution operation. Verify the output of the convolution using two tests.

3.1.1

$$3x^{-2} + 2x^{-1} - 2 + x^2, \qquad 2x^{-3} + x^{-1} + 4 + x$$

***3.1.2**

$$x^{-3} - 2x^{-1} + 5 + x, \qquad x^{-1} + x^2$$

3.1.3

$$3x^{-2} - 2 + x^2, \qquad 2x^{-1} - 3 + x^3$$

3.2 Find the linear convolution of the sequences $x(n)$ and $h(n)$. Verify the output of the convolution using two tests.

3.2.1

$$\{x(n), n = 0, 1, 2, 3, 4\} = \{2, 3, 4, 2, 1\}, \qquad \{h(n), n = 0, 1\} = \{1, 1\}$$

3.2.2

$$\{x(n), n = 0, 1, 2, 3, 4\} = \{2, 3, 1, 4, -1\}, \qquad \{h(n), n = 0, 1\} = \{-1, 1\}$$

***3.2.3**

$$\{x(n), n = 0, 1, 2, 3, 4\} = \{2, 3, 1, 4, -1\}, \qquad \{h(n), n = 0, 1, 2\} = \{1, -2, 1\}$$

3.3 Find the double-shifted convolution of the sequences $x(n)$ and $h(n)$ using the following equation.

$$y(n) = \sum_{k=-\infty}^{\infty} h(k)x(2n - k)$$

3.3.1

$$\{x(n), n = 0, 1, 2, 3, 4\} = \{3, 1, 2, 4, -2\}, \qquad \{h(n), n = 0, 1\} = \{1, 1\}$$

***3.3.2**

$$\{x(n), n = 0, 1, 2, 3, 4\} = \{3, 1, 2, 4, -2\}, \qquad \{h(n), n = 0, 1\} = \{-1, 1\}$$

3.3.3

$$\{x(n), n = 0, 1, 2, 3, 4\} = \{3, 1, 2, 4, -2\}, \qquad \{h(n), n = 0, 1\} = \{-3, 2\}$$

3.3.4

$$\{x(n), n = 0, 1, 2, 3, 4\} = \{3, 1, 2, 4, -2\}, \qquad \{h(n), n = 0, 1\} = \{1, -2, 1\}$$

3.4 Find the linear convolution of the sequences $x(n)$ and $h(n)$. How to get the first convolution output from the rest?

3.4.1

$$\{x(n), n = 0, 1, 2, 3, 4, 5, 6, 7, 8, 9\} = \{2, 0, 3, 0, 4, 0, 2, 0, 1, 0\},$$
$$\{h(n), n = 0, 1\} = \{1, 1\}$$
$$\{x(n), n = 0, 1, 2, 3, 4\} = \{2, 3, 4, 2, 1\}, \qquad \{h(n), n = 0\} = \{1\}$$

3.4.2

$$\{x(n), n = 0, 1, 2, 3, 4, 5, 6, 7, 8, 9\} = \{2, 0, 1, 0, 3, 0, -2, 0, 4, 0\},$$
$$\{h(n), n = 0, 1\} = \{1, -1\}$$
$$\{x(n), n = 0, 1, 2, 3, 4\} = \{2, 1, 3, -2, 4\}, \qquad \{h(n), n = 0\} = \{1\}$$
$$\{x(n), n = 0, 1, 2, 3, 4\} = \{2, 1, 3, -2, 4\}, \qquad \{h(n), n = 0\} = \{-1\}$$

***3.4.3**

$$\{x(n), n = 0, 1, 2, 3, 4, 5, 6, 7, 8\} = \{2, 0, 3, 0, 4, 0, 2, 0, 1\},$$
$$\{h(n), n = 0, 1, 2, 3, 4\} = \{1, 2, -6, 2, 1\}$$
$$\{x(n), n = 0, 1, 2, 3, 4\} = \{2, 3, 4, 2, 1\}, \qquad \{h(n), n = 0, 1, 2\} = \{1, -6, 1\}$$
$$\{x(n), n = 0, 1, 2, 3, 4\} = \{2, 3, 4, 2, 1\}, \qquad \{h(n), n = 0, 1\} = \{2, 2\}$$

3.4.4

$$\{x(n), n = 0, 1, 2, 3, 4, 5, 6, 7, 8\} = \{2, 0, 1, 0, 3, 0, -2, 0, 4\},$$
$$\{h(n), n = 0, 1, 2\} = \{1, 2, 1\}$$
$$\{x(n), n = 0, 1, 2, 3, 4\} = \{2, 1, 3, -2, 4\}, \qquad \{h(n), n = 0, 1\} = \{1, 1\}$$
$$\{x(n), n = 0, 1, 2, 3, 4\} = \{2, 1, 3, -2, 4\}, \qquad \{h(n), n = 0\} = \{2\}$$

3.5 Find the linear convolution of the sequences $x(n)$ and $h(n)$ using the following equation:

$$y(n) = \sum_{k=-\infty}^{\infty} x_d(k)h(n - 2k)$$

where $x_d(k)$ is $x(n)$ downsampled by a factor of 2. Verify the answer by direct convolution of $x(n)$ and $h(n)$.

3.5.1

$$\{x(n), n = 0, 1, 2, 3, 4, 5, 6, 7\} = \{2, 0, 4, 0, 1, 0, 3, 0\},$$
$$\{h(n), n = 0, 1, 2\} = \{1, -2, 1\}$$

***3.5.2**

$$\{x(n), n = 0, 1, 2, 3, 4, 5, 6, 7\} = \{2, 0, 4, 0, 1, 0, 3, 0\},$$
$$\{h(n), n = 0, 1, 2, 3\} = \{-1, 3, 3, -1\}$$

3.5.3

$$\{x(n), n = 0, 1, 2, 3, 4, 5, 6, 7\} = \{2, 0, 4, 0, 1, 0, 3, 0\},$$
$$\{h(n), n = 0, 1, 2, 3, 4\} = \{-1, 2, 6, 2, -1\}$$

3.6 Using the three methods of signal extension, extend $x(n)$ at either end by two samples so that we get a signal with eight data values. Using the equation

$$y(n) = \sum_{k=-\infty}^{\infty} x(k)h(2n - k)$$

find the three values of $y(n)$ found with complete overlap of the extended $x(n)$ and $h(n)$ in each case.

3.6.1

$$\{x(n), n = 0, 1, 2, 3\} = \{2, 3, 4, 1\},$$
$$\{h(n), n = 0, 1, 2, 3\} = \{1 + \sqrt{3}, 3 + \sqrt{3}, 3 - \sqrt{3}, 1 - \sqrt{3}\}$$

3.6.2

$$\{x(n), n = 0, 1, 2, 3\} = \{2, 1, 4, 3\},$$
$$\{h(n), n = 0, 1, 2, 3\} = \{-1, 3, 3, -1\}$$

***3.6.3**

$$\{x(n), n = 0, 1, 2, 3\} = \{5, 1, 3, 4\},$$
$$\{h(n), n = 0, 1, 2, 3\} = \{-1, 3, -3, 1\}$$

***3.7** Find the eight values of the linear convolution of the sequences $x(n)$ and $h(n)$ obtained by the complete overlap of $x(n)$ and $h(n)$. Assuming whole-point extension at the borders, prefix one value to $x(n)$ and suffix one value.

$$\{x(n), n = 0, 1, 2, 3, 4, 5, 6, 7\} = \{2, 3, 1, 2, 4, 1, 4, -1\},$$
$$\{h(n), n = 0, 1, 2\} = \{1, -2, 1\}$$

3.8 Find the eight values of the linear convolution of the sequences $x(n)$ and $h(n)$ obtained by the complete overlap of $x(n)$ and $h(n)$. Assuming whole-point extension at the borders, prefix two values to $x(n)$ and suffix two values.

$$\{x(n), n = 0, 1, 2, 3, 4, 5, 6, 7\} = \{2, 3, 1, 2, 4, 1, 4, -1\},$$
$$\{h(n), n = 0, 1, 2, 3, 4\} = \{-1, 2, 6, 2, -1\}$$

3.9 Find the cross-correlation of $\{x(n), n = 0, 1, 2, 3, 4\} = \{2, 3, 1, 4, -1\}$ and $\delta(n)$, the unit-impulse. Find the convolution of $\{x(n), n = 0, 1, 2, 3, 4\} = \{2, 3, 1, 4, -1\}$ and $\delta(n)$, the unit-impulse.

3.10 Find the cross-correlation $r_{xy}(k)$ of $\{x(n), n = 0, 1, 2, 3, 4\} = \{2, 3, 1, 4, -1\}$ and $\delta(n - 3)$, the shifted unit-impulse. Find the convolution of $\{x(n), n = 0, 1, 2, 3, 4\} = \{2, 3, 1, 4, -1\}$ and $\delta(n + 3)$, the shifted unit-impulse.

3.11 Find the autocorrelation $r_{xx}(k)$ of $\{x(n), n = 0, 1, 2, 3, 4\} = \{2, 3, 1, 4, -1\}$.

3.12 Find the cross-correlation $r_{xy}(k)$ of $\{x(n), n = 0, 1, 2, 3, 4\} = \{2, 3, 1, 4, -1\}$ and $\{y(n), n = 0, 1, 2, 3, 4\} = \{2, 1, -2, 3, 1\}$.

3.13 Compute the double-shifted cross-correlation of $x(n)$ and $p(n)$ using the equation

$$\sum_{n} x(n)p(n - 2m), \quad m = 0, \pm 1, \pm 2, \ldots$$

3.13.1

$$\{x(n), n = 0, 1\} = \{1, 1\}$$
$$\{p(n), n = 0, 1\} = \{1, -1\}$$

3.13.2

$$\{x(n), n = 0, 1, 2, 3, 4, 5\} = \{0.0352, -0.0854, -0.1350, 0.4599, 0.8069, 0.3327\}$$
$$\{p(n), n = 0, 1, 2, 3, 4, 5\} = \{-0.3327, 0.8069, -0.4599, -0.1350, 0.0854, 0.0352\}$$

3.13.3

$$\{x(n), n = -1, 0, 1, 2\} = \{1, 3, 3, 1\}$$
$$\{p(n), n = -1, 0, 1, 2\} = \{-2, 6, 6, -2\}$$

3.13.4

$$\{x(n), n = -1, 0, 1, 2\} = \{-1, 3, -3, 1\}$$
$$\{p(n), n = -1, 0, 1, 2\} = \{2, 6, -6, -2\}$$

3.13.5

$$\{x(n), n = -1, 0, 1, 2\} = \{1, 3, 3, 1\}$$
$$\{p(n), n = -1, 0, 1, 2\} = \{2, 6, -6, -2\}$$

3.13.6

$$\{x(n), n = -1, 0, 1, 2\} = \{-1, 3, -3, 1\}$$
$$\{p(n), n = -1, 0, 1, 2\} = \{-2, 6, 6, -2\}$$

***3.13.7**

$$\{x(n), n = -2, -1, 0, 1, 2\} = \{-1, 2, 6, 2, -1\}$$
$$\{p(n), n = -1, 0, -1\} = \{2, 4, 2\}$$

3.13.8

$$\{x(n), n = -1, 0, 1\} = \{2, -4, 2\}$$
$$\{p(n), n = -2, -1, 0, 1, 2\} = \{1, 2, -6, 2, 1\}$$

3.13.9

$$\{x(n), n = -1, 0, 1\} = \{1, 2, 1\}$$
$$\{p(n), n = 0, 1, 2\} = \{1, -2, 1\}$$

***3.13.10**

$$\{x(n), n = -2, -1, 0, 1, 2\} = \{-1, 2, 6, 2, -1\}$$
$$\{p(n), n = -1, 0, 1, 2, 3\} = \{1, 2, -6, 2, 1\}$$

4

Fourier Analysis of Discrete Signals

In general, the processing of a signal is facilitated if the processing is carried out after representing it in terms of a set of simple and well-defined basis functions. For example, convolution becomes multiplication when the signals are represented by their spectra. The basis functions are usually well defined, whereas the amplitude profile of practical signals is arbitrary. Representing an entity by a set of basis functions is used in describing many things. There are infinite points in the plane. All of them are specified by their two coordinates along x-axis and y-axis. Any of the infinite places on earth can be located by its latitude and longitude. An arbitrary color can be specified in terms of a set basis colors, for example, red, green, and blue. In all cases, different sets of basis functions exist. The choice of the set to be selected depends on the ease of representation and the suitability of the basis for the particular application.

4.1 Transform Analysis

The basis of most of the transforms is to change the representation of a signal into a form that is more suitable for solving a problem. An arbitrary 1-D signal $x(n)$ can be represented by a series

$$x(n) = c_0 b_0(n) + c_1 b_1(n) + c_2 b_2(n) + \cdots \tag{4.1}$$

in which $x(n)$ and the basis signals $b_k(n)$ are known and c_k are the coefficients to be determined so that the equation is satisfied. For each transform, the basis functions are different. The basis signal used in Fourier analysis is the sinusoidal waveform. For convenience of analysis, the sinusoidal waveform is expressed in an equivalent exponential form that is obtained from Euler's formula

$$A\cos(\omega n + \theta) = \frac{A}{2}\left(e^{j(\omega n+\theta)} + e^{-j(\omega n+\theta)}\right)$$

The essence of the DWT is the representation of the signals in terms of their components of transient nature corresponding to a set of continuous group of frequency components of their

Discrete Wavelet Transform: A Signal Processing Approach, First Edition. D. Sundararajan.

Companion Website: www.wiley.com/go/sundararajan/wavelet

spectra. Therefore, the spectrum (the representation of a signal with frequency as the independent variable) of a signal is the central object in the DWT, as is the case in most of the frequently used transforms in signal and system analysis. The easiest way to understand the spectrum is through the DFT, the practically most often used version of the Fourier analysis. This is due to the fact that its input and output values are finite and discrete and fast numerical algorithms are readily available for its computation. Therefore, it is capable of approximating other versions adequately. In this book, we use the DFT to approximate the DTFT. Furthermore, the transform concepts and the orthogonality property are also introduced. In this chapter, we describe the DFT and the DTFT, which are two of the four versions of the Fourier analysis for discrete signal analysis. In the DWT, the signals of interest are aperiodic discrete signals. The DTFT is primarily used to analyze this type of signals. As the wavelet transform is a generalized version of the Fourier analysis, Fourier analysis is required in the formulation and the understanding of the DWT.

4.2 The Discrete Fourier Transform

Consider the discrete periodic waveform, $x(n) = 1 + \sin\left(\frac{2\pi}{4}n\right) + 0\ \cos\left(2\frac{2\pi}{4}n\right)$, with period four samples, shown in Figure 4.1(a). The samples are

$$\{x(0) = 1, x(1) = 2, x(2) = 1, x(3) = 0\}$$

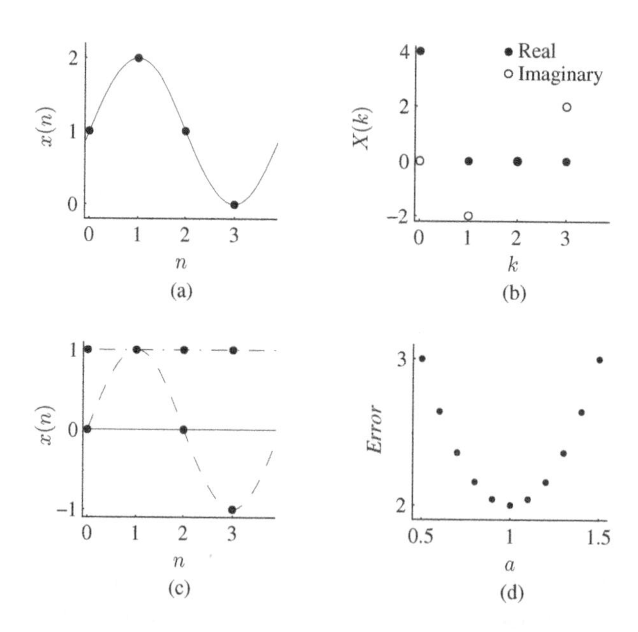

Figure 4.1 (a) A periodic waveform, $x(n) = 1 + \sin\left(\frac{2\pi}{4}n\right) + 0\ \cos\left(2\frac{2\pi}{4}n\right)$, with period four samples and (b) its frequency-domain representation; (c) the frequency components of the waveform in (a); (d) the square error in approximating the waveform in (a) using only the DC component with different amplitudes

The independent variable n represents time, and the dependent variable is amplitude. Figure 4.1(b) shows the frequency-domain representation of the waveform in (a).

$$\{X(0) = 4, X(1) = -j2, X(2) = 0, X(3) = j2\}$$

It shows the complex amplitude, multiplied by the factor 4, of its constituent complex exponentials. To find the real sinusoids, shown in Figure 4.1(c), that constitute the signal, we add up the complex exponentials.

$$\begin{aligned} x(n) &= \frac{1}{4}\left(4e^{j0\frac{2\pi}{4}n} - j2e^{j\frac{2\pi}{4}n} + 0e^{j2\frac{2\pi}{4}n} + j2e^{j3\frac{2\pi}{4}n}\right) \\ &= 1 + \sin\left(\frac{2\pi}{4}n\right) + 0\cos\left(2\frac{2\pi}{4}n\right) \end{aligned}$$

As can be seen from this example, Fourier analysis represents a signal as a linear combination of sinusoids or, equivalently, complex exponentials with pure imaginary exponents.

The Fourier reconstruction of a waveform is with respect to the least square error criterion. That is, the sum of the squared magnitude of the error between the given waveform and the corresponding Fourier reconstructed waveform is guaranteed to be the minimum if part of the constituent sinusoids of a waveform is used in the reconstruction and will be zero if all the constituent sinusoids are used. The reason this criterion, based on signal energy or power, is used rather than a minimum uniform deviation criterion is that: (i) it is acceptable for most applications and (ii) it leads to closed-form formulas for the analytical determination of the Fourier coefficients.

Let $x_a(n)$ be an approximation to a given waveform $x(n)$ of period N, using fewer harmonics than that is required. The square error between $x(n)$ and $x_a(n)$ is defined as

$$\text{error} = \sum_{n=0}^{N-1} |x(n) - x_a(n)|^2$$

For a given number of harmonics, there is no better approximation for the signal than that provided by the Fourier analysis when the least square error criterion is applied. Assume that we are constrained to use only the DC component to approximate the waveform in Figure 4.1(a). Let the optimal value of the DC component be a. To minimize the square error,

$$(1-a)^2 + (2-a)^2 + (1-a)^2 + (0-a)^2$$

must be minimum. Differentiating this expression with respect to a and equating it to zero, we get

$$2(1-a)(-1) + 2(2-a)(-1) + 2(1-a)(-1) + 2(0-a)(-1) = 0$$

Solving this equation, we get $a = 1$ as given by the Fourier analysis. The square error, for various values of a, is shown in Figure 4.1(d).

For two complex exponentials $e^{j\frac{2\pi}{N}ln}$ and $e^{j\frac{2\pi}{N}kn}$ over a period of N samples, the orthogonality property is defined as

$$\sum_{n=0}^{N-1} e^{j\frac{2\pi}{N}(l-k)n} = \begin{cases} N & \text{for } l = k \\ 0 & \text{for } l \neq k \end{cases}$$

where $l, k = 0, 1, \ldots, N-1$. If $l = k$, the summation is equal to N as $e^{j\frac{2\pi}{N}(l-k)n} = e^0 = 1$. Otherwise, by using the closed-form expression for the sum of a geometric progression, we get

$$\sum_{n=0}^{N-1} e^{j\frac{2\pi}{N}(l-k)n} = \frac{1 - e^{j2\pi(l-k)}}{1 - e^{j\frac{2\pi(l-k)}{N}}} = 0, \quad \text{for } l \neq k$$

This is also obvious from the fact that the sum of the samples of the complex exponential over an integral number of cycles is zero. That is, in order to find the coefficient, with a scale factor N, of a complex exponential, we multiply the samples of a signal with the corresponding samples of the complex conjugate of the complex exponential. Using each complex exponential in turn, we get the frequency coefficients of all the components of a signal as

$$X(k) = \sum_{n=0}^{N-1} x(n) W_N^{nk}, \quad k = 0, 1, \ldots, N-1 \tag{4.2}$$

where $W_N = e^{-j\frac{2\pi}{N}}$. This is the DFT equation analyzing a waveform with harmonically related discrete complex sinusoids. $X(k)$ is the coefficient, scaled by N, of the complex sinusoid $e^{j\frac{2\pi}{N}kn}$ with a specific frequency index k (frequency $\frac{2\pi}{N}k$ radians per sample). DFT computation is based on assumed periodicity. That is, the N input values are considered as one period of a periodic waveform with period N. The summation of the sample values of the N complex sinusoids multiplied by their respective frequency coefficients $X(k)$ is the IDFT operation. The N-point IDFT of the frequency coefficients $X(k)$ is defined as

$$x(n) = \frac{1}{N} \sum_{k=0}^{N-1} X(k) W_N^{-nk}, \quad n = 0, 1, \ldots, N-1 \tag{4.3}$$

The sum of the sample values is divided by N in Equation (4.3) as the coefficients $X(k)$ have been scaled by the factor N in the DFT computation.

The DFT and IDFT definitions can be expressed in matrix form. Expanding the DFT definition with $N = 4$, we get

$$\begin{bmatrix} X(0) \\ X(1) \\ X(2) \\ X(3) \end{bmatrix} = \begin{bmatrix} e^{-j\frac{2\pi}{4}(0)(0)} & e^{-j\frac{2\pi}{4}(0)(1)} & e^{-j\frac{2\pi}{4}(0)(2)} & e^{-j\frac{2\pi}{4}(0)(3)} \\ e^{-j\frac{2\pi}{4}(1)(0)} & e^{-j\frac{2\pi}{4}(1)(1)} & e^{-j\frac{2\pi}{4}(1)(2)} & e^{-j\frac{2\pi}{4}(1)(3)} \\ e^{-j\frac{2\pi}{4}(2)(0)} & e^{-j\frac{2\pi}{4}(2)(1)} & e^{-j\frac{2\pi}{4}(2)(2)} & e^{-j\frac{2\pi}{4}(2)(3)} \\ e^{-j\frac{2\pi}{4}(3)(0)} & e^{-j\frac{2\pi}{4}(3)(1)} & e^{-j\frac{2\pi}{4}(3)(2)} & e^{-j\frac{2\pi}{4}(3)(3)} \end{bmatrix} \begin{bmatrix} x(0) \\ x(1) \\ x(2) \\ x(3) \end{bmatrix}$$

$$= \begin{bmatrix} 1 & 1 & 1 & 1 \\ 1 & -j & -1 & j \\ 1 & -1 & 1 & -1 \\ 1 & j & -1 & -j \end{bmatrix} \begin{bmatrix} x(0) \\ x(1) \\ x(2) \\ x(3) \end{bmatrix}$$

Using vector and matrix quantities, the DFT definition is given by

$$\boldsymbol{X} = \boldsymbol{W}\boldsymbol{x}$$

where $\boldsymbol{x}$ is the input vector, $\boldsymbol{X}$ is the coefficient vector, and $\boldsymbol{W}$ is the transform matrix, defined as

$$\boldsymbol{W} = \begin{bmatrix} 1 & 1 & 1 & 1 \\ 1 & -j & -1 & j \\ 1 & -1 & 1 & -1 \\ 1 & j & -1 & -j \end{bmatrix}$$

Expanding the IDFT definition with $N = 4$, we get

$$\begin{bmatrix} x(0) \\ x(1) \\ x(2) \\ x(3) \end{bmatrix} = \frac{1}{4} \begin{bmatrix} e^{j\frac{2\pi}{4}(0)(0)} & e^{j\frac{2\pi}{4}(0)(1)} & e^{j\frac{2\pi}{4}(0)(2)} & e^{j\frac{2\pi}{4}(0)(3)} \\ e^{j\frac{2\pi}{4}(1)(0)} & e^{j\frac{2\pi}{4}(1)(1)} & e^{j\frac{2\pi}{4}(1)(2)} & e^{j\frac{2\pi}{4}(1)(3)} \\ e^{j\frac{2\pi}{4}(2)(0)} & e^{j\frac{2\pi}{4}(2)(1)} & e^{j\frac{2\pi}{4}(2)(2)} & e^{j\frac{2\pi}{4}(2)(3)} \\ e^{j\frac{2\pi}{4}(3)(0)} & e^{j\frac{2\pi}{4}(3)(1)} & e^{j\frac{2\pi}{4}(3)(2)} & e^{j\frac{2\pi}{4}(3)(3)} \end{bmatrix} \begin{bmatrix} X(0) \\ X(1) \\ X(2) \\ X(3) \end{bmatrix}$$

$$= \frac{1}{4} \begin{bmatrix} 1 & 1 & 1 & 1 \\ 1 & j & -1 & -j \\ 1 & -1 & 1 & -1 \\ 1 & -j & -1 & j \end{bmatrix} \begin{bmatrix} X(0) \\ X(1) \\ X(2) \\ X(3) \end{bmatrix}$$

Concisely,

$$\boldsymbol{x} = \frac{1}{4}\boldsymbol{W}^{-1}\boldsymbol{X} = \frac{1}{4}(\boldsymbol{W}^*)^T\boldsymbol{X}$$

The inverse and forward transform matrices are orthogonal. That is,

$$\frac{1}{4} \begin{bmatrix} 1 & 1 & 1 & 1 \\ 1 & j & -1 & -j \\ 1 & -1 & 1 & -1 \\ 1 & -j & -1 & j \end{bmatrix} \begin{bmatrix} 1 & 1 & 1 & 1 \\ 1 & -j & -1 & j \\ 1 & -1 & 1 & -1 \\ 1 & j & -1 & -j \end{bmatrix} = \begin{bmatrix} 1 & 0 & 0 & 0 \\ 0 & 1 & 0 & 0 \\ 0 & 0 & 1 & 0 \\ 0 & 0 & 0 & 1 \end{bmatrix}$$

The DFT is defined for sequences of any length. However, it is assumed that the length N of a column sequence

$$\{x(0), x(1), x(2), \ldots, x(N-1)\}$$

is a power of 2 in most applications. The reason is that fast algorithms for the computation of the DFT are available only for these lengths.

Some examples of 4-point DFT computation are given as follows.

$$\{x(0) = 1, x(1) = 2, x(2) = 1, x(3) = 0\}$$

The DFT of this set of data is computed as

$$\begin{bmatrix} X(0) \\ X(1) \\ X(2) \\ X(3) \end{bmatrix} = \begin{bmatrix} 1 & 1 & 1 & 1 \\ 1 & -j & -1 & j \\ 1 & -1 & 1 & -1 \\ 1 & j & -1 & -j \end{bmatrix} \begin{bmatrix} 1 \\ 2 \\ 1 \\ 0 \end{bmatrix} = \begin{bmatrix} 4 \\ -j2 \\ 0 \\ j2 \end{bmatrix}$$

The DFT spectrum is $\{X(0) = 4, X(1) = -j2, X(2) = 0, X(3) = j2\}$, as shown in Figure 4.1(b). Using the IDFT, we get back the input $x(n)$.

$$\begin{bmatrix} x(0) \\ x(1) \\ x(2) \\ x(3) \end{bmatrix} = \frac{1}{4} \begin{bmatrix} 1 & 1 & 1 & 1 \\ 1 & j & -1 & -j \\ 1 & -1 & 1 & -1 \\ 1 & -j & -1 & j \end{bmatrix} \begin{bmatrix} 4 \\ -j2 \\ 0 \\ j2 \end{bmatrix} = \begin{bmatrix} 1 \\ 2 \\ 1 \\ 0 \end{bmatrix}$$

$$\{x(0) = 1, x(1) = 0, x(2) = 0, x(3) = 0\}$$

The DFT of this set of data is computed as

$$\begin{bmatrix} X(0) \\ X(1) \\ X(2) \\ X(3) \end{bmatrix} = \begin{bmatrix} 1 & 1 & 1 & 1 \\ 1 & -j & -1 & j \\ 1 & -1 & 1 & -1 \\ 1 & j & -1 & -j \end{bmatrix} \begin{bmatrix} 1 \\ 0 \\ 0 \\ 0 \end{bmatrix} = \begin{bmatrix} 1 \\ 1 \\ 1 \\ 1 \end{bmatrix}$$

The DFT spectrum is $\{X(0) = 1, X(1) = 1, X(2) = 1, X(3) = 1\}$. Using the IDFT, we get back the input $x(n)$.

$$\begin{bmatrix} x(0) \\ x(1) \\ x(2) \\ x(3) \end{bmatrix} = \frac{1}{4} \begin{bmatrix} 1 & 1 & 1 & 1 \\ 1 & j & -1 & -j \\ 1 & -1 & 1 & -1 \\ 1 & -j & -1 & j \end{bmatrix} \begin{bmatrix} 1 \\ 1 \\ 1 \\ 1 \end{bmatrix} = \begin{bmatrix} 1 \\ 0 \\ 0 \\ 0 \end{bmatrix}$$

$$\{x(0) = 1, x(1) = 1, x(2) = 1, x(3) = 1\}$$

The DFT of this set of data is computed as

$$\begin{bmatrix} X(0) \\ X(1) \\ X(2) \\ X(3) \end{bmatrix} = \begin{bmatrix} 1 & 1 & 1 & 1 \\ 1 & -j & -1 & j \\ 1 & -1 & 1 & -1 \\ 1 & j & -1 & -j \end{bmatrix} \begin{bmatrix} 1 \\ 1 \\ 1 \\ 1 \end{bmatrix} = \begin{bmatrix} 4 \\ 0 \\ 0 \\ 0 \end{bmatrix}$$

The DFT spectrum is $\{X(0) = 4, X(1) = 0, X(2) = 0, X(3) = 0\}$. Using the IDFT, we get back the input $x(n)$.

$$\begin{bmatrix} x(0) \\ x(1) \\ x(2) \\ x(3) \end{bmatrix} = \frac{1}{4} \begin{bmatrix} 1 & 1 & 1 & 1 \\ 1 & j & -1 & -j \\ 1 & -1 & 1 & -1 \\ 1 & -j & -1 & j \end{bmatrix} \begin{bmatrix} 4 \\ 0 \\ 0 \\ 0 \end{bmatrix} = \begin{bmatrix} 1 \\ 1 \\ 1 \\ 1 \end{bmatrix}$$

Figure 4.2 shows some standard signals and their DFT spectra. Figures 4.2(a) and (b) shows the unit-impulse signal $\delta(n)$ and its DFT spectrum. One of the reasons for the importance of the impulse signal is that its transform is a constant. The DC signal $u(n)$ and its DFT spectrum, shown in Figures 4.2(c) and (d), are almost the reversal of those of the impulse. Figures 4.2(e) and (f) shows the sinusoid $\cos\left(\frac{2\pi}{16}2n + \frac{\pi}{3}\right)$ and its DFT spectrum. The sinusoid

$$\cos\left(\frac{2\pi}{16}2n + \frac{\pi}{3}\right) = \frac{1}{2}\cos\left(\frac{2\pi}{16}2n\right) - \frac{\sqrt{3}}{2}\sin\left(\frac{2\pi}{16}2n\right)$$

Therefore, the real parts of the spectrum are $\{4, 4\}$ at $k = 2, 14$, as the frequency index is 2. The imaginary parts are $\sqrt{3}\{4, -4\}$ at $k = 2, 14$.

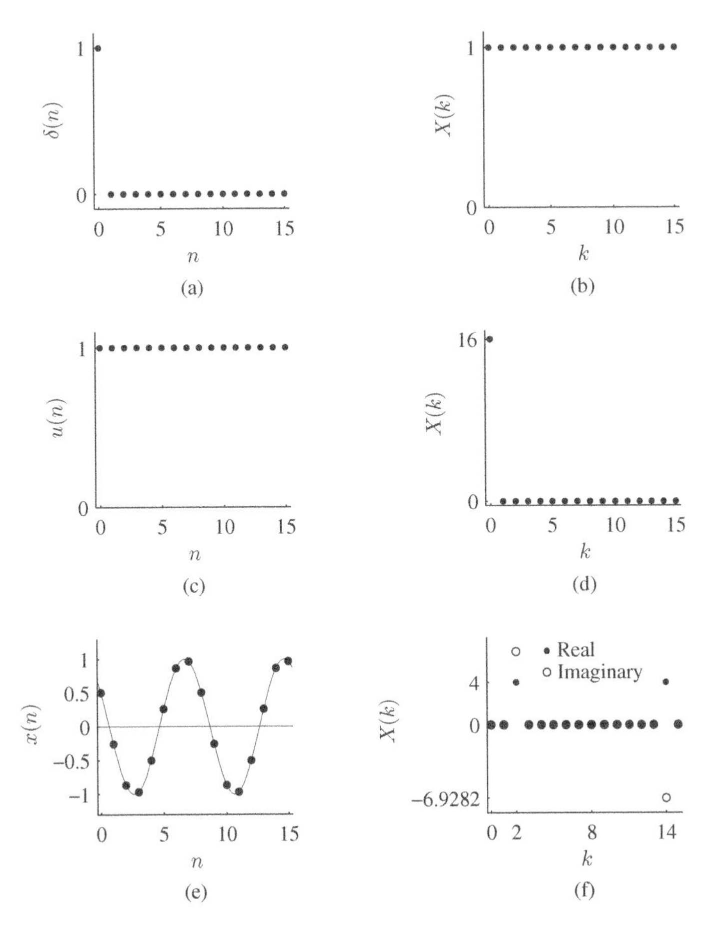

Figure 4.2 (a) The unit-impulse signal $\delta(n)$ and (b) its DFT spectrum; (c) the DC signal $u(n)$ and (d) its DFT spectrum; (e) the sinusoid $\cos\left(\frac{2\pi}{16}2n + \frac{\pi}{3}\right)$ and (f) its DFT spectrum

4.2.1 Parseval's Theorem

This theorem expresses the power of a signal in terms of its DFT spectrum. The orthogonal transforms have the energy preservation property. Let $x(n) \leftrightarrow X(k)$ with sequence length N. The double-headed arrow indicates that $X(k)$ is the representation of $x(n)$ in the transform domain. The sum of the squared magnitude of the samples of a complex exponential with amplitude one, over the period N, is N. Remember that these samples occur on the unit circle. The DFT decomposes a signal in terms of complex exponentials with coefficients $X(k)/N$. Therefore, the power of a complex exponential is $\frac{|X(k)|^2}{N^2}N = \frac{|X(k)|^2}{N}$. The power of a signal is the sum of the powers of its constituent complex exponentials and is given as

$$\sum_{n=0}^{N-1} |x(n)|^2 = \frac{1}{N}\sum_{k=0}^{N-1} |X(k)|^2$$

Example 4.1 Verify the Parseval's theorem for the DFT pair

$$\{3, 1, 3, 4\} \leftrightarrow \{11, j3, 1, -j3\}$$

Solution

The sum of the squared magnitude of the data sequence is 35 and that of the DFT coefficients divided by 4 is also 35. ■

4.3 The Discrete-Time Fourier Transform

A continuum of discrete sinusoids over a finite frequency range is used as the basis signals in the DTFT to analyze aperiodic discrete signals. Compared to the DFT, as the discrete aperiodic time-domain waveform contains an infinite number of samples, the frequency increment of the periodic spectrum of the DFT tends to zero, and the spectrum becomes continuous. The period is not affected, as it is determined by the sampling interval in the time domain. An alternate view of the DTFT is that it is the same as the Fourier Series with the roles of time- and frequency-domain functions interchanged.

The DTFT $X(e^{j\omega})$ of the signal $x(n)$ is defined as

$$X(e^{j\omega}) = \sum_{n=-\infty}^{\infty} x(n)e^{-j\omega n} \tag{4.4}$$

The inverse DTFT $x(n)$ of $X(e^{j\omega})$ is defined as

$$x(n) = \frac{1}{2\pi}\int_{-\pi}^{\pi} X(e^{j\omega})e^{j\omega n}\, d\omega, \quad n = 0, \pm 1, \pm 2, \ldots \tag{4.5}$$

The frequency-domain function $X(e^{j\omega})$ is periodic of period $\omega = 2\pi$, as

$$e^{-jn\omega} = e^{-jn(\omega+2\pi)} = e^{-jn\omega}e^{-jn2\pi}$$

Therefore, the integration in Equation (4.5) can be evaluated over any interval of length 2π. The spectral density $X(e^{j\omega})$, which is proportional to the spectral amplitude, represents the frequency content of a signal and is called the spectrum. The amplitude, $\frac{1}{2\pi}X(e^{j\omega})d\omega$, of the constituent sinusoids of a signal is infinitesimal. The DTFT spectrum is a relative amplitude spectrum.

The convergence properties are similar to those for Fourier series. The summation in Equation (4.4) converges uniformly to $X(e^{j\omega})$ for all ω, if $x(n)$ is absolutely summable.

$$\sum_{n=-\infty}^{\infty} |x(n)e^{-j\omega n}| = \sum_{n=-\infty}^{\infty} |x(n)||e^{-j\omega n}| = \sum_{n=-\infty}^{\infty} |x(n)| < \infty$$

That is, $X(e^{j\omega})$ is a continuous function. The summation converges in the least square error sense, if $x(n)$ is square summable. That is

$$\sum_{n=-\infty}^{\infty} |x(n)|^2 = \frac{1}{2\pi}\int_{-\pi}^{\pi} |X(e^{j\omega})|^2 e^{j\omega n}\, d\omega < \infty$$

The DTFT exists for all energy signals. Gibbs phenomenon occurs in the spectrum. That is, uniform convergence is not possible in the vicinity of a discontinuity of the spectrum.

The fact that the DTFT and its inverse form a transform pair is as follows. Multiplying both sides of Equation (4.4) by $e^{jk\omega}$ and integrating over the limits $-\pi$ to π, we get

$$\int_{-\pi}^{\pi} X(e^{j\omega})e^{jk\omega}d\omega = \sum_{n=-\infty}^{\infty} x(n)\int_{-\pi}^{\pi} e^{-j\omega n}e^{jk\omega}d\omega$$

$$= \sum_{n=-\infty}^{\infty} x(n)\int_{-\pi}^{\pi} e^{j(k-n)\omega}d\omega = \int_{-\pi}^{\pi} d\omega = 2\pi x(k), \quad k = n$$

For $k \neq n$,

$$\sum_{n=-\infty}^{\infty} x(n)\int_{-\pi}^{\pi} e^{j(k-n)\omega}d\omega = \sum_{n=-\infty}^{\infty} x(n)\left.\frac{e^{j(k-n)\omega}}{j(k-n)}\right|_{-\pi}^{\pi} = \sum_{n=-\infty}^{\infty} x(n)(0) = 0$$

The sampling interval is assumed to be 1 second in Equations (4.4) and (4.5). However, the DTFT and its inverse are defined for arbitrary sampling interval T_s.

$$X(e^{j\omega T_s}) = \sum_{n=-\infty}^{\infty} x(nT_s)e^{-jn\omega T_s} \tag{4.6}$$

$$x(nT_s) = \frac{1}{\omega_s}\int_{-\frac{\omega_s}{2}}^{\frac{\omega_s}{2}} X(e^{j\omega T_s})e^{jn\omega T_s}d\omega, \quad n = 0, \pm 1, \pm 2, \ldots \tag{4.7}$$

where $\omega_s = \frac{2\pi}{T_s}$.

Example 4.2 Find the DTFT of the unit-impulse signal $\delta(n)$, shown in Figure 4.3(a).

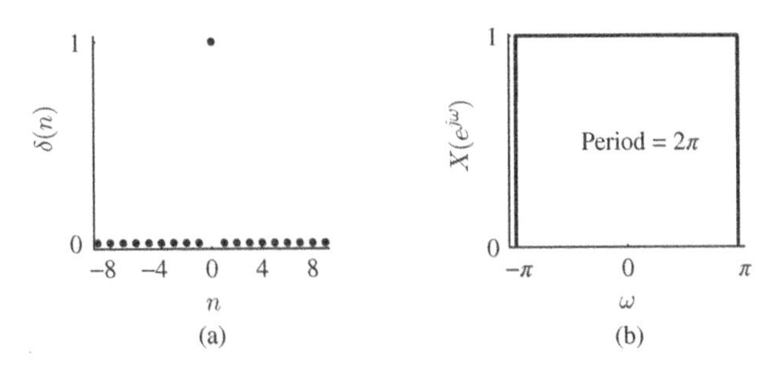

Figure 4.3 (a) The unit-impulse signal $\delta(n)$; (b) one period of the DTFT spectrum of $\delta(n)$

Solution

$$X(e^{j\omega}) = \sum_{n=-\infty}^{\infty} \delta(n)e^{-j\omega n} = 1 \quad \textit{and} \quad \delta(n) \leftrightarrow 1$$

That is, the unit-impulse signal is composed of complex sinusoids of all frequencies from $\omega = -\pi$ to $\omega = \pi$ in equal proportion, as shown in Figure 4.3(b). ■

Example 4.3 Find the DTFT of the sequence $x(n)$, whose nonzero values are given as

$$x(0) = -\frac{1}{4}, x(1) = \frac{1}{2}, x(2) = -\frac{1}{4}$$

Solution

The DTFT of $x(n)$ is the frequency response given by

$$X(e^{j\omega}) = \sum_{n=0}^{2} x(n)e^{-j\omega n}$$

$$= -\frac{1}{4} + \frac{1}{2}e^{-j\omega} - \frac{1}{4}e^{-j2\omega} = e^{-j\omega}\left(-\frac{1}{4}e^{j\omega} + \frac{1}{2} - \frac{1}{4}e^{-j\omega}\right) = \frac{1}{2}e^{-j\omega}(1 - \cos(\omega))$$

The plot of the magnitude of the frequency response is shown in Figure 4.4. The magnitude response is even-symmetric, which is always the case for real signals. Therefore, the frequency response is plotted only for the range from 0 to π radians. The frequency response indicates how the spectrum of a signal is modified when filtered with the specific filter. For example, the frequency response shown in Figure 4.4 is that of a highpass filter. The low-frequency components of a signal are attenuated more than the high-frequency components, when the signal passes through this filter. Verify that the magnitude of the frequency response is 0 at $\omega = 0$ and 1 at $\omega = \pi$. The phase response $\theta(\omega) = -\omega$ is a straight line. This is called linear-phase response, which is important in many applications in that the filtered signal has no phase distortion except for a delay. The linear-phase response is produced by filters whose coefficients are even- or odd-symmetric.

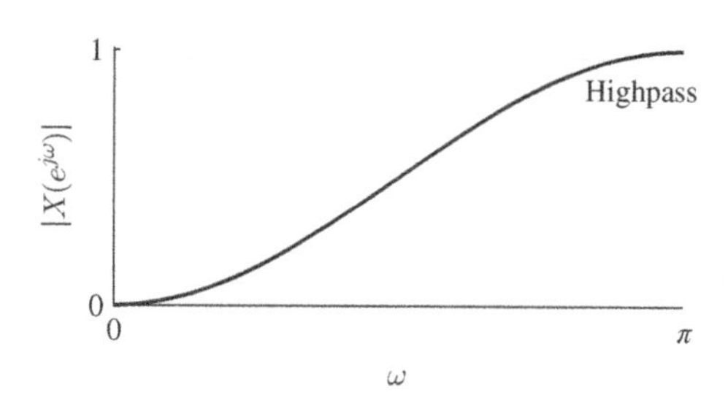

Figure 4.4 The magnitude of the frequency response of a highpass filter

■

Let the spectrum of a signal $x(n)$ be $X(e^{j\omega})$. Then, $x(-n) \leftrightarrow X(e^{-j\omega})$. That is, the time reversal of a signal results in its spectrum also reflected about the vertical axis at the origin. This result is obtained if we replace n by $-n$ and ω by $-\omega$ in the DTFT definition.

Example 4.4 Find the DTFT of the sequence $l(n)$, whose nonzero values are given as

$$\{l(0) = (1 - \sqrt{3}), l(1) = (3 - \sqrt{3}), l(2) = (3 + \sqrt{3}), l(3) = (1 + \sqrt{3})\}$$

Find also the DTFT of $h(n) = (-1)^n l(n)$ and $hb(n) = (-1)^n l(-n)$.

Solution

$$L(e^{j\omega}) = \sum_{n=0}^{3} l(n)e^{-j\omega n}$$
$$= (1-\sqrt{3}) + (3-\sqrt{3})e^{-j\omega} + (3+\sqrt{3})e^{-j2\omega} + (1+\sqrt{3})e^{-j3\omega}$$

The magnitude of the frequency response, shown in Figure 4.5 by thick lines, is that of a lowpass filter. Verify that the magnitude of the frequency response is 8 at $\omega = 0$ and 0 at $\omega = \pi$.

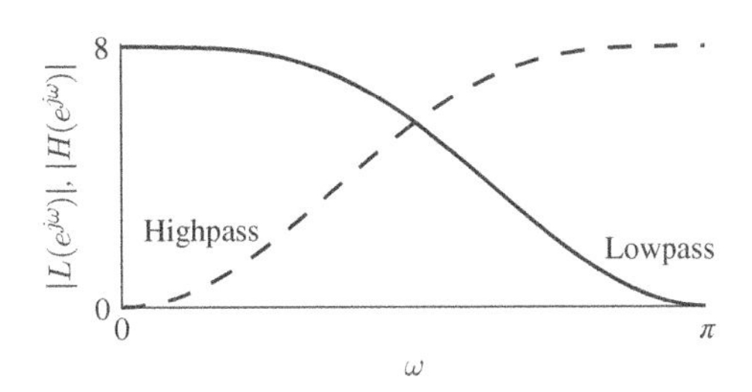

Figure 4.5 The magnitude of the frequency response

Let us form the sequence $h(n) = (-1)^n l(n)$. Then,

$$\{h(0) = (1-\sqrt{3}), h(1) = -(3-\sqrt{3}), h(2) = (3+\sqrt{3}), h(3) = -(1+\sqrt{3})\}$$

The DTFT of $h(n)$ is the frequency response given by

$$H(e^{j\omega}) = \sum_{n=0}^{3} h(n)e^{-j\omega n}$$
$$= (1-\sqrt{3}) - (3-\sqrt{3})e^{-j\omega} + (3+\sqrt{3})e^{-j2\omega} - (1+\sqrt{3})e^{-j3\omega}$$

The magnitude of the frequency response, shown in Figure 4.5 by dashed lines, is that of a highpass filter. The shift of the frequency response of the lowpass filter by π radians results in a highpass frequency response. Verify that the magnitude of the frequency response is 0 at $\omega = 0$ and 8 at $\omega = \pi$.

Let us form the sequence $hb(n) = (-1)^n l(-n)$. Then,

$$\{hb(0) = (1-\sqrt{3}), hb(-1) = -(3-\sqrt{3}), hb(-2) = (3+\sqrt{3}), hb(-3) = -(1+\sqrt{3})\}$$

The DTFT of $hb(n)$ is the frequency response given by

$$Hb(e^{j\omega}) = \sum_{n=0}^{3} hb(-n)e^{j\omega n}$$
$$= (1-\sqrt{3}) - (3-\sqrt{3})e^{j\omega} + (3+\sqrt{3})e^{j2\omega} - (1+\sqrt{3})e^{j3\omega}$$

The time-reversal operation does not change the magnitude of the frequency response $|H(e^{j\omega})|$. ■

4.3.1 Convolution

The convolution of signals $x(n)$ and $h(n)$ is defined, in Chapter 3, as

$$y(n) = \sum_{k=-\infty}^{\infty} x(k)h(n-k)$$

The convolution of $h(n)$ with a complex exponential $e^{j\omega_0 n}$ is given as

$$\sum_{k=-\infty}^{\infty} h(k)e^{j\omega_0(n-k)} = e^{j\omega_0 n}\sum_{k=-\infty}^{\infty} h(k)e^{-j\omega_0 k} = H(e^{j\omega_0})e^{j\omega_0 n}$$

As an arbitrary $x(n)$ is reconstructed by the inverse DTFT as $x(n) = \frac{1}{2\pi}\int_{-\pi}^{\pi} X(e^{j\omega})e^{j\omega n}\,d\omega$, the convolution of $x(n)$ and $h(n)$ is given by $y(n) = \frac{1}{2\pi}\int_{-\pi}^{\pi} X(e^{j\omega})H(e^{j\omega})e^{j\omega n}\,d\omega$, where $X(e^{j\omega})$ and $H(e^{j\omega})$ are, respectively, the DTFT of $x(n)$ and and $h(n)$. The inverse DTFT of $X(e^{j\omega})H(e^{j\omega})$ is the convolution of $x(n)$ and $h(n)$. Therefore, we get the transform pair

$$\sum_{k=-\infty}^{\infty} x(k)h(n-k) = \frac{1}{2\pi}\int_{-\pi}^{\pi} X(e^{j\omega})H(e^{j\omega})e^{j\omega n}\,d\omega \leftrightarrow X(e^{j\omega})H(e^{j\omega})$$

Example 4.5 Find the convolution of the rectangular signal $x(n)$ with itself.

$$x(n) = \begin{cases} 1 & \text{for } n = 0, 1 \\ 0 & \text{otherwise} \end{cases}$$

Solution

The rectangular signal and its magnitude spectrum are shown, respectively, in Figures 4.6(a) and (b).The DTFT of the rectangular signal is $2e^{-j\omega/2}\cos(\omega/2)$. The DTFT of the convolution of this signal with itself is, due to the convolution property, $2e^{-j\omega}(1 + \cos(\omega))$ (the product of the DTFT of the rectangular signal with itself). As the convolution of a rectangular signal with itself is a triangular signal, this DTFT is that of a triangular signal. The triangular signal and its magnitude spectrum are shown, respectively, in Figures 4.6(c) and (d). ■

The frequency response of the triangular signal is more closer to an ideal lowpass frequency response compared to that of the first signal. The price that is paid for a better frequency response is that the impulse response is longer, requiring more computation.

4.3.2 Convolution in the DWT

Apart from its major use in the implementation of the DWT, convolution is also used in the design of filters. We present an example of the use of the convolution operation in the DWT. Given a set of four filters, the following two conditions must be satisfied for the filters to decompose and reconstruct the input to a filter bank perfectly.

$$\tilde{L}(e^{j\omega})L(e^{j\omega}) + \tilde{H}(e^{j\omega})H(e^{j\omega}) = 2e^{-jK\omega}$$

$$\tilde{L}(e^{j\omega})L(e^{j(\omega+\pi)}) + \tilde{H}(e^{j\omega})H(e^{j(\omega+\pi)}) = 0$$

where K is a constant. These conditions are called perfect reconstruction conditions.

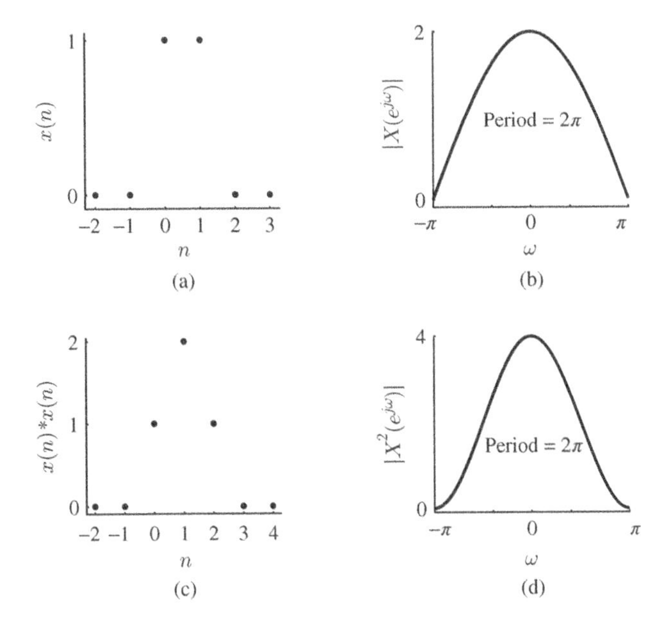

Figure 4.6 (a) The rectangular signal and (b) its spectrum; (c) the triangular signal, which is the convolution of the signal in (a) with itself, and (d) its spectrum

Example 4.6 The impulse response of a set of four filters are

$$\{l(n), n = 0, 1, 2, 3, 4, 5\} = \frac{1}{(8\sqrt{2})}\{-1, 1, 8, 8, 1, -1\}$$

$$\{h(n), n = 0, 1, 2, 3, 4, 5\} = \frac{1}{(8\sqrt{2})}\{0, 0, -8, 8, 0, 0\}$$

$$\{\tilde{l}(n), n = 0, 1, 2, 3, 4, 5\} = \frac{1}{(8\sqrt{2})}\{0, 0, 8, 8, 0, 0\}$$

$$\{\tilde{h}(n), n = 0, 1, 2, 3, 4, 5\} = \frac{1}{(8\sqrt{2})}\{-1, -1, 8, -8, 1, 1\}$$

Verify that they satisfy the perfect reconstruction conditions.

Solution

The frequency response of a filter is the DTFT of its impulse response. For example, the frequency response of the first filter is

$$L(e^{j\omega}) = \frac{1}{(8\sqrt{2})}(-1 + e^{-j\omega} + 8e^{-j2\omega} + 8e^{-j3\omega} + e^{-j4\omega} - e^{-j5\omega})$$

As polynomial multiplication is the same as convolution, the coefficients of the product polynomial can be found by the convolution of their coefficients. That is,

$$l(n) * \tilde{l}(n) = \{-8, 0, 72, 128, 72, 0, -8\}/128$$

$$h(n) * \tilde{h}(n) = \{8, 0, -72, 128, -72, 0, 8\}/128$$

The sum of the two convolutions is

$$l(n) * \tilde{l}(n) + h(n) * \tilde{h}(n) = \{0, 0, 0, 2, 0, 0, 0\}$$

The DTFT of this sequence is of the form $2e^{-jK\omega}$. The first of the two conditions is met.

Regarding the second condition, the term $L(e^{j(\omega+\pi)})$ corresponds to $(-1)^n l(n)$ in the time domain. Therefore,

$$((-1)^n l(n)) * \tilde{l}(n) = \{-8, -16, 56, 0, -56, 16, 8\}/128$$

$$((-1)^n h(n)) * \tilde{h}(n) = \{8, 16, -56, 0, 56, -16, -8\}/128$$

The sum of the two convolutions is

$$((-1)^n l(n)) * \tilde{l}(n) + ((-1)^n h(n)) * \tilde{h}(n) = \{0, 0, 0, 0, 0, 0, 0\}$$

The second of the two conditions is also met. ■

4.3.3 *Correlation*

The convolution of two signals $a(n)$ and $b(n)$ in the time domain corresponds to multiplication of their transforms in the frequency domain, $A(e^{j\omega})B(e^{j\omega})$. The correlation operation is the same as that of the convolution if one of the appropriate signal is time reversed. Therefore, the correlation of the two signals in the frequency domain is $A(e^{j\omega})B^*(e^{j\omega})$. If $a(n) = b(n)$, the autocorrelation of $a(n)$ in the frequency domain is

$$A(e^{j\omega})A^*(e^{j\omega}) = |A(e^{j\omega})|^2$$

4.3.4 *Correlation in the DWT*

We present an example of the use of the correlation operation in the DWT. Given a set of four filters, four relations, called double-shift orthogonality conditions, must be satisfied for the filters to be used in the DWT.

Example 4.7 The impulse response of a set of four filters is

$$\{l(n), n = -1, 0, 1, 2\} = \sqrt{2}\{-2, 6, 6, -2\}/8$$

$$\{h(n), n = -1, 0, 1, 2\} = \sqrt{2}\{-1, 3, -3, 1\}/8$$

$$\{\tilde{l}(n), n = -1, 0, 1, 2\} = \sqrt{2}\{1, 3, 3, 1\}/8$$

$$\{\tilde{h}(n), n = -1, 0, 1, 2\} = \sqrt{2}\{2, 6, -6, -2\}/8$$

Very that they satisfy the double-shift orthogonality conditions.

Solution

$$\tilde{L}(e^{j\omega})L^*(e^{j\omega}) + \tilde{L}(e^{j(\omega+\pi)})L^*(e^{j(\omega+\pi)}) = 2 \tag{4.8}$$

The term $\tilde{L}(e^{j\omega})L^*(e^{j\omega})$ corresponds to correlation of $\tilde{l}(n)$ and $l(n)$ in the time domain or convolution of $\tilde{l}(n)$ and $l(-n)$. Instead of using another symbol, we just write the correlation of $\tilde{l}(n)$ and $l(n)$ as $\tilde{l}(n) * l(-n)$. Therefore,

$$\tilde{l}(n) * l(-n) + ((-1)^n \tilde{l}(n)) * ((-1)^{-n} l(-n)) = 2$$

The convolution of the two terms is

$$\{-2, 0, 18, 32, 18, 0, -2\}/32$$

$$\{2, 0, -18, 32, -18, 0, 2\}/32$$

The sum of the two convolutions is

$$\{0, 0, 0, 2, 0, 0, 0\}$$

$$\tilde{H}(e^{j\omega})H^*(e^{j\omega}) + \tilde{H}(e^{j(\omega+\pi)})H^*(e^{j(\omega+\pi)}) = 2 \quad (4.9)$$

The convolution of the two terms is

$$\{2, 0, -18, 32, -18, 0, 2\}/32$$

$$\{-2, 0, 18, 32, 18, 0, -2\}/32$$

The sum of the two convolutions is

$$\{0, 0, 0, 2, 0, 0, 0\}$$

$$\tilde{H}(e^{j\omega})L^*(e^{j\omega}) + \tilde{H}(e^{j(\omega+\pi)})L^*(e^{j(\omega+\pi)}) = 0 \quad (4.10)$$

The convolution of the two terms is

$$\{-4, 0, 60, 0, -60, 0, 4\}/32$$

$$\{4, 0, -60, 0, 60, 0, -4\}/32$$

The sum of the two convolutions is

$$\{0, 0, 0, 0, 0, 0, 0\}$$

$$\tilde{L}(e^{j\omega})H^*(e^{j\omega}) + \tilde{L}(e^{j(\omega+\pi)})H^*(e^{j(\omega+\pi)}) = 0 \quad (4.11)$$

The convolution of the two terms is

$$\{1, 0, -3, 0, 3, 0, -1\}/32$$

$$\{-1, 0, 3, 0, -3, 0, 1\}/32$$

The sum of the two convolutions is

$$\{0, 0, 0, 0, 0, 0, 0\}$$

■

4.3.5 *Time Expansion*

As we have seen in Chapter 2, a signal is compressed or expanded by the scaling operation. Consider the case of signal expansion by a factor of 2. Let the spectrum of a signal $x(n)$ be $X(e^{j\omega})$. The upsampled signal $x_u(n)$ is defined as

$$x_u(2n) = \begin{cases} x(n) & \text{for} - \infty < \text{n} < \infty \\ 0 & \text{otherwise} \end{cases}$$

Then,

$$X_u(e^{j\omega}) = X(e^{j2\omega})$$

The DTFT of the sequence $x_u(n)$ is given by

$$X_u(e^{j\omega}) = \sum_{n=-\infty}^{\infty} x_u(n)e^{-j\omega n}$$

As we have nonzero input values only if $n = 2k, \quad k = 0, \pm 1, \pm 2, \ldots,$ we get

$$X_u(e^{j\omega}) = \sum_{k=-\infty}^{\infty} x_u(2k)e^{-j\omega 2k} = \sum_{k=-\infty}^{\infty} x(k)e^{-j\omega 2k} = X(e^{j2\omega})$$

Therefore,

$$x_u(n) \leftrightarrow X(e^{j2\omega})$$

The spectrum is compressed. That is, the spectral value at ω in the spectrum of the signal occurs at $\frac{\omega}{2}$ in the spectrum of its expanded version.

Example 4.8 The DTFT of the signal $x(n)$, shown in Figure 4.7(a) with its only nonzero values given as $x(-1) = 1$ and $x(1) = 1$, is

$$X(e^{j\omega}) = e^{j\omega} + e^{-j\omega} = 2\,\cos(\omega)$$

shown in Figure 4.7(b). Find the upsampled signal $x_u(n)$ and its spectrum.

Solution

Using the theorem, we get the DTFT of $x_u(n) = x(n/2)$, shown in Figure 4.7(c), as

$$X_u(e^{j\omega}) = X(e^{j2\omega}) = 2\,\cos(2\omega)$$

shown in Figure 4.7(d). This result can be verified from the DTFT definition. As the signal is expanded by a factor of 2, the spectrum is compressed by a factor of 2. The spectrum of the expanded signal is the concatenation of two copies of the spectrum of the original signal in the range $0 < \omega < 2\pi$. Therefore, the inverse DTFT of one-half of the spectrum of the expanded signal yields the original signal. ■

4.3.6 *Sampling Theorem*

Consider the sampling theorem in the frequency domain. Let the Fourier transform (FT) of the continuous signal $x(t)$ be $X(j\omega)$. If a signal $x(t)$ is bandlimited to f_m Hz ($|X(j\omega)| = 0$,

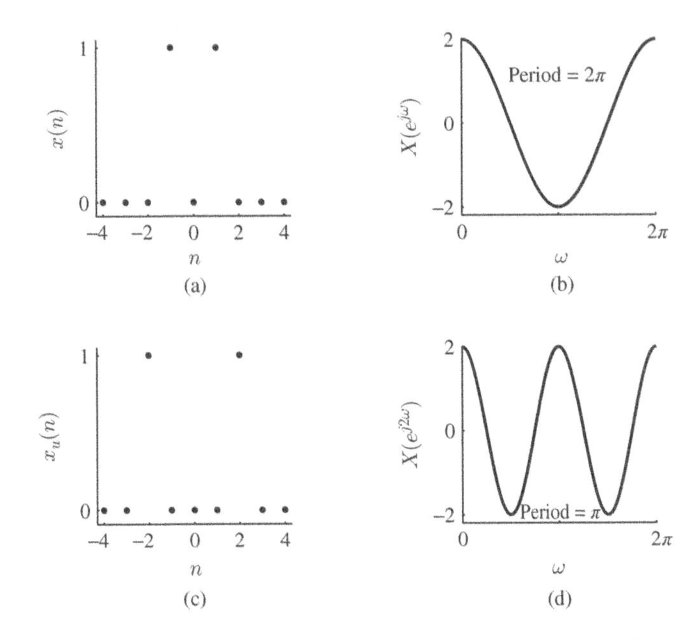

Figure 4.7 (a) Signal $x(n)$ and (b) its DTFT spectrum; (c) expanded version $x_u(n) = x(n/2)$ and (d) its DTFT spectrum

$|\omega| > 2\pi f_m$), then it can be reconstructed perfectly from its samples if sampled at a frequency $f_s > 2f_m$. Consider the FT pair

$$\frac{\sin(\omega_0 t)}{\pi t} \leftrightarrow u(\omega + \omega_0) - u(\omega - \omega_0)$$

With $\omega_0 = \pi/2$, the continuous sinc signal $\frac{\sin((\pi/2)t)}{\pi t}$ and its rectangular spectrum are shown, respectively, in Figures 4.8(a) and (b). Let the DTFT of the discrete signal $x(n)$, obtained by sampling $x(t)$ with a sampling interval of 1 second, be $X(e^{j\omega})$. Sampling a signal in the time domain makes its spectrum periodic in the frequency domain, and the effective frequency range of the spectrum is reduced. Consider the DTFT pair

$$\frac{\sin(\omega_0 n)}{\pi n}, \quad 0 < \omega_0 \leq \pi \leftrightarrow \begin{cases} 1 & \text{for } |\omega| < \omega_0 \\ 0 & \text{for } \omega_0 < |\omega| \leq \pi \end{cases}$$

Note the restriction on ω_0 in the DTFT pair. With $\omega_0 = \pi/2$, the discrete signal $\frac{\sin((\pi/2)n)}{\pi n}$ and its spectrum are shown, respectively, in Figures 4.8(c) and (d). With $\omega_0 = 2\pi/3$, the discrete signal and its spectrum are shown, respectively, in Figures 4.8(e) and (f). The point is, with a sampling interval of 1 second, the highest frequency component that can be represented unambiguously in the spectrum is π radians, as the spectrum is periodic of period 2π radians. Any frequency component higher than π radians will have aliased representation. A sampling interval of 1 second corresponds to sampling frequency 2π radians. Therefore, the sampling theorem states that the sampling frequency must be greater than two times that of the highest frequency component of a signal.

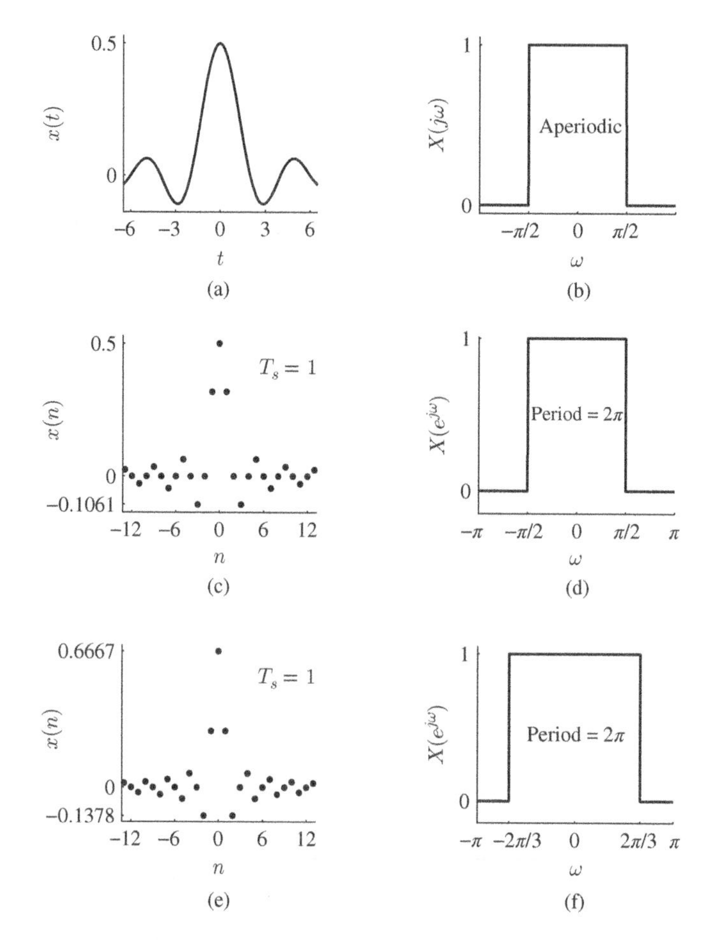

Figure 4.8 (a) The sinc function $\frac{\sin((\pi/2)t)}{\pi t}$; (b) its rectangular FT spectrum; (c) samples of (a) with sampling interval $T_s = 1$ seconds; (d) its DTFT spectrum; (e) samples of the sinc function $\frac{\sin((2\pi/3)t)}{\pi t}$ with sampling interval $T_s = 1$ seconds; (f) its DTFT spectrum

4.3.7 Parseval's Theorem

As the frequency-domain representation of a signal is an equivalent representation, the energy of a signal can also be expressed in terms of its spectrum.

$$E = \sum_{n=-\infty}^{\infty} |x(n)|^2 = \frac{1}{2\pi}\int_0^{2\pi} |X(e^{j\omega})|^2 d\omega$$

As $x(n)$ can be considered as the FS coefficients of $X(e^{j\omega})$, this expression is the same as that corresponding to the FS with the roles of the domains interchanged. The quantity $|X(e^{j\omega})|^2$ is called the energy spectral density of the signal, as $\frac{1}{2\pi}|X(e^{j\omega})|^2 d\omega$ is the signal energy over the infinitesimal frequency band ω to $\omega + d\omega$.

Example 4.9 Verify the Parseval's theorem for the triangular signal

$$x(n) = \begin{cases} 1 & \text{for } n = 0, 2 \\ 2 & \text{for } n = 1 \\ 0 & \text{otherwise} \end{cases}$$

and its DTFT $1 + 2e^{-j\omega} + e^{-j2\omega}$.

Solution

The energy of the signal, from its time-domain representation, is $1^2 + 2^2 + 1^2 = 6$. The energy of the signal, from its frequency-domain representation, is

$$\begin{aligned} E &= \frac{1}{2\pi}\int_0^{2\pi} |1 + 2e^{-j\omega} + e^{-j2\omega}|^2 d\omega \\ &= \frac{1}{2\pi}\int_0^{2\pi} ((1)(1) + (2e^{-j\omega})(2e^{j\omega}) + e^{-j2\omega}e^{j2\omega})d\omega \\ &= \frac{1}{2\pi}\int_0^{2\pi} (1^2 + 2^2 + 1^2)d\omega = 6 \end{aligned}$$

■

4.4 Approximation of the DTFT

In studying any transform, we derive the transform of some well-defined signals analytically. That gives us understanding. In practice, the amplitude profile of signals is arbitrary. Therefore, the transform has to be approximated using numerical methods. The DWT is an example of a transform that can be computed by a digital computer. In studying the wavelet transform, as is the case in almost all aspects of signals and systems, we have to use the Fourier analysis, and it is approximated by the DFT. We use the DFT to approximate the DTFT in some chapters. The samples of the continuous DTFT spectrum can be approximated by taking the DFT of the time-domain samples. Let

$$\{x(0) = 3, x(1) = 2, x(2) = 1, x(3) = 4\}$$

The DTFT of the signal is

$$X(e^{j\omega}) = 3 + 2e^{-j\omega} + e^{-j2\omega} + 4e^{-j3\omega}$$

The samples of the spectrum at $\omega = 0, \pi/2, \pi, 3\pi/2$ are $\{10, 2 + j2, -2, 2 - j2\}$. We can get these samples by taking the 4-point DFT of the input. The IDFT of these samples yields the time-domain input. Consider finding the inverse DTFT of the spectrum

$$\frac{1}{2}\left(X\left(e^{j\frac{\omega}{2}}\right) + X\left(e^{j\left(\frac{\omega}{2}+\pi\right)}\right)\right)$$

where

$$X(e^{j\omega}) = 3 + 2e^{-j\omega} + e^{-j2\omega} + 4e^{-j3\omega}$$

Now,

$$X\left(e^{j\frac{\omega}{2}}\right) = 3 + 2e^{-j\frac{\omega}{2}} + e^{-j\omega} + 4e^{-j3\frac{\omega}{2}}$$

and

$$X\left(e^{j\left(\frac{\omega}{2}+\pi\right)}\right) = 3 + 2e^{-j\left(\frac{\omega}{2}+\pi\right)} + e^{-j2\left(\frac{\omega}{2}+\pi\right)} + 4e^{-j3\left(\frac{\omega}{2}+\pi\right)}$$
$$= 3 - 2e^{-j\frac{\omega}{2}} + e^{-j\omega} - 4e^{-j3\frac{\omega}{2}}$$

Then,

$$\frac{1}{2}\left(X\left(e^{j\frac{\omega}{2}}\right) + X\left(e^{j\left(\frac{\omega}{2}+\pi\right)}\right)\right) = 3 + e^{-j\omega}$$

The inverse DTFT of this expression is $\{x(0) = 3, x(1) = 1\}$, which is the downsampled version of $x(n)$. The samples of $3 + e^{-j\omega}$ at $\omega = 0, \pi$ are $\{4, 2\}$. The IDFT of these samples is $\{x(0) = 3, x(1) = 1\}$. These simple problems are presented for understanding. In practice, the DFT is required to approximate arbitrary DTFT spectra.

4.5 The Fourier Transform

The limit of the Fourier series, as the period T tends to infinity, is the FT. The periodic time-domain waveform becomes aperiodic and the line spectrum becomes continuous as the fundamental frequency tends to zero. The FT $X(j\omega)$ of $x(t)$ is defined as

$$X(j\omega) = \int_{-\infty}^{\infty} x(t)e^{-j\omega t}\, dt \tag{4.12}$$

The inverse FT $x(t)$ of $X(j\omega)$ is defined as

$$x(t) = \frac{1}{2\pi}\int_{-\infty}^{\infty} X(j\omega)e^{j\omega t}\, d\omega \tag{4.13}$$

The FT spectrum is composed of components of all frequencies ($-\infty < \omega < \infty$). The amplitude of any component is $X(j\omega)\ d\omega/(2\pi)$, which is infinitesimal. The FT is a relative amplitude spectrum.

4.6 Summary

- Fourier analysis, along with its generalizations, forms the core of the transform analysis of signals and systems.
- Fourier analysis is a tool to transform a time-domain signal to the frequency domain and back. An arbitrary signal is decomposed into a sum of sinusoids in Fourier analysis.
- It is easier to interpret the characteristics of signals and systems in the frequency domain.
- Convolution in the time domain becomes multiplication in the frequency domain.
- The DFT version of Fourier analysis is most often used in practice, due to its discrete nature in both the domains and the availability of fast algorithms for its computation.
- The DTFT version of Fourier analysis is primarily used in the analysis and design of DWT filters.
- The FT version of Fourier analysis is used in the analysis and design of continuous aperiodic signals. Usually, both the signal and the spectrum are continuous and aperiodic.
- The FT and DTFT can be approximated by the DFT.

Exercises

4.1 Find the value of a so that $x(n)$ is approximated in the least square error sense, if it is reconstructed by the DC component alone with coefficient a. Compute the DFT of $x(n)$ and verify the result.

4.1.1 $\{x(0) = 1, x(1) = -2, x(2) = 2, x(3) = 4\}$.
***4.1.2** $\{x(0) = 1, x(1) = 3, x(2) = 2, x(3) = 1\}$.
4.1.3 $\{x(0) = 2, x(1) = -4, x(2) = 3, x(3) = 2\}$.

4.2 Compute the DFT of $x(n)$ and verify the Parseval's theorem.

4.2.1 $\{x(0) = 1, x(1) = 3, x(2) = 1, x(3) = -1\}$.
***4.2.2** $\{x(0) = 2, x(1) = 4, x(2) = 1, x(3) = 3\}$.
4.2.3 $\{x(0) = 4, x(1) = -1, x(2) = 3, x(3) = 2\}$.

4.3 Find the DTFT of $x(n)$. Compute four samples of the DTFT spectrum of $x(n)$ and verify that they are the same as that of the 4-point DFT of $x(n)$. Compute the IDFT of the four spectral samples and verify that we get back $x(n)$.

4.3.1 $\{x(0) = 1, x(1) = 1, x(2) = 1, x(3) = 1\}$.
4.3.2 $\{x(0) = 1, x(1) = 0, x(2) = 0, x(3) = 0\}$.
4.3.3 $\{x(0) = 1, x(1) = -1, x(2) = 1, x(3) = -1\}$.
***4.3.4** $\{x(0) = 0, x(1) = 1, x(2) = 0, x(3) = -1\}$.
4.3.5 $\{x(0) = 1, x(1) = 0, x(2) = -1, x(3) = 0\}$.
4.3.6 $\{x(0) = 1, x(1) = 3, x(2) = 3, x(3) = 4\}$.
4.3.7 $\{x(0) = 2, x(1) = 4, x(2) = -1, x(3) = 3\}$.

4.4 Find the DTFT of $x(n)$. Using the time expansion property, find the DTFT of $x(n/2)$. Use the IDFT to find the inverse of the spectrum and verify that $x(n/2)$ is obtained.

4.4.1 $\{x(0) = 1, x(1) = 2\}$.
4.4.2 $\{x(0) = 3, x(1) = -1\}$.
***4.4.3** $\{x(0) = 2, x(1) = 3\}$.
4.4.4 $\{x(0) = 1, x(1) = -4\}$.

4.5 Given the impulse response of a set of four filters, verify that the following two identities hold.

$$\tilde{L}(e^{j\omega})L(e^{j\omega}) + \tilde{H}(e^{j\omega})H(e^{j\omega}) = 2e^{-jK\omega}$$

$$\tilde{L}(e^{j\omega})L(e^{j(\omega+\pi)}) + \tilde{H}(e^{j\omega})H(e^{j(\omega+\pi)}) = 0$$

where K is a constant.

4.5.1

$$\{l(n), n = 0, 1, 2, 3\} = \{-0.3536, 1.0607, 1.0607, -0.3536\}$$

$$\{h(n), n = 0, 1, 2, 3\} = \{-0.1768, 0.5303, -0.5303, 0.1768\}$$

$$\{\tilde{l}(n), n = 0, 1, 2, 3\} = \{0.1768, 0.5303, 0.5303, 0.1768\}$$

$$\{\tilde{h}(n), n = 0, 1, 2, 3\} = \{-0.3536, -1.0607, 1.0607, 0.3536\}.$$

4.5.2

$$\{l(n), n = 0, 1, \ldots, 7\} = \{0.0663, -0.1989, -0.1547, 0.9944, 0.9944, -0.1547, -0.1989, 0.0663\}$$

$$\{h(n), n = 0, 1, \ldots, 7\} = \{0, 0, -0.1768, 0.5303, -0.5303, 0.1768, 0, 0\}$$

$$\{\tilde{l}(n), n = 0, 1, \ldots, 7\} = \{0, 0, 0.1768, 0.5303, 0.5303, 0.1768, 0, 0\}$$

$$\{\tilde{h}(n), n = 0, 1, \ldots, 7\} = \{0.0663, 0.1989, -0.1547, -0.9944, 0.9944, 0.1547, -0.1989, -0.0663\}.$$

4.6 Given the impulse response of a set of four filters, verify that the following four identities hold.

$$\tilde{L}(e^{j\omega})L^*(e^{j\omega}) + \tilde{L}(e^{j(\omega+\pi)})L^*(e^{j(\omega+\pi)}) = 2$$

$$\tilde{H}(e^{j\omega})H^*(e^{j\omega}) + \tilde{H}(e^{j(\omega+\pi)})H^*(e^{j(\omega+\pi)}) = 2$$

$$\tilde{H}(e^{j\omega})L^*(e^{j\omega}) + \tilde{H}(e^{j(\omega+\pi)})L^*(e^{j(\omega+\pi)}) = 0$$

$$\tilde{L}(e^{j\omega})H^*(e^{j\omega}) + \tilde{L}(e^{j(\omega+\pi)})H^*(e^{j(\omega+\pi)}) = 0$$

$$l(n), n = -1, 0, 1\} = \{0.3536, 0.7071, 0.3536\}$$

$$\{h(n), n = -1, 0, 1, 2, 3\} = \{0.1768, 0.3536, -1.0607, 0.3536, 0.1768\}$$

$$\{\tilde{l}(n), n = -2, -1, 0, 1, 2\} = \{-0.1768, 0.3536, 1.0607, 0.3536, -0.1768\}$$

$$\{\tilde{h}(n), n = 0, 1, 2\} = \{0.3536, -0.7071, 0.3536\}.$$

4.7 Find the DTFT of $x(n)$ and verify the Parseval's theorem.

4.7.1 $\{x(0) = 1, x(1) = 3, x(2) = 1, x(3) = -1\}$.
***4.7.2** $\{x(0) = 1, x(1) = -2, x(2) = -1, x(3) = 2\}$.
4.7.3 $\{x(0) = 4, x(1) = -2, x(2) = -1, x(3) = -4\}$.

5

The z-Transform

The z-transform is a generalization of the Fourier analysis in that it uses a larger set of basis functions. The set becomes larger because the basis functions include exponentially varying amplitude sinusoids. Obviously, the Fourier analysis, which uses constant-amplitude sinusoids as basis functions, is a special case of the z-transform, if the z-transform exists on the unit circle. We can analyze a broader class of signals using the z-transform. In addition, it is easier to manipulate the resulting expressions involving the variable z. It is particularly useful in discrete system analysis and design. As the signals analyzed with the DWT are of finite length, we present only a small part of the z-transform analysis that is essential for our purposes.

5.1 The z-Transform

The z-transform $X(z)$ of a discrete signal $x(n)$ is defined as

$$\begin{aligned} X(z) &= \sum_{n=-\infty}^{\infty} x(n)z^{-n} \\ &= \cdots + x(-2)z^2 + x(-1)z + x(0) + x(1)z^{-1} + x(2)z^{-2} + \cdots \end{aligned} \tag{5.1}$$

where z is a complex variable. The z-transform is defined in the region of its convergence of the complex plane. Constant-amplitude sinusoids correspond to the independent variable $z = e^{j\omega}$ on the unit circle. Decaying-amplitude sinusoids correspond to z inside the unit circle. Growing-amplitude sinusoids correspond to z outside the unit circle. Writing the complex variable z in various forms, we get

$$z^n = r^n e^{j\omega n} = r^n(\cos(\omega n) + j\ \sin(\omega n))$$

The basis functions are the same as those used in DTFT with an exponential term r^n. On the unit circle ($r = 1$), the z-transform, if it exists, reverts to the DTFT, as

$$z^{-n} = 1^{-n}e^{-j\omega n} = e^{-j\omega n}$$

Discrete Wavelet Transform: A Signal Processing Approach, First Edition. D. Sundararajan.

Companion Website: www.wiley.com/go/sundararajan/wavelet

Therefore, the DTFT is obtained from the z-transform simply by replacing z by $e^{-j\omega}$, if the z-transform converges on the unit circle. Depending on the context, we use the z-transform or the DTFT representation of a signal. The region of convergence of the z-transform of finite-duration signals, which are of interest in DWT, is the entire z-plane, except possibly at the points $z = 0$ and/or $z = \infty$.

Example 5.1 Find the z-transform of the unit-impulse signal, $\delta(n)$.

Solution

Using the definition, we get

$$X(z) = 1, \text{ for all } z \quad \text{and} \quad \delta(n) \leftrightarrow 1, \text{ for all } z$$

The transform pair for a shifted impulse $\delta(n-m)$ is

$$\delta(n-m) \leftrightarrow z^{-m}, \quad \begin{cases} \text{entire } z\text{-plane except } z = 0 & \text{for } m > 0 \\ \text{entire } z\text{-plane except } z = \infty & \text{for } m < 0 \end{cases}$$

■

Example 5.2 Find the z-transform of the finite sequence with its only nonzero samples specified as $\{x(0) = 1, x(1) = 2, x(2) = -6, x(3) = 2, x(4) = 1\}$.

Solution

Using the definition, we get

$$X(z) = 1 + 2z^{-1} - 6z^{-2} + 2z^{-3} + z^{-4} = \frac{z^4 + 2z^3 - 6z^2 + 2z + 1}{z^4}, \quad |z| > 0$$

The time-domain sequence $x(n)$ corresponding to a z-transform can be readily obtained by inspection for finite-duration sequences, as can be seen from the right side of the last expression. ■

5.2 Properties of the z-Transform

Properties present the z-domain effect of time-domain characteristics and operations on signals and vice versa. In addition, they are used to find new transform pairs more easily.

5.2.1 *Linearity*

It is often advantageous to decompose a complex sequence into a linear combination of simpler sequences in the manipulation of sequences and their transforms. If $x(n) \leftrightarrow X(z)$ and $y(n) \leftrightarrow Y(z)$, then

$$ax(n) + by(n) \leftrightarrow aX(z) + bY(z)$$

where a and b are arbitrary constants. The z-transform of a linear combination of sequences is the same linear combination of the z-transforms of the individual sequences. This property is due to the linearity of the defining summation operation of the transform.

5.2.2 *Time Shift of a Sequence*

The time shift property is used to express the transform of the shifted version, $x(n+m)$, of a sequence $x(n)$ in terms of its transform $X(z)$. If $x(n) \leftrightarrow X(z)$ and m is a positive or negative integer, then

$$x(n+m) \leftrightarrow z^m X(z)$$

Consider the sequence $x(n)$ with $x(-2) = 1, x(-1) = 2, x(0) = 6, x(1) = 2, x(2) = 1$, and $x(n) = 0$ otherwise. The transform of $x(n)$ is

$$X(z) = z^2 + 2z + 6 + 2z^{-1} + z^{-2}$$

The transform of $x(n+1)$ is

$$z^3 + 2z^2 + 6z + 2 + z^{-1} = zX(z)$$

The transform of $x(n-2)$ is

$$1 + 2z^{-1} + 6z^{-2} + 2z^{-3} + z^{-4} = z^{-2}X(z)$$

5.2.3 *Convolution*

If $x(n) \leftrightarrow X(z)$ and $h(n) \leftrightarrow H(z)$, then

$$y(n) = \sum_{m=-\infty}^{\infty} h(m)x(n-m) \leftrightarrow Y(z) = H(z)X(z)$$

The DTFT of $x(n)r^{-n}$ is the z-transform $X(z)$ of $x(n)$. The convolution of $x(n)r^{-n}$ and $h(n)r^{-n}$ corresponds to $X(z)H(z)$ in the frequency domain. The inverse DTFT of $X(z)H(z)$, therefore, is the convolution of $x(n)r^{-n}$ and $h(n)r^{-n}$, given by

$$\sum_{m=-\infty}^{\infty} x(m)r^{-m}h(n-m)r^{(-n+m)} = r^{-n} \sum_{m=-\infty}^{\infty} x(m)h(n-m) = r^{-n}(x(n) * h(n))$$

As finding the inverse z-transform is the same as finding the inverse DTFT in addition to multiplying the signal by r^n, we get the convolution of $x(n)$ and $h(n)$ by finding the inverse z-transform of $X(z)H(z)$.

Consider the two sequences and their transforms $x(n) = \{1, -1\} \leftrightarrow X(z) = 1 - z^{-1}$ and $h(n) = \{1, -1\} \leftrightarrow H(z) = 1 - z^{-1}$. The convolution of the sequences, in the z-domain, is given by the product of their transforms,

$$X(z)H(z) = (1 - z^{-1})(1 - z^{-1}) = (1 - 2z^{-1} + z^{-2})$$

The inverse transform of $X(z)H(z)$ is the convolution of the sequences in the time domain, and it is $\{1, -2, 1\}$.

For 2-D signals, the convolution relation in the z-domain is given by

$$Y(z_1, z_2) = X(z_1, z_2)H(z_1, z_2)$$

If $H(z_1, z_2)$ is separable,

$$H(z_1, z_2) = H(z_1)H(z_2)$$

then

$$Y(z_1, z_2) = X(z_1, z_2)H(z_1, z_2) = X(z_1, z_2)H(z_1)H(z_2)$$

As convolution is commutative, either the row transform of the signal can be carried out first followed by the column transform of the resulting signal or vice versa.

5.3 Summary

- The z-transform is a generalization of the Fourier analysis.
- The basis signals include exponentially varying amplitude sinusoids, in contrast to the constant-amplitude sinusoids used in Fourier analysis.
- The z-transform is particularly useful in the analysis and design of discrete systems.
- It is easier to manipulate expressions involving the variable z compared to those involving the variable $e^{j\omega}$.

Exercises

5.1 The nonzero values of a sequence $x(n)$ are given. Find the z-transform of

$$x(n), x(n+2), x(n-3)$$

5.1.1 $\{x(-3) = 1, x(0) = 1, x(3) = -1\}$.
5.1.2 $\{x(-1) = -1, x(0) = -2, x(2) = -4\}$.
***5.1.3** $\{x(-2) = 4, x(0) = 3, x(1) = 3\}$.
5.1.4 $\{x(-3) = 3, x(0) = 4, x(4) = -2\}$.
5.1.5 $\{x(-2) = -2, x(0) = -5, x(1) = -4\}$.

5.2 Find the inverse z-transform of
5.2.1 $X(z) = z^2 - 1 - 2z^{-2} + 3z^{-4}$.
5.2.2 $X(z) = 2z^3 + 2 - 2z^{-5}$.
5.2.3 $X(z) = -2z - 3 - 3z^{-1} - z^{-8}$.
***5.2.4** $X(z) = 3z^5 + 1 + z^{-1} - 4z^{-3}$.
5.2.5 $X(z) = z^4 + z^{-2} + z^{-3}$.

5.3 The nonzero values of two sequences $x(n)$ and $h(n)$ are given. Using the z-transform, find the convolution of the sequences $y(n) = x(n) * h(n)$.
5.3.1 $\{x(-2) = 2, x(0) = 1, x(4) = -2\}$ and $\{h(-1) = 3, h(2) = -4\}$.
5.3.2 $\{x(-1) = 3, x(0) = -2\}$ and $\{h(0) = -1, h(2) = 2\}$.
5.3.3 $\{x(1) = 3, x(2) = -1\}$ and $\{h(-1) = -2, h(1) = 1\}$.
5.3.4 $\{x(0) = -4, x(2) = 3\}$ and $\{h(-2) = 1, h(0) = -2\}$.
***5.3.5** $\{x(-3) = 3, x(0) = -4\}$ and $\{h(-3) = 3, h(-1) = 2\}$.

6

Finite Impulse Response Filters

In everyday usage, a filter is a device that removes the undesirable part of something that passes through it. A coffee filter removes coffee grounds. A water filter destroys bacteria and removes salts. In signal processing, a filter modifies the spectrum of a signal in a desired manner. Filters can be designed to any specified frequency response. However, there are some standard types. In particular, the lowpass and highpass filters are of primary interest in the implementation of the DWT. In the time domain, a linear system is characterized by its response to an impulse, assuming that the system is initially relaxed (zero initial conditions). One way of classifying linear time-invariant filters is by the duration of their impulse response $h(n)$. If the impulse response is of a finite duration, then it is called a finite impulse response (FIR) filter. Causal FIR filters, characterized by

$$h(n) = 0, \quad n < 0 \text{ and } n \geq N$$

are of primary interest in DWT implementation. This type of filters has the advantages of linear phase response, inherent stability, and less susceptibility to coefficient quantization. In Section 6.1, the characterization of ideal digital filters is presented. In Section 6.2, the required symmetry of the impulse response of linear-phase FIR filters is studied.

6.1 Characterization

A causal FIR filter is characterized by a difference equation that is a linear combination of the products of the impulse response (filter coefficients) with present and past input samples only. The output of this type of filter of length N (order $N-1$) is given by the convolution sum or the difference equation

$$y(n) = \sum_{k=0}^{N-1} h(k)x(n-k) = \sum_{k=0}^{N-1} x(k)h(n-k)$$

where $y(n)$, $h(n)$, and $x(n)$ are, respectively, the output, the impulse response, and the input of the filter. The difference equation of a second-order FIR filter is given as

$$y(n) = h(0)x(n) + h(1)x(n-1) + h(2)x(n-2)$$

Discrete Wavelet Transform: A Signal Processing Approach, First Edition. D. Sundararajan.

Companion Website: www.wiley.com/go/sundararajan/wavelet

At each instant n, the weighted sum of the most recent three input values by the filter coefficients is the output. As the output is related only by the input and the impulse response, the difference equation is equivalent to convolution-summation. By the proper design of the set of coefficients of a filter, the weighting of the input samples results in an output with a desired spectrum. A digital filter is made up of three basic components. They are multipliers, adders, and unit delays. The delays are used to create delayed versions of the signal. These signals are multiplied by the impulse response values, called the coefficients, of the filter and summed to yield the output. The design of a filter is the determination of its impulse response so that the sum of the weighted input samples produces the desired response.

The impulse response is an important characterization of a filter. If the input to the filter is the unit-impulse, then

$$y(n) = \sum_{k=0}^{N-1} h(k)\delta(n-k) = h(n)$$

The transform of the impulse response of the filter is its transfer function. As convolution in the time domain corresponds to multiplication in the frequency domain, it is easier to interpret the operation of a filter in the frequency domain. With the z-transforms of the input $x(n)$, output $y(n)$, and the impulse response $h(n)$ denoted, respectively, by $X(z)$, $Y(z)$, and $H(z)$, the output of the filter is given by

$$Y(z) = X(z)H(z)$$

in the z-domain and

$$Y(e^{j\omega}) = X(e^{j\omega})H(e^{j\omega})$$

in the frequency domain. With the transform of the unit-impulse $X(z) = 1$, the z-transform transfer function of a second-order FIR filter is given as

$$H(z) = \frac{Y(z)}{X(z)} = h(0) + h(1)z^{-1} + h(2)z^{-2}$$

The transfer function has no poles except at $z = 0$, and hence, FIR filters are inherently stable. The transfer function with respect to the DTFT or the frequency response of a second-order FIR filter is given as

$$H(e^{j\omega}) = \frac{Y(e^{j\omega})}{X(e^{j\omega})} = h(0) + h(1)e^{-j\omega} + h(2)e^{-j2\omega}$$

Filters are classified according to their frequency responses, such as lowpass, highpass, and bandpass filter. The frequency response characterizes the effect of a system on an input sinusoid of any frequency. The magnitude response indicates the change in the amplitude of the sinusoid. The phase response indicates the change in phase. While it is not physically realizable, these types of filters are introduced through their ideal frequency-response characteristics.

6.1.1 Ideal Lowpass Filters

The magnitude of the frequency response, which is even-symmetric and periodic with period 2π, of an ideal digital lowpass filter is shown in Figure 6.1. Due to the symmetry, the specification of the response over the period 0 to π, shown by thick lines, characterizes a filter. The

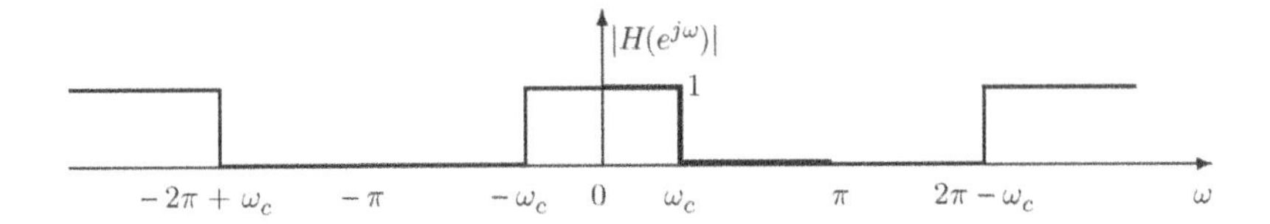

Figure 6.1 Magnitude of the periodic frequency response of an ideal digital lowpass filter

response is specified as

$$|H(e^{j\omega})| = \begin{cases} 1 & \text{for } 0 \le \omega \le \omega_c \\ 0 & \text{for } \omega_c < \omega \le \pi \end{cases}$$

The filter passes frequency components of a signal with unity gain (or a constant gain) in the range $\omega = 0$ to $\omega = \omega_c$ (the passband of the filter) and rejects everything else. It is obvious, as the output of the filter, in the frequency domain, is given by $Y(e^{j\omega}) = H(e^{j\omega})X(e^{j\omega})$. The magnitudes of the frequency components of the signal, $X(e^{j\omega})$, with frequencies up to ω_c are multiplied by 1, and the rest are multiplied by zero. The range of frequencies from 0 to ω_c is called the passband and that from ω_c to π is called the stopband. This ideal filter model is physically unrealizable because it requires a noncausal system. Realizable filters approximate this model to satisfy the required specifications. Typical frequency responses of practical filters are shown in this and other chapters. There is a transition band, between the passband and the stopband, in the frequency response of practical filters. Note that, due to periodicity, the digital lowpass filter, unlike the analog filter, rejects signal components in the frequency range from ω_c to π (or half the sampling frequency) only. It is assumed that the signal has components in the frequency range from 0 to π radians only. Otherwise, the signal will be distorted due to aliasing effect.

6.1.2 *Ideal Highpass Filters*

The magnitude of the periodic frequency response of an ideal digital highpass filter is shown in Figure 6.2. The response is specified as

$$|H(e^{j\omega})| = \begin{cases} 0 & \text{for } 0 \le \omega < \omega_c \\ 1 & \text{for } \omega_c \le \omega \le \pi \end{cases}$$

The filter passes frequency components of a signal with a gain of 1 from $\omega = \omega_c$ to $\omega = \pi$, called the passband, and rejects everything else in the range from $\omega = 0$ to $\omega = \omega_c$, called the stopband. Note that the shift of the frequency response of a lowpass filter by π radians is the frequency response of a highpass filter.

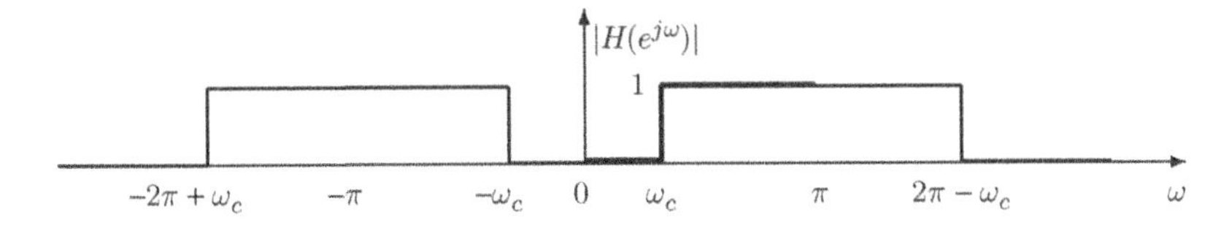

Figure 6.2 Magnitude of the periodic frequency response of an ideal digital highpass filter

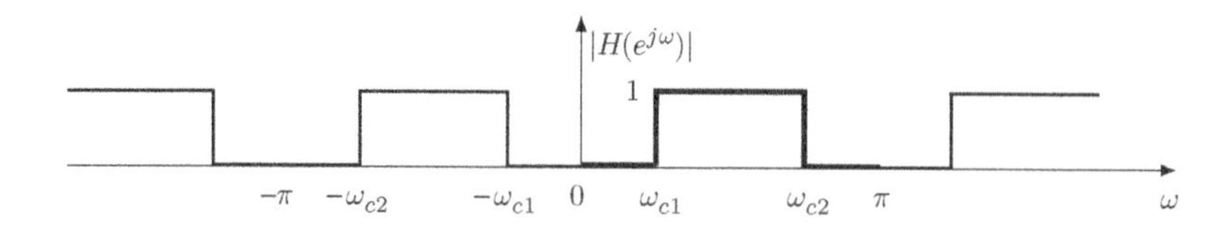

Figure 6.3 Magnitude of the periodic frequency response of an ideal digital bandpass filter

6.1.3 *Ideal Bandpass Filters*

The magnitude of the periodic frequency response of an ideal digital bandpass filter is shown in Figure 6.3. The response is specified as

$$|H(e^{j\omega})| = \begin{cases} 1 & \text{for} \quad \omega_{c1} \leq \ \omega \leq \omega_{c2} \\ 0 & \text{for} \quad 0 \leq \omega < \omega_{c1} \quad \text{and} \quad \omega_{c2} < \omega \leq \pi \end{cases}$$

The filter passes frequency components with unity gain in the range ω_{c1} to ω_{c2}, called the passband, and rejects everything else in the two stopbands.

6.2 Linear Phase Response

The requirement of linear phase response is essential for major applications such as signal transmission and image processing. Linear phase response ensures that the processed signal has no phase distortion except for a delay. The phase response is linear over the passband and is given by

$$\angle H(e^{j\omega}) = e^{-j\omega k_0}$$

The time-domain response is a delay, $y(n) = x(n - k_0)$ with k_0 positive. The phase response need not be defined over the stopband, as the magnitude response is negligible. It is relatively easy to meet the requirements of linear phase response in FIR filter design. The transfer function of an FIR filter with N coefficients is given as

$$H(z) = h(0) + h(1)z^{-1} + \cdots + h(N-1)z^{-(N-1)} = \sum_{n=0}^{N-1} h(n)z^{-n}$$

The frequency response is obtained, by replacing z by $e^{j\omega}$, as

$$H(e^{j\omega}) = \sum_{n=0}^{N-1} h(n)e^{-j\omega n}$$

If the coefficients $h(n)$ have even or odd symmetry, then the frequency response can be expressed in terms of summation of the products of $h(n)$ and cosine or sine functions, in addition to constant and linear phase factors. The symmetry also reduces the number of multiplications by about one-half in the implementation. There are four cases, depending on whether the symmetry is even or odd and N is even or odd, resulting in four types of FIR filters with linear phase response. In this and other chapters, we have presented typical frequency responses of filters used in the DWT.

6.2.1 *Even-Symmetric FIR Filters with Odd Number of Coefficients*

The number of coefficients of the filter or the terms of the impulse response N is odd. The even symmetry of the coefficients is given by $h(n) = h(N-1-n)$ for $0 \le n \le (N-1)/2$. Typical impulse response of this type of filters is shown in Figure 6.4, with $N = 5$. The center of symmetry occurs at a sample point (at $n = \frac{N-1}{2} = 2$ in Figure 6.4), and the filter delay, $\frac{N-1}{2}$, is an integral number of samples. The frequency response can be expanded as

$$H(e^{j\omega}) = \sum_{n=0}^{N-1} h(n)e^{-j\omega n} = \sum_{n=0}^{\frac{N-3}{2}} h(n)e^{-j\omega n} + h\left(\frac{N-1}{2}\right) e^{-j\omega\left(\frac{N-1}{2}\right)} + \sum_{n=\frac{N+1}{2}}^{N-1} h(n)e^{-j\omega n}$$

By making the substitution $n = N-1-n$ in the last term, we get, due to the symmetry of the coefficients,

$$\begin{aligned}
H(e^{j\omega}) &= \sum_{n=0}^{\frac{N-3}{2}} h(n)e^{-j\omega n} + h\left(\frac{N-1}{2}\right) e^{-j\omega\left(\frac{N-1}{2}\right)} + \sum_{n=0}^{\frac{N-3}{2}} h(N-1-n)e^{-j\omega(N-1-n)} \\
&= \sum_{n=0}^{\frac{N-3}{2}} h(n)e^{-j\omega n} + h\left(\frac{N-1}{2}\right) e^{-j\omega\left(\frac{N-1}{2}\right)} + \sum_{n=0}^{\frac{N-3}{2}} h(n)e^{-j\omega(N-1-n)}
\end{aligned}$$

Taking the factor $e^{-j\omega\left(\frac{N-1}{2}\right)}$ out of the summation, we get

$$\begin{aligned}
H(e^{j\omega}) &= e^{-j\omega\left(\frac{N-1}{2}\right)} \left(h\left(\frac{N-1}{2}\right) + \sum_{n=0}^{\frac{N-3}{2}} h(n) \left(e^{j\omega\left(\frac{N-1}{2}-n\right)} + e^{-j\omega\left(\frac{N-1}{2}-n\right)} \right) \right) \\
&= e^{-j\omega\left(\frac{N-1}{2}\right)} \left(h\left(\frac{N-1}{2}\right) + \sum_{n=0}^{\frac{N-3}{2}} 2h(n) \cos\left(\omega\left(\frac{N-1}{2} - n\right)\right) \right)
\end{aligned}$$

Let $k = \frac{N-1}{2} - n$. Then,

$$H(e^{j\omega}) = e^{-j\omega\left(\frac{N-1}{2}\right)} \sum_{k=0}^{\frac{N-1}{2}} c(k)\cos(\omega k), \quad c(k) = \begin{cases} h\left(\frac{N-1}{2}\right) & \text{for } k = 0 \\ 2h\left(\frac{N-1}{2} - k\right) & \text{for } k = 1, 2, \dots, \frac{N-1}{2} \end{cases} \tag{6.1}$$

The phase of the frequency response of the filter, proportional to an integer, is a linear function of ω as the summation value is real (with a phase of 0 or π radians). The filter shown in Figure 6.4 has the impulse response

$$\{h(0) = -1, h(1) = 2, h(2) = 6, h(3) = 2, h(4) = -1\}$$

With $c(0) = 6$, $c(1) = 4$, and $c(2) = -2$, the frequency response of the filter, from Equation (6.1), is

$$H(e^{j\omega}) = e^{-j2\omega}(6 + 4\cos(\omega) - 2\cos(2\omega))$$

The magnitude of the frequency response of the filter is

$$|H(e^{j\omega})| = (6 + 4\cos(\omega) - 2\cos(2\omega))$$

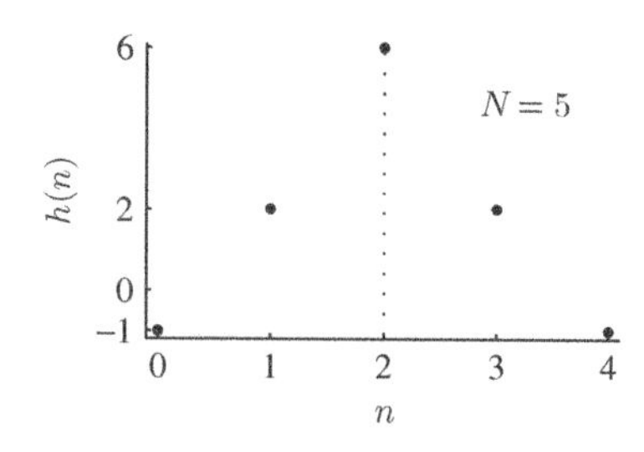

Figure 6.4 Typical FIR filter with even-symmetric impulse response and odd number of coefficients

The phase response is

$$\angle H(e^{j\omega}) = -2\omega \text{ radians}$$

Example 6.1 Derive an expression for the frequency response of the filter with the impulse response

$$\{h(0) = 1, h(1) = 2, h(2) = 1\}$$

What are the magnitudes and phases in degrees of the filter at $\omega = \{0, \pi/2, \pi\}$ radians? Is the filter lowpass or highpass type?

Solution

The frequency response is obtained using Equation (6.1) as

$$H(e^{j\omega}) = 2e^{-j\omega}(1 + \cos(\omega))$$

The magnitudes are 4, 2, and 0. It is a lowpass filter. The phases are $0°$, $-90°$ and $-180°$. The magnitude of the frequency response is shown in Figure 6.5(a). The linear phase response of the filter is shown in Figure 6.5(b). ■

6.2.2 *Even-Symmetric FIR Filters with Even Number of Coefficients*

The number of coefficients of the filter or the terms of the impulse response, N, is even. The even symmetry of the coefficients is given by $h(n) = h(N - 1 - n)$ for $0 \leq n < (N/2) - 1$.

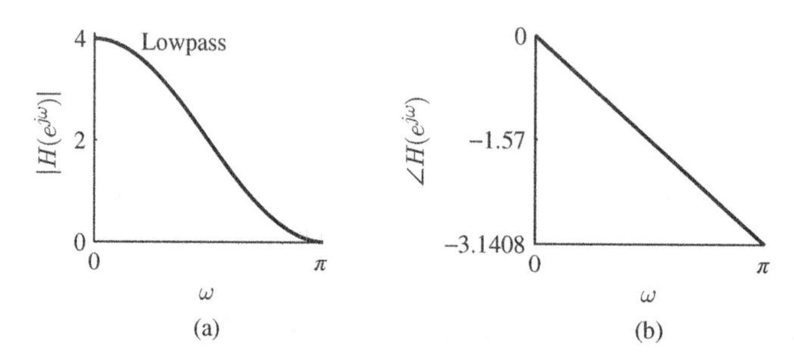

Figure 6.5 (a) The magnitude of the frequency response of the filter; (b) the phase response of the filter

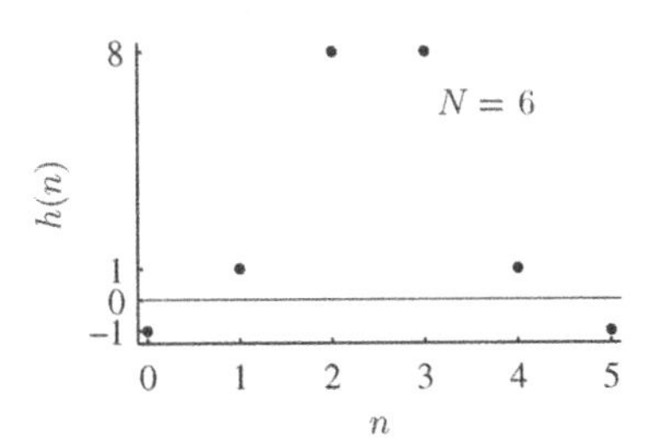

Figure 6.6 Typical FIR filter with even-symmetric impulse response and even number of coefficients

Typical impulse response of this type of filters is shown in Figure 6.6, with $N = 6$. With $M = (N-1)/2$, the frequency response is given by

$$|H(e^{j\omega})| = 2\sum_{n=0}^{\frac{N}{2}-1} h(n)\cos((M-n)\omega)$$

$$\angle(H(e^{j\omega})) = -M\omega \text{ radians}$$

Let $\{l(0) = 1/\sqrt{2}, l(1) = 1/\sqrt{2}\}$ be the impulse response. Then, the magnitude and the phase of the frequency response are $\sqrt{2}\cos(\omega/2)$ and $-\omega/2$, respectively.

6.3 Summary

- Digital filter is a linear time-invariant system. It modifies its input signal to produce an output signal that is more desirable in some sense.
- Two standard types of filters, lowpass and highpass filters, are an essential part of the DWT implementation.
- A lowpass filter passes the low-frequency components of a signal and suppresses the high-frequency components.
- A highpass filter passes the high-frequency components of a signal and suppresses the low-frequency components.
- FIR filters are characterized by their impulse response of finite duration.
- FIR filter can produce linear phase response that is desirable in several applications.

Exercises

6.1 Derive an expression for the frequency response of the filter with the impulse response $h(n)$. What are the magnitudes and phases in degrees of the filter at $\omega = \{10^{-7}, \pi/2, (\pi - 10^{-7})\}$ radians? Is the filter lowpass or highpass type?

6.1.1

$$\frac{1}{\sqrt{2}}\{h(0) = 1, h(1) = 1\}$$

6.1.2

$$\frac{1}{\sqrt{2}}\{h(0) = -1, h(1) = 1\}$$

6.1.3

$$\frac{\sqrt{2}}{8}\{h(0) = -1, h(1) = 2, h(2) = 6, h(3) = 2, h(4) = -1\}$$

***6.1.4**

$$\frac{\sqrt{2}}{8}\{h(0) = 1, h(1) = 2, h(2) = -6, h(3) = 2, h(4) = 1\}$$

6.1.5

$$\frac{1}{8\sqrt{2}}\{h(0) = -1, h(1) = 1, h(2) = 8, h(3) = 8, h(4) = 1, h(5) = -1\}$$

***6.1.6**

$$\frac{\sqrt{2}}{4}\{h(0) = -1, h(1) = 3, h(2) = 3, h(3) = -1\}$$

6.1.7

$$\frac{\sqrt{2}}{8}\{h(0) = -1, h(1) = 3, h(2) = -3, h(3) = 1\}$$

6.2 Derive the linear-phase FIR filter frequency response with four coefficients $\{h(0), h(1), h(1), h(0)\}$. Generalize the result to an N-coefficient filter, where N is even.

6.3 Derive an expression for the linear-phase FIR filter frequency response with five coefficients $\{h(0), h(1), h(2), h(1), h(0)\}$.

7

Multirate Digital Signal Processing

Most naturally occurring signals are of continuous type. That is, the signal is defined at every instant of time. However, due to the fact that all practical signals can be considered as bandlimited with adequate accuracy and due to the advantages of digital systems and fast numerical algorithms, continuous signals are mostly converted into digital form, processed, and converted back (if required). The conversion of a continuous signal into a digital signal requires sampling in time and quantizing its amplitude. While a signal can be processed with a fixed sampling rate satisfying the sampling theorem, it is found that the processing is more efficient if the sampling rate is changed to suit the stage of processing. For example, a higher sampling rate is desired in reconstructing a continuous signal from its samples, while a lower sampling rate reduces the execution time of algorithms. If different sampling rates are used in processing a signal, then the process is called multirate digital signal processing. Sampling rate can be increased or reduced as long as the sampling theorem is not violated, or in situations where it is possible to recover the original signal from its aliased versions. The process of changing the sampling rate of a signal is called the sampling rate conversion. It is found that this conversion can be efficiently carried out in the digital domain itself.

In this chapter, downsampling and upsampling operations, which reduce and increase the sampling rate, respectively, are first presented. Combinations of these two operations and convolution (filtering) result in decimation and interpolation operations. Convolution followed by downsampling is the basic step in the computation of the DWT. In addition to adders, multipliers, and delays, two more devices, decimators and interpolators, are used in implementing multirate digital signal processing systems. The two-channel filter bank, which is made of decimators, interpolators, and delays, is then described. The filter bank is the implementation structure of the DWT. Basically, a filter bank decomposes an input signal into subband components and reconstructs the input from these components. Finally, an efficient implementation of the filter bank, called the polyphase form, is derived. When we decompose the spectrum of a signal into subbands with smaller bandwidths, the corresponding signals can be processed at a reduced sampling rate. Filtering first and then reducing the sampling rate are inefficient, as the filter works at a higher sampling rate than that is required. The polyphase implementation of the filters enables the reduction of the sampling rate before filtering. A similar advantage is achieved when a signal is reconstructed from its subband components.

Discrete Wavelet Transform: A Signal Processing Approach, First Edition. D. Sundararajan.

Companion Website: www.wiley.com/go/sundararajan/wavelet

7.1 Decimation

A signal $x(n)$, assuming properly sampled, can be composed of frequency components with frequencies up to $f_s/2$, where f_s is the sampling frequency. Retaining one out of every M samples and discarding the rest is called the downsampling of a signal by a factor of M, a positive integer. The nth sample of the downsampled signal $x_d(n)$ is the (Mn)th sample of $x(n)$

$$x_d(n) = x(Mn)$$

The sampling interval in the downsampled signal becomes M times longer than that of the original signal. Therefore, the sampling frequency is reduced to f_s/M. This implies that frequency components with frequencies only up to $f_s/(2M)$ can be unambiguously represented. To prevent aliasing in the downsampled signal, the original signal must be filtered by a lowpass filter with cutoff frequency $f_s/(2M)$ before downsampling. Lowpass filtering followed by downsampling is called the decimation of a signal. The decimation of a signal $x(n)$ by a factor of 2 is shown in Figure 7.1. Filtering is achieved by the convolution of the input $x(n)$ with the filter (cutoff frequency $f_s/4$) impulse response $h(n)$. The resulting output $v(n)$ is downsampled by a factor of 2 to get the decimated output $y(n) = v_d(n) = v(2n)$. Note that the odd-indexed values of the filter output $v(n)$ are also a valid representation of the decimated signal.

Assuming that the length of $x(n)$ is N and that of $h(n)$ is M, the convolution output of $x(n)$ and $h(n)$ is a sequence of length $N + M - 1$. In the DWT computation, the first output corresponds to the value with index $M - 1$. That is, the filter coefficients must completely overlap the input sequence. In the case of symmetric filters with more than two coefficients, the filters are centered in the first row of the transform matrix in order to reduce the edge effects. Symmetric filters are considered in later chapters. Furthermore, the convolution output is downsampled. Combining the convolution and downsampling operations, the output of the double-shifted convolution is given by

$$y(n) = \sum_k h(k)x(2n - k) = \sum_k x(k)h(2n - k)$$

This is a double-shifted convolution equation, where either the data or the coefficients (not both) are shifted by two sample intervals rather than one. While the filtering operation is time invariant, the downsampling operation is time variant. The convolution of the input $\{x(0) = 3, x(1) = 4, x(2) = 3, x(3) = -2\}$ with the impulse response $\{h(0) = -1, h(1) = 1\}$ is shown in Figure 7.2(a). The computation of the DWT coefficients alone is shown in Figure 7.2(b).

7.1.1 *Downsampling in the Frequency-Domain*

Consider the signal $x(n)$ with its DTFT $X(e^{j\omega})$. We express the spectrum $X_d(e^{j\omega})$ of the downsampled signal (by a factor of 2) in terms of $X(e^{j\omega})$. The DTFT of the signal

$x(n)$ → [$*h(n)$] (f_s) → $v(n) = x(n)*h(n)$ (f_s) → [$2\downarrow$] ($\frac{f_s}{2}$) → $vd(n) = v(2n) = y(n)$

Figure 7.1 Decimation of a signal $x(n)$ by a factor of 2

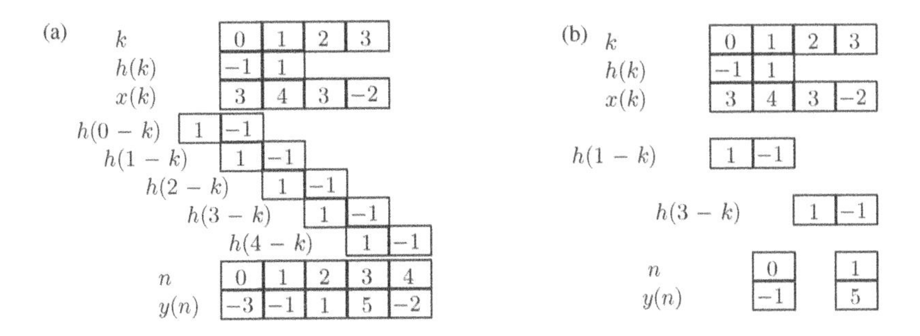

Figure 7.2 Convolution of two sequences (a), and computation of the DWT coefficients (b)

$e^{-j\pi n}x(n) = (-1)^n x(n)$ is $X(e^{j(\omega+\pi)})$. The even-indexed samples of the two signals are the same, while the odd-indexed samples are the negatives of the other. For example,

$$x(n) = \{x(0) = 1, x(1) = 3, x(2) = 2, x(3) = 4\} \leftrightarrow X(e^{j\omega}) = 1 + 3e^{-j\omega} + 2e^{-j2\omega} + 4e^{-j3\omega}$$

$$(-1)^n x(n) = \{x(0) = 1, x(1) = -3, x(2) = 2, x(3) = -4\} \leftrightarrow X(e^{j(\omega+\pi)}) = 1 - 3e^{-j\omega} + 2e^{-j2\omega} - 4e^{-j3\omega}$$

The inverse DTFT of the spectrum

$$\frac{1}{2}(X(e^{j\omega}) + X(e^{j(\omega+\pi)})) \tag{7.1}$$

is the signal

$$\frac{1}{2}(x(n) + (-1)^n x(n))$$

The sum is $x(n)$ with zero-valued odd-indexed samples. The first-half and the second-half of the spectrum are identical, as by circularly shifting the spectrum $X(e^{j\omega})$ by π radians and adding it with itself amount to summing the same values in the first-half and the second-half. Note that the spectrum of a real-valued discrete signal is periodic and conjugate symmetric. For the specific example, we get

$$\begin{aligned}\frac{1}{2}(x(n) + (-1)^n x(n)) &= \{x(0) = 1, x(1) = 0, x(2) = 2, x(3) = 0\} \\ &\leftrightarrow \frac{1}{2}(X(e^{j\omega}) + X(e^{j(\omega+\pi)})) = 1 + 2e^{-j2\omega}\end{aligned}$$

The term $2e^{-j2\omega}$ is periodic with period $\pi(2e^{-j2(\omega+\pi)} = 2e^{-j2\pi}e^{-j2\omega} = 2e^{-j2\omega})$, and 1 is periodic with any period. Therefore, the spectrum is periodic with period π. By taking the samples of the spectrum, we get

$$\{1 + 2e^{-j2\omega},\ \ \omega = 0, \pi/2, \pi, 3\pi/2\} = \{3, -1, 3, -1\}$$

Another confirmation that the spectrum is periodic with period π. Now, if we take the spectral values over one period $\{3, -1\}$ and compute the IDFT, we get $\{1, 2\}$, which is the down-sampled version of $x(n)$. The spectrum of $\{1, 2\}$, by definition, is $1 + 2e^{-j\omega}$, which can also be obtained from $1 + 2e^{-j2\omega}$ by replacing ω by $\omega/2$. According to a theorem (proved in Chapter 4), if the DTFT of $\{a, b\}$ is $P(e^{j\omega})$, then the DTFT of the upsampled signal

$\{a, 0, b, 0\}$ is the concatenation and compression of $P(e^{j\omega})$ and $P(e^{j\omega})$ in the range $0 < \omega < 2\pi$, $P(e^{j2\omega})$. The spectrum is replicated. Therefore, replacing ω by $\omega/2$ in Equation (7.1), we take the values of the spectrum of the upsampled signal in the frequency range from 0 to π only and expand it in forming the spectrum $X_d(e^{j\omega})$ of the downsampled signal

$$x_d(n) = x(2n) \leftrightarrow X_d(e^{j\omega}) = \frac{1}{2}\left(X\left(e^{j\frac{\omega}{2}}\right) + X\left(e^{j\left(\frac{\omega}{2}+\pi\right)}\right)\right), \quad 0 < \omega < 2\pi \tag{7.2}$$

The inverse DTFT of this spectrum gives the downsampled signal. Formally,

$$\begin{aligned}
X_d(e^{j\omega}) &= \sum_n x(2n)e^{-j\omega n} \\
&= \frac{1}{2}\sum_k \left(x(k)\, e^{-j\frac{\omega}{2}k} + x(k)e^{-j\left(\frac{\omega}{2}+\pi\right)k}\right) \\
&= \frac{1}{2}\left(X\left(e^{j\frac{\omega}{2}}\right) + X\left(e^{j\left(\frac{\omega}{2}+\pi\right)}\right)\right)
\end{aligned}$$

Example 7.1 Find the DTFT $X(e^{j\omega})$ of $x(n)$.

$$x(n) = \{x(-2) = 1, x(-1) = 1, x(1) = 1, x(2) = 1\}$$

Express the DTFT $X_d(e^{j\omega})$ of $x(2n)$ in terms of $X(e^{j\omega})$.

Solution

Figure 7.3 shows the process of finding the DTFT of $x(2n)$. The signal $x(n)$ is shown in Figure 7.3(a). Its DTFT

$$X(e^{j\omega}) = e^{j2\omega} + e^{j\omega} + e^{-j\omega} + e^{-j2\omega} = 2\cos(2\omega) + 2\cos(\omega)$$

is shown in Figure 7.3(b). The signal $(-1)^n x(n)$

$$\{x(-2) = 1, x(-1) = -1, x(0) = 0, x(1) = -1, x(2) = 1\}$$

is shown in Figure 7.3(c). Its DTFT, which is the shifted (by π radians) version of that in Figure 7.3(b),

$$X(e^{j(\omega+\pi)}) = 2\cos(2(\omega + \pi)) + 2\cos(\omega + \pi) = 2\cos(2\omega) - 2\cos(\omega)$$

is shown in Figure 7.3(d). The signal $(x(n) + (-1)^n x(n))/2$

$$\{x(-2) = 1, x(-1) = 0, x(0) = 0, x(1) = 0, x(2) = 1\}$$

is shown in Figure 7.3(e). Its DTFT

$$\frac{1}{2}(X(e^{j\omega}) + X(e^{j(\omega+\pi)})) = \cos(2\omega) + \cos(\omega) + \cos(2\omega) - \cos(\omega) = 2\cos(2\omega)$$

is shown in Figure 7.3(f). The spectrum is periodic with period π. The downsampled signal $x_d(n) = x(2n)$

$$\{x(-2) = 0, x(-1) = 1, x(0) = 0, x(1) = 1, x(2) = 0\}$$

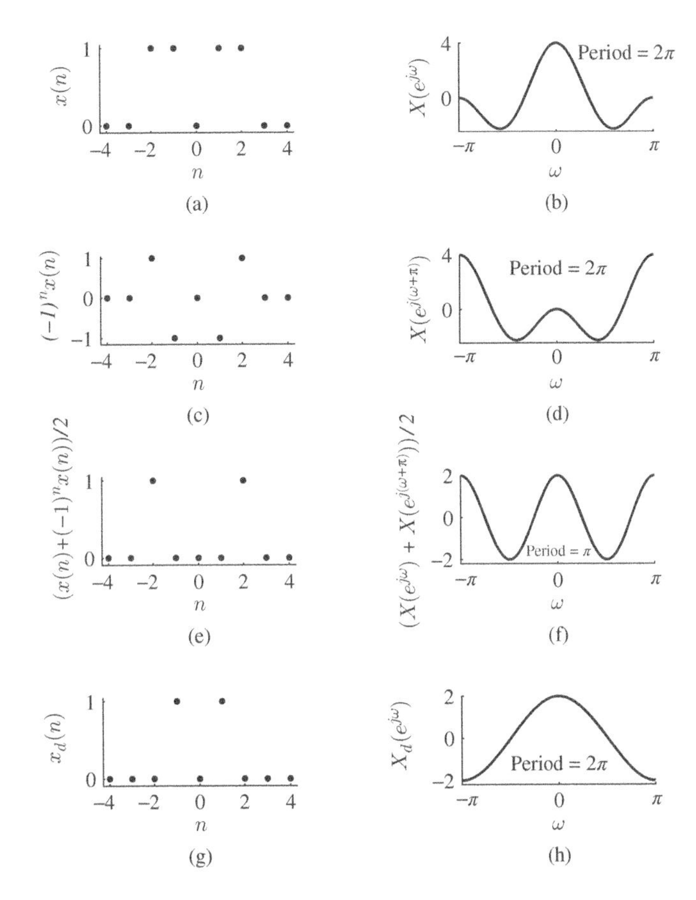

Figure 7.3 (a) Signal $x(n)$; (b) its DTFT $X(e^{j\omega})$; (c) signal $(-1)^n x(n)$; (d) its DTFT $X(e^{j(\omega+\pi)})$; (e) signal $(x(n) + (-1)^n x(n))/2$; (f) its DTFT $(X(e^{j\omega}) + X(e^{j(\omega+\pi)}))/2$; (g) downsampled signal $x_d(n) = x(2n)$; (h) its DTFT $X_d(e^{j\omega}) = \left(X\left(e^{j\frac{\omega}{2}}\right) + X\left(e^{j\left(\frac{\omega}{2}+\pi\right)}\right)\right)/2$

is shown in Figure 7.3(g). Its DTFT, which is the expanded (by a factor of 2) version of that shown in Figure 7.3(f),

$$X_d(e^{j\omega}) = 2\cos(2(\omega/2)) = 2\cos(\omega)$$

is shown in Figure 7.3(h). ■

7.1.2 Downsampling Followed by Filtering

In Figure 7.1, filtering is followed by downsampling. In this process, half of the filter output values are discarded in downsampling. Obviously, it is more efficient not to compute those values in the first place. It is possible to interchange the downsampler with the filter under the condition that the filter impulse response is of the form

$$h(n) = \{h(0), 0, h(1), 0, \dots, h(N-1), 0\}$$

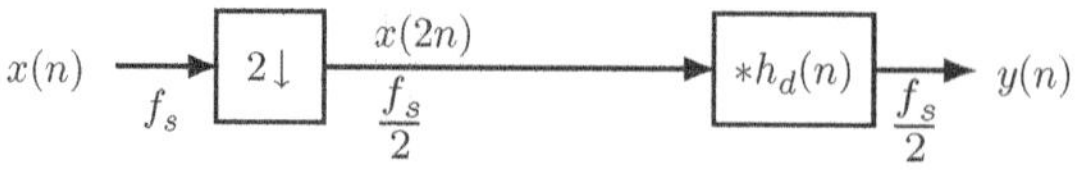

$x(n)$ f_s $2\downarrow$ $x(2n)$ $\frac{f_s}{2}$ $*h_d(n)$ $\frac{f_s}{2}$ $y(n)$

Figure 7.4 Decimation of a signal $x(n)$ by a factor of 2 with downsampling followed by filtering

Then, the downsampled version of $h(n)$ is

$$h_d(n) = \{h(0), h(1), \ldots, h(N-1)\}$$

and

$$H_d(e^{j2\omega}) = H(e^{j\omega})$$

The response $h(n)$ is the upsampled version of $h_d(n)$. The output $v_d(n)$ in Figure 7.1 can also be generated using the more efficient configuration shown in Figure 7.4.With reference to Figure 7.1, the output after convolution of the input with the impulse response is

$$X(e^{j\omega})H_d(e^{j2\omega}) = X(e^{j\omega})H(e^{j\omega}) \tag{7.3}$$

After downsampling, we get

$$Y(e^{j\omega}) = \frac{1}{2}\left(H_d\left(e^{j\frac{2\omega}{2}}\right)X\left(e^{j\frac{\omega}{2}}\right) + H_d\left(e^{j2\left(\frac{\omega}{2}+\pi\right)}\right)X\left(e^{j\left(\frac{\omega}{2}+\pi\right)}\right)\right) \tag{7.4}$$

$$= \frac{1}{2}\left(H_d\left(e^{j\omega}\right)X\left(e^{j\frac{\omega}{2}}\right) + H_d(e^{j\omega})X\left(e^{j\left(\frac{\omega}{2}+\pi\right)}\right)\right) \tag{7.5}$$

With reference to Figure 7.4, the output is

$$Y(e^{j\omega}) = H_d(e^{j\omega})\frac{\left(X\left(e^{j\frac{\omega}{2}}\right) + X\left(e^{j\left(\frac{\omega}{2}+\pi\right)}\right)\right)}{2} \tag{7.6}$$

Both the outputs are the same.

Example 7.2 Let

$$x(n) = \{2, 1, 3\} \quad \text{and} \quad h(n) = \{1, 0, 3, 0, -2, 0\}$$

Verify that the outputs of decimator configurations shown in Figures 7.1 and 7.4 are the same.

Solution

$$h_d(n) = \{1, 3, -2\}$$

From Figure 7.1, the convolution output of $x(n)$ and $h(n)$ is

$$x(n) * h(n) = \{2, 1, 3\} * \{1, 0, 3, 0, -2, 0\} = \{2, 1, 9, 3, 5, -2, -6, 0\}$$

The downsampling of the convolution output gives

$$\{2, 9, 5, -6\}$$

From Figure 7.4 also, we get the same decimated output.

$$x_d(n) * h_d(n) = \{2, 3\} * \{1, 3, -2\} = \{2, 9, 5, -6\}$$

■

7.2 Interpolation

In upsampling a signal $x(n)$ by a factor of M, $M-1$ zeros are inserted between every two adjacent samples. The upsampled signal $x_u(n)$ is defined as

$$x_u(n) = \begin{cases} x\left(\frac{n}{M}\right) & \text{for } n = 0,\ \pm M,\ \pm 2M,\ \dots, \\ 0 & \text{otherwise} \end{cases}$$

The sampling interval is reduced by a factor of M. The sampling frequency f_s is increased to Mf_s. The spectrum of the original signal is replicated $M-1$ times. New high-frequency components, called image frequencies, are required to make up the upsampled signal. In order to get back the original signal with proper intermediate values, the high-frequency components have to be filtered out by a filter with cutoff frequency $f_s/(2M)$. Lowpass filtering preceded by upsampling is called the interpolation operation. The interpolation of a signal $x(n)$ by a factor of 2 is shown in Figure 7.5. The input is upsampled by a factor of 2 to get the upsampled signal $x_u(n)$. The output $y(n)$ of the interpolation operation is the convolution of the filter impulse response $h(n)$ with $x_u(n)$.

The convolution of the upsampled input $x_u(n)$ with the filter impulse response $h(n)$ is given by

$$y(n) = \sum_k x_u(k)h(n-k)$$

As the odd-indexed values of the upsampled sequence $x_u(n)$ are zero,

$$y(n) = \sum_{\text{even } k} x\left(\frac{k}{2}\right) h(n-k) = \sum_m x(m)h(n-2m)$$

The coefficients are right shifted by one sample interval after computation of each output value. That is, when the input $x(n)$ is convolved alternately with even- and odd-indexed coefficients, the output is the same. The convolution of the upsampled input

$$\{x_u(0) = 4, x_u(1) = 0, x_u(2) = 2, x_u(3) = 0\}$$

with the impulse response $\{h(0) = -1, h(1) = 2\}$ is shown in Figure 7.6(a). Convolution of

$$\{x(0) = 4, x(1) = 2\}$$

alternately with $h_e(n) = -1$ and $h_o(n) = 2$ is shown in Figure 7.6(b).

7.2.1 *Upsampling in the Frequency-Domain*

Consider the signal $x(n)$ with its DTFT $X(e^{j\omega})$. The replication of the spectrum, due to upsampling, creates high-frequency components. Insertion of zeros in upsampling a signal

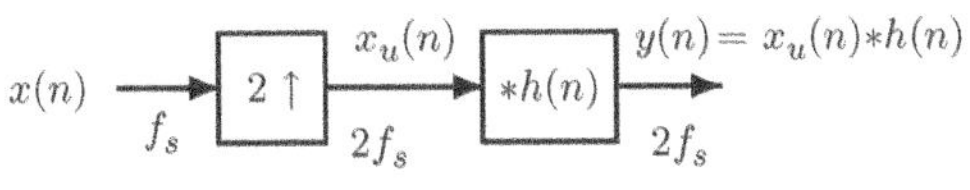

Figure 7.5 Interpolation of a signal $x(n)$ by a factor of 2

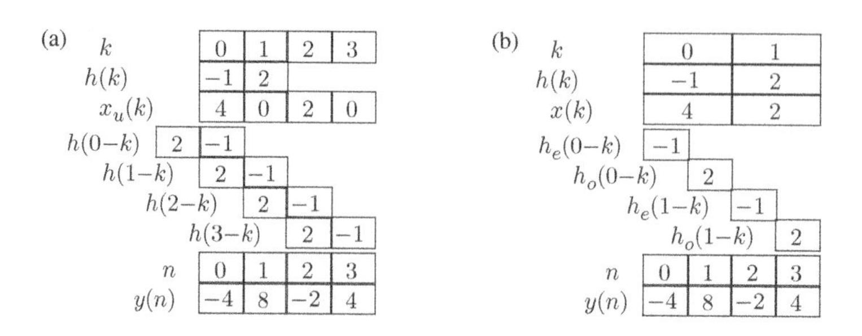

Figure 7.6 Convolution of an upsampled sequence $x_u(n)$ and $h(n)$ (a), and convolution of $x(n)$ alternately with $h_e(n)$ and $h_o(n)$ (b)

makes it bumpy requiring high-frequency components, called image frequencies, in its composition. For example, with an upsampling factor of 2, the spectrum of the upsampled signal $x_u(n)$ is given by replacing ω by 2ω in $X(e^{j\omega})$.

$$x_u(n) = \begin{cases} x\left(\frac{n}{2}\right) & \text{for } n = 0, \ \pm 2, \ \pm 4, \ldots \\ 0 & \text{otherwise} \end{cases} \leftrightarrow X_u(e^{j\omega}) = X(e^{j2\omega}), \quad 0 < \omega < 2\pi$$

This result is proved in Chapter 4. Consider the transform pair

$$\{x(0) = 3, x(1) = -4\} \leftrightarrow 3 - 4e^{-j\omega}$$

Then,

$$\{x(0) = 3, x(1) = 0, x(2) = -4, x(3) = 0\} \leftrightarrow 3 - 4e^{-j2\omega}$$

The term $-4e^{-j2\omega}$ is periodic with period π, and 3 is periodic with any period. Therefore, the spectrum is periodic with period π. The spectrum of the original signal is compressed into $|\omega| \leq \pi/2$, and an image of the compressed spectrum appears next to it. Upsampling results in imaging. In upsampling by a factor of 2, one component with frequency ω is the source of two components with frequencies ω and $\pi - \omega$. Downsampling leads to aliasing. In downsampling by a factor of 2, two input components with frequencies ω and $\pi - \omega$ produce the same sequence corresponding to frequency ω.

7.2.2 Filtering Followed by Upsampling

In Figure 7.5, filtering is preceded by upsampling. In this process, half of the filter input values are zeros. Obviously, it is more efficient not to process those values in the first place. It is possible to interchange the upsampler with the filter under the condition that the filter impulse response is of the form

$$h(n) = \{h(0), 0, h(1), 0, \ldots, h(N-1), 0\}$$

Then, the downsampled version of $h(n)$ is

$$h_d(n) = \{h(0), h(1), \ldots, h(N-1)\}$$

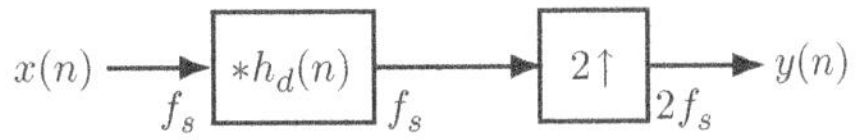

$x(n) \xrightarrow{f_s} \boxed{*h_d(n)} \xrightarrow{f_s} \boxed{2\uparrow} \xrightarrow{2f_s} y(n)$

Figure 7.7 Interpolation of a signal $x(n)$ by a factor of 2 with filtering followed by upsampling

and

$$H_d(e^{j2\omega}) = H(e^{j\omega})$$

The response $h(n)$ is the upsampled version of $h_d(n)$. The output $y(n)$ in Figure 7.5 can also be generated using the more efficient configuration shown in Figure 7.7. From Figure 7.5, we get

$$Y(e^{j\omega}) = X(e^{j2\omega})H_d(e^{j2\omega}) = X(e^{j2\omega})H(e^{j\omega}) \tag{7.7}$$

From Figure 7.7 also, we get the same interpolated output. Firstly, the filtering operation produces the intermediate signal

$$X(e^{j\omega})H_d(e^{j\omega}) \tag{7.8}$$

the upsampling of which yields the output.

$$Y(e^{j\omega}) = X(e^{j2\omega})H_d(e^{j2\omega}) = X(e^{j2\omega})H(e^{j\omega}) \tag{7.9}$$

Example 7.3 Let

$$x(n) = \{1, 2\} \quad \text{and} \quad h(n) = \{2, 0, 3\}$$

Verify that the outputs of interpolator configurations shown in Figures 7.5 and 7.7 are the same.

Solution

$$h_d(n) = \{2, 3\}$$

The convolution of the upsampled input $x_u(n)$ with the filter impulse response $h(n)$ yields

$$\{1, 0, 2\} * \{2, 0, 3\} = \{2, 0, 7, 0, 6\}$$

The convolution of the input and the decimated impulse response yields

$$\{1, 2\} * \{2, 3\} = \{2, 7, 6\}$$

The upsampled version $\{2, 0, 7, 0, 6\}$ of this output is also the same. ■

7.3 Two-Channel Filter Bank

Assuming a sampling interval of 1 second, the effective frequency range of a real discrete signal is $0 - \pi$ radians. Frequencies close to zero are low frequencies, and those near π are high frequencies. In the DWT representation of a signal, the signal is decomposed into a number of components corresponding to a set of nonoverlapping subbands of its spectrum. This representation, called the time-frequency representation, is between time-domain and frequency-domain representations of a signal. Obviously, a set of bandpass filters is required to do the task. The subbands are, usually, of different lengths. The requirement is that lower frequency components are to be represented over longer time intervals (higher frequency resolution), and higher frequency components are to be represented over shorter time intervals (higher time resolution). It turns out that the decomposition of the spectrum into subbands can be carried out efficiently using a set of lowpass and highpass filters recursively in the analysis filter bank rather than using bandpass filters. The spectrum is divided into two equal parts using a lowpass filter and a highpass filter. Now, the highpass filter output is downsampled and left alone. Note that as each of the subbands is half the width of the spectrum of the original signal, downsampling operation does not destroy any information and prevents doubling the size of the data. The subband spectrum of the downsampled lowpass filter output is again divided into two subbands of smaller size using the same set of lowpass and highpass filters. This recursive process is repeated until the lowpass filter output meets the required resolution. In the signal reconstruction process, the outputs of the decomposition process are upsampled first, filtered by the respective synthesis filters, and then added in the synthesis filter bank. This process is also recursive.

Lowpass and highpass filters, samplers, and delays (not shown) constitute a two-channel filter bank, shown in Figure 7.8. The efficient design and implementation of a two-channel filter bank are most important in the computation of the DWT. There is an analysis filter bank and a synthesis filter bank. If the transform matrices of the two filter banks are transposes as well as inverses, then the filter bank is called orthogonal. In a biorthogonal filter bank, the transpose of one transform matrix is the inverse of the other. The problem reduces to the design of a set of lowpass and highpass filters for analysis and another set for reconstruction. The filters are typically FIR type and are characterized by the impulse response in the time domain and by the frequency response (the DTFT of the impulse response) in the frequency domain, as is the case with conventional FIR filters.

The filter bank is represented in the frequency domain. It can also be represented in the time domain or z-domain. Basically, a two-channel filter bank decomposes an input signal into a pair of components in the analysis section and reconstructs the input signal from this pair in the synthesis section. The input $x(n)$ is represented by its DTFT $X(e^{j\omega})$. The input goes through a lowpass channel and a highpass channel. The lowpass component of the input is obtained by filtering with the lowpass filter. The impulse response is $l(n)$ (frequency response $L(e^{j\omega})$). This component is downsampled by a factor of 2 by the downsampler yielding the output $X_\phi(e^{j\omega})$ of the lowpass channel of the analysis filter bank. The input data are usually much longer than the filter impulse response. In the convolution operation, the outputs are the weighted averages of sections of the input with the filter coefficients. A similar process produces the output $X_\psi(e^{j\omega})$ of the highpass channel of the analysis filter bank.

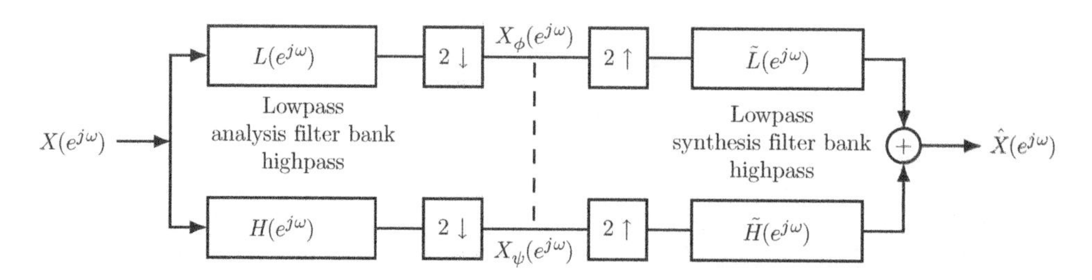

Figure 7.8 A frequency-domain representation of a two-channel analysis and synthesis filter bank

Using the two outputs of the analysis filter bank, the input can be perfectly reconstructed by the synthesis filter bank. The flow-graphs of the analysis and synthesis filter banks are the transposed forms of each other. The transposed form is obtained by (i) reversing the directions of all the signal flow paths, (ii) replacing adders by junction points or vice versa, (iii) replacing upsamplers by downsamplers or vice versa, (iv) interchanging the filters, and (v) interchanging the input and output points. To reconstruct the input from the analysis filter bank output, we have to make the output of double the length. This is carried out by the upsamplers. The upsampling operation creates some unwanted components, which are filtered out by the synthesis bank filters with impulse responses $\widetilde{l}(n)$ (frequency response $\widetilde{L}(e^{j\omega})$) and $\widetilde{h}(n)$ (frequency response $\widetilde{H}(e^{j\omega})$). The sum of the output of the lowpass and highpass channels of the synthesis filter bank constitutes the reconstructed input $\widehat{x}(n)$ ($\widehat{X}(e^{j\omega})$ in the frequency domain). The reason we are decomposing and reconstructing the signal is that the decomposed signal components are more suitable to process (such as compression and denoising) than the input.

The lowpass and highpass filters of the analysis filter bank decompose the input signal into components corresponding to lower and higher frequency subbands. However, they are not individually invertible, as the frequency response $L(e^{j\pi}) = 0$ for the lowpass filter and $H(e^{j0}) = 0$ for the highpass filter. In combination, inversion is possible despite the zeros of the frequency response. The inverse of an FIR filter is usually not an FIR filter. In order to invert an FIR filter with another FIR filter, we need a filter bank. The zeros in the frequency response are essential for the decomposition of the spectrum into various subbands and to reconstruct the signal back, even though the cutoff frequencies of the filters are not sharp.

7.3.1 Perfect Reconstruction Conditions

As practical filters are nonideal, there is aliasing in each channel of the filter bank. There is also amplitude and phase distortion. The combined effect of the analysis and synthesis filters should be such that the signal is reconstructed from its unaltered decomposed components with a constant gain and a constant delay. Such a system is called a perfect reconstruction (PR) multirate system. Without samplers, PR implies that

$$\widetilde{L}(e^{j\omega})L(e^{j\omega}) + \widetilde{H}(e^{j\omega})H(e^{j\omega}) = Me^{-jK\omega} \tag{7.10}$$

where M and K are constants. The term $e^{-jK\omega}$ corresponds to a delayed output ($\widehat{x}(n) = x(n-K)$) for positive K. Note that practical filters are causal. The filter bank without samplers is two systems connected in parallel, each one of them being a cascade of two systems. The term $\widetilde{L}(e^{j\omega})L(e^{j\omega})$ is a cascade of two systems with frequency responses $\widetilde{L}(e^{j\omega})$ and $L(e^{j\omega})$. The term $\widetilde{H}(e^{j\omega})H(e^{j\omega})$ is a cascade of two systems with frequency responses $\widetilde{H}(e^{j\omega})$ and $H(e^{j\omega})$. These two systems connected in parallel are represented by Equation 7.10. If the overall frequency response is $Me^{-jK\omega}$, then an input to such a system will not be distorted except for a constant gain factor and a constant delay.

Let the input signal to the downsampler be

$$x(n) = \{\ldots, x(-1), x(0), x(1), \ldots\}$$

The output of the downsampler is

$$x(2n) = \{\ldots, x(-2), x(0), x(2), \ldots\}$$

The downsampled and upsampled output is

$$\{\ldots, x(-2), 0, x(0), 0, x(2), 0, \ldots\}$$

That is, downsampling followed by upsampling makes the odd-indexed values of the given signal zero. The DTFT of this signal is

$$0.5(X(e^{j\omega}) + X(e^{j(\omega+\pi)}))$$

The term $X(e^{j\omega})$ corresponds to $x(n)$, and the term $X(e^{j(\omega+\pi)})$ corresponds to $(-1)^n x(n)$.

In the filter bank, the DTFT of the input to the downsampler of the lowpass channel is

$$L(e^{j\omega})X(e^{j\omega})$$

Therefore, the outputs $X_\phi(e^{j\omega})$ and $X_\psi(e^{j\omega})$ of the analysis filter banks are given by

$$X_\phi(e^{j\omega}) = 0.5\left(L\left(e^{j\frac{\omega}{2}}\right)X\left(e^{j\frac{\omega}{2}}\right) + L\left(e^{j\left(\frac{\omega}{2}+\pi\right)}\right)X\left(e^{j\left(\frac{\omega}{2}+\pi\right)}\right)\right)$$
$$X_\psi(e^{j\omega}) = 0.5\left(H\left(e^{j\frac{\omega}{2}}\right)X\left(e^{j\frac{\omega}{2}}\right) + H\left(e^{j\left(\frac{\omega}{2}+\pi\right)}\right)X\left(e^{j\left(\frac{\omega}{2}+\pi\right)}\right)\right)$$

The DTFT of the output of the lowpass channel is

$$0.5\widetilde{L}(e^{j\omega})(L(e^{j\omega})X(e^{j\omega}) + L(e^{j(\omega+\pi)})X(e^{j(\omega+\pi)}))$$

Similarly, the DTFT of the output of the highpass channel is

$$0.5\widetilde{H}(e^{j\omega})(H(e^{j\omega})X(e^{j\omega}) + H(e^{j(\omega+\pi)})X(e^{j(\omega+\pi)}))$$

Now, the sum of the two DTFT expressions, which is the DTFT of the approximation $\widehat{x}(n)$ of the input signal $x(n)$, is given by

$$\begin{aligned}\widehat{X}(e^{j\omega}) &= 0.5\widetilde{L}(e^{j\omega})(L(e^{j\omega})X(e^{j\omega}) + L(e^{j(\omega+\pi)})X(e^{j(\omega+\pi)}))\\ &\quad + 0.5\widetilde{H}(e^{j\omega})(H(e^{j\omega})X(e^{j\omega}) + H(e^{j(\omega+\pi)})X(e^{j(\omega+\pi)}))\end{aligned}$$

This expression can be rearranged as

$$\begin{aligned}\widehat{X}(e^{j\omega}) &= 0.5(\widetilde{L}(e^{j\omega})L(e^{j\omega}) + \widetilde{H}(e^{j\omega})H(e^{j\omega}))X(e^{j\omega})\\ &\quad + 0.5(L(e^{j(\omega+\pi)})\widetilde{L}(e^{j\omega}) + H(e^{j(\omega+\pi)})\widetilde{H}(e^{j\omega})X(e^{j(\omega+\pi)}))\end{aligned}$$

The required output corresponds to $X(e^{j\omega})$, and the other is aliased. Therefore, the conditions for PR are

$$\widetilde{L}(e^{j\omega})L(e^{j\omega}) + \widetilde{H}(e^{j\omega})H(e^{j\omega}) = 2e^{-jK\omega}$$
$$\widetilde{L}(e^{j\omega})L(e^{j(\omega+\pi)}) + \widetilde{H}(e^{j\omega})H(e^{j(\omega+\pi)}) = 0$$

The first condition ensures no distortion, and the second guarantees alias cancellation. As the terms making up the PR conditions are product of transforms, they correspond to convolution in the time domain. It is so because the problem is concerned with the input and output of a system (the filter bank). Assuming that analysis filters $L(e^{j\omega})$ and $H(e^{j\omega})$ have been designed, the synthesis filters are given by the relations

$$\widetilde{L}(e^{j\omega}) = \pm H(e^{j(\omega+\pi)}) \quad \text{and} \quad \widetilde{H}(e^{j\omega}) = \mp L(e^{j(\omega+\pi)})$$

PR is not strictly possible, in practice, due to finite wordlength, as it introduces errors in the coding process. Typical error in the reconstructed input signal is in the range 10^{-12}–10^{-15} for floating point representation of numbers. Furthermore, the processing of the DWT coefficients, such as quantization and thresholding, makes the PR conditions difficult to hold and leads to artifacts in the reconstructed signal.

Example 7.4 Let

$$L(e^{j\omega}) = \frac{1}{\sqrt{2}}(1 + e^{-j\omega}) \quad \text{and} \quad H(e^{j\omega}) = \frac{1}{\sqrt{2}}(-1 + e^{-j\omega}) \tag{7.11}$$

Find the reconstruction filters using the PR conditions.

Solution

One possible choice for synthesis filters is

$$\widetilde{L}(e^{j\omega}) = \frac{1}{\sqrt{2}}(1 + e^{-j\omega}) \quad \text{and} \quad \widetilde{H}(e^{j\omega}) = \frac{1}{\sqrt{2}}(1 - e^{-j\omega}) \tag{7.12}$$

For this set of filters, the PR conditions hold as

$$\frac{1}{\sqrt{2}}(1 + e^{-j\omega})\frac{1}{\sqrt{2}}(1 + e^{-j\omega}) + \frac{1}{\sqrt{2}}(1 - e^{-j\omega})\frac{1}{\sqrt{2}}(-1 + e^{-j\omega}) = 2e^{-j\omega}$$

$$\frac{1}{\sqrt{2}}(1 + e^{-j\omega})\frac{1}{\sqrt{2}}(1 - e^{-j\omega}) + \frac{1}{\sqrt{2}}(1 - e^{-j\omega})\frac{1}{\sqrt{2}}(-1 - e^{-j\omega}) = 0$$

In the time domain, the first equation of the PR condition corresponds to the sum of two convolutions.

$$\left(\frac{1}{\sqrt{2}}, \frac{1}{\sqrt{2}}\right) * \left(\frac{1}{\sqrt{2}}, \frac{1}{\sqrt{2}}\right) + \left(\frac{1}{\sqrt{2}}, -\frac{1}{\sqrt{2}}\right) * \left(-\frac{1}{\sqrt{2}}, \frac{1}{\sqrt{2}}\right) = \left(\frac{1}{2}, 1, \frac{1}{2}\right)$$
$$+ \left(-\frac{1}{2}, 1, -\frac{1}{2}\right) = (0, 2, 0)$$

The transform of $(0, 2, 0)$ is $2e^{-j\omega}$. The second equation of the PR condition can be interpreted similarly.

$$\left(\frac{1}{\sqrt{2}}, \frac{1}{\sqrt{2}}\right) * \left(\frac{1}{\sqrt{2}}, -\frac{1}{\sqrt{2}}\right) + \left(\frac{1}{\sqrt{2}}, -\frac{1}{\sqrt{2}}\right) * \left(-\frac{1}{\sqrt{2}}, -\frac{1}{\sqrt{2}}\right) = \left(\frac{1}{2}, 0, -\frac{1}{2}\right)$$
$$+ \left(-\frac{1}{2}, 0, \frac{1}{2}\right) = (0, 0, 0)$$

■

7.4 Polyphase Form of the Two-Channel Filter Bank

7.4.1 Decimation

In the straightforward implementation of the decimation operation, the filter works at a sampling rate of f_s, whereas the sampling rate of the decimated output is $f_s/2$. That is, the filter has to produce an output in each sampling interval, which is $1/f_s$. By partitioning the input signal and the filter coefficients into even- and odd-indexed groups, the filter can work at a lower sampling rate of $f_s/2$. Each of the set of two smaller filters is called a polyphase filter. Each filter produces an output sample every $2/f_s$ time intervals. The word polyphase indicates that there is a delay between the odd-indexed and the even-indexed input values.

The difference equation of an FIR filter with impulse response $h(n), n = 0, 1, \ldots, N-1$, relating the input $x(n)$ and the output $y(n)$, is given by

$$y(n) = h(0)x(n) + h(1)x(n-1) + \cdots + h(N-1)x(n-N+1)$$

The output of a four-coefficient FIR filter, for some consecutive samples, is given by

$$y(3) = h(0)x(3) + h(1)x(2) + h(2)x(1) + h(3)x(0)$$
$$y(4) = h(0)x(4) + h(1)x(3) + h(2)x(2) + h(3)x(1)$$
$$y(5) = h(0)x(5) + h(1)x(4) + h(2)x(3) + h(3)x(2)$$
$$y(6) = h(0)x(6) + h(1)x(5) + h(2)x(4) + h(3)x(3)$$

The downsampled output of the FIR filter, consisting of the even-indexed output samples $y_e(n)$, is given by

$$y(4) = h(0)x(4) + h(1)x(3) + h(2)x(2) + h(3)x(1)$$
$$y(6) = h(0)x(6) + h(1)x(5) + h(2)x(4) + h(3)x(3)$$

The product terms are formed using the even-indexed input samples $x_e(n)$ and the even-indexed coefficients $h_e(n)$ or the odd-indexed input samples $x_o(n)$ and the odd-indexed coefficients $h_o(n)$.

$$y(4) = (h(0)x(4) + h(2)x(2)) + (h(1)x(3) + h(3)x(1))$$
$$y(6) = (h(0)x(6) + h(2)x(4)) + (h(1)x(5) + h(3)x(3))$$

Therefore, the even-indexed input samples can be convolved with the even-indexed coefficients. Simultaneously, the odd-indexed input samples can be convolved with the odd-indexed coefficients. The output can be formed from the two partial results. Noting that the odd-indexed samples are delayed by a sampling interval and taking the z-transform, we get

$$Y_e(z) = X_e(z)H_e(z) + z^{-1}X_o(z)H_o(z) \tag{7.13}$$

The operation of a decimator with a four-coefficient filter

$$H(z) = h(0) + h(1)z^{-1} + h(2)z^{-2} + h(3)z^{-3} \tag{7.14}$$

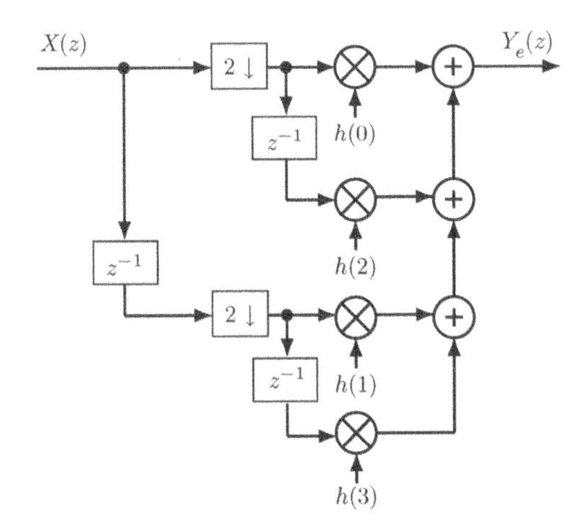

Figure 7.9 Polyphase form of an FIR decimation filter

is characterized by

$$Y_e(z) = \left[h(0) + h(2)z^{-1} \;\; h(1) + h(3)z^{-1}\right] \begin{bmatrix} X_e(z) \\ z^{-1}X_0(z) \end{bmatrix} \tag{7.15}$$

The polyphase implementation of a four-coefficient FIR decimation filter is shown in Figure 7.9. The even-indexed input $X_e(z)$ is derived from the input $X(z)$ by downsampling by a factor of 2. The odd-indexed input $X_o(z)$ is derived from the input $X(z)$ by delaying by one sampling interval and then downsampling by a factor of 2. The even-indexed input is convolved with the even-indexed filter coefficients, and odd-indexed input is convolved with the odd-indexed filter coefficients simultaneously. The sum of the outputs of the two sections is the output of the filter.

Example 7.5 Let the input signal be

$$x(n) = \{2, -1, 3, 0, 4, -2, 3, 2\}$$

Let the impulse response of the filter be

$$h(n) = \{1, 3, -2, 4\}$$

Find the even-indexed convolution output of $x(n)$ and $h(n)$ directly and by the polyphase approach.

Solution

The output $y(n)$ of convolving $h(n)$ and $x(n)$ is

$$y(n) = x(n) * h(n) = \{2, 5, -4, 19, -6, 22, -11, 31, -8, 8, 8\}$$

The downsampled output $y_e(n)$ is

$$y_e(n) = \{2, -4, -6, -11, -8, 8\}$$

The outputs of convolving the even- and odd-indexed coefficients of $h(n)$ with those of $x(n)$ are, respectively,

$$\{1, -2\} * \{2, 3, 4, 3\} = \{2, -1, -2, -5, -6\}$$
$$\{3, 4\} * \{-1, 0, -2, 2\} = \{-3, -4, -6, -2, 8\}$$

The output $y_e(n)$ is

$$\{2, -1, -2, -5, -6, 0\} + \{0, -3, -4, -6, -2, 8\} = \{2, -4, -6, -11, -8, 8\}$$

In the z-domain (from Equation (7.15)), we get

$$\begin{aligned} Y_e(z) &= \begin{bmatrix} 1 - 2z^{-1} & 3 + 4z^{-1} \end{bmatrix} \begin{bmatrix} 2 + 3z^{-1} + 4z^{-2} + 3z^{-3} \\ z^{-1}\left(-1 - 2z^{-2} + 2z^{-3}\right) \end{bmatrix} \\ &= (2 - z^{-1} - 2z^{-2} - 5z^{-3} - 6z^{-4}) + z^{-1}(-3 - 4z^{-1} - 6z^{-2} - 2z^{-3} + 8z^{-4}) \\ &= 2 - 4z^{-1} - 6z^{-2} - 11z^{-3} - 8z^{-4} + 8z^{-5} \end{aligned}$$

■

The analysis filter bank output can be written as

$$X_\phi(z) = L_e(z)X_e(z) + z^{-1}L_o(z)X_o(z)$$
$$X_\psi(z) = H_e(z)X_e(z) + z^{-1}H_o(z)X_o(z)$$

Using matrix notation, we get

$$\begin{bmatrix} X_\phi(z) \\ X_\psi(z) \end{bmatrix} = \begin{bmatrix} L_e(z) & L_o(z) \\ H_e(z) & H_o(z) \end{bmatrix} \begin{bmatrix} X_e(z) \\ z^{-1}X_o(z) \end{bmatrix} = P(z) \begin{bmatrix} X_e(z) \\ z^{-1}X_o(z) \end{bmatrix} \tag{7.16}$$

where $P(z)$ is a polyphase matrix of the analysis filter bank. The polyphase matrix is the z-transform of a block of filters. There are four filters from the two phases of the two original filters. As two parallel filters of half the size work simultaneously, the execution time is reduced by one-half if the same processors that are used without polyphase are used. Otherwise, processors with half the speed of those used without polyphase can be used with the same execution time.

A general proof for polyphase decomposition is as follows. Let the input be $x(n)$, and the impulse response of the filter be $h(n)$. Then, the DTFT of the even part of the sequences is given by

$$\frac{1}{2}(X(e^{j\omega}) + X(e^{j(\omega+\pi)})) \quad \text{and} \quad \frac{1}{2}(H(e^{j\omega}) + H(e^{j(\omega+\pi)})) \tag{7.17}$$

Similarly, the DTFT of the odd part of the sequences is given by

$$\frac{1}{2}(X(e^{j\omega}) - X(e^{j(\omega+\pi)})) \quad \text{and} \quad \frac{1}{2}(H(e^{j\omega}) - H(e^{j(\omega+\pi)})) \tag{7.18}$$

The DTFT of $x(n)$ and $h(n)$ is, respectively, the sum of their even and odd parts. The even part of the product of $X(e^{j\omega})$ and $H(e^{j\omega})$, $X(e^{j\omega})H(e^{j\omega})$, is given by

$$\begin{aligned} &\tfrac{1}{2}\left(X(e^{j\omega})H(e^{j\omega}) + X(e^{j(\omega+\pi)})H(e^{j(\omega+\pi)})\right) \\ &= \frac{1}{4}(X(e^{j\omega}) + X(e^{j(\omega+\pi)}))(H(e^{j\omega}) + H(e^{j(\omega+\pi)})) \\ &\quad + \frac{1}{4}(X(e^{j\omega}) - X(e^{j(\omega+\pi)}))(H(e^{j\omega}) - H(e^{j(\omega+\pi)})) \end{aligned}$$

On the right side, the product is formed with powers of even and even or odd and odd of the DTFT series. Adding the even and even or odd and odd exponents results in even-indexed exponents. The two terms on the right side can be computed simultaneously.

Example 7.6 Consider the transform pair

$$x(n) = \{x(0) = 1, x(1) = 2, x(2) = 1, x(3) = 3\} \leftrightarrow X(e^{j\omega}) = 1 + 2e^{-j\omega} + e^{-j2\omega} + 3e^{-j3\omega}$$
$$h(n) = \{x(0) = 2, x(1) = 1, x(2) = 3, x(3) = 4\} \leftrightarrow H(e^{j\omega}) = 2 + e^{-j\omega} + 3e^{-j2\omega} + 4e^{-j3\omega}$$

Find the even-indexed output of convolving $x(n)$ and $h(n)$.

Solution

The even and odd components of $x(n)$ and their DTFT are

$$x_e(n) = \{x(0) = 1, x(2) = 1\} \leftrightarrow X_e(e^{j\omega}) = 1 + e^{-j2\omega}$$
$$x_o(n) = \{x(1) = 2, x(3) = 3\} \leftrightarrow X_o(e^{j\omega}) = 2e^{-j\omega} + 3e^{-j3\omega}$$

The even and odd components of $h(n)$ and their DTFT are

$$h_e(n) = \{x(0) = 2, x(2) = 3\} \leftrightarrow H_e(e^{j\omega}) = 2 + 3e^{-j2\omega}$$
$$h_o(n) = \{x(1) = 1, x(3) = 4\} \leftrightarrow H_o(e^{j\omega}) = e^{-j\omega} + 4e^{-j3\omega}$$

The convolution of $x(n)$ and $h(n)$ and the product of $X(e^{j\omega})$ and $H(e^{j\omega})$ are

$$y(n) = \{y(0) = 2, y(1) = 5, y(2) = 7, y(3) = 17, y(4) = 14, y(5) = 13, y(6) = 12\} \leftrightarrow$$
$$X(e^{j\omega})H(e^{j\omega}) = 2 + 5e^{-j\omega} + 7e^{-j2\omega} + 17e^{-j3\omega} + 14e^{-j4\omega} + 13e^{-j5\omega} + 12e^{-j6\omega}$$

The sum of the products formed with powers of even and even and odd and odd of the two DTFT series is

$$(1 + e^{-j2\omega})(2 + 3e^{-j2\omega}) + (2e^{-j\omega} + 3e^{-j3\omega})(e^{-j\omega} + 4e^{-j3\omega}) = (2 + 5e^{-j2\omega} + 3e^{-j4\omega})$$
$$+(2e^{-j2\omega} + 11e^{-j4\omega} + 12e^{-j6\omega}) = 2 + 7e^{-j2\omega} + 14e^{-j4\omega} + 12e^{-j6\omega}$$

which is the even part of the product $X(e^{j\omega})H(e^{j\omega})$. ■

7.4.2 Interpolation

In the polyphase implementation of interpolation, filtering is implemented at a lower rate, and the output is upsampled. This is in contrast to upsampling the signal first and filtering the signal at a higher rate. Assume that the interpolation factor is $M = 2$. Then, one sample with zero value is inserted between every two adjacent samples of the given signal $x(n)$. The output of a four-coefficient FIR filter, for some consecutive samples, is given by

$$y(4) = h(0)x(2) + h(1)0 \quad + h(2)x(1) + h(3)0$$
$$y(5) = h(0)0 \quad + h(1)x(2) + h(2)0 \quad + h(3)x(1)$$
$$y(6) = h(0)x(3) + h(1)0 \quad + h(2)x(2) + h(3)0$$
$$y(7) = h(0)0 \quad + h(1)x(3) + h(2)0 \quad + h(3)x(2)$$

Discarding the terms corresponding to zero inputs, we get

$$y(4) = h(0)x(2) + h(2)x(1)$$
$$y(5) = h(1)x(2) + h(3)x(1)$$
$$y(6) = h(0)x(3) + h(2)x(2)$$
$$y(7) = h(1)x(3) + h(3)x(2)$$

That is, the even- and odd-indexed samples of the output $y(n)$ can be obtained by the convolution of $x(n)$ separately with the even- and odd-indexed coefficients of $h(n)$, respectively. Inserting zeros between the samples of the two output sets and summing, we get

$$\{\dots, y(4), 0, y(6), 0, \dots\} + \{\dots, 0, y(5), 0, y(7), \dots\} = \{\dots, y(4), y(5), y(6), y(7), \dots\}$$

In practice, we just interleave the two partial outputs to get the interpolated signal.

Example 7.7 Let the input signal be

$$x_u(n) = \{4, 0, 2, 0, 1, 0, 3, 0\}$$

Let the impulse response of the filter be

$$h(n) = \{1, 2, 3, 4\}$$

Find the convolution of the signals using the direct and polyphase method.

Solution

The output of convolving $h(n)$ and $x_u(n)$ is

$$x_u(n) * h(n) = \{4, 8, 14, 20, 7, 10, 6, 10, 9, 12, 0\}$$

The downsampled input signal, by a factor of 2, is

$$x(n) = \{4, 2, 1, 3\}$$

In the polyphase form of interpolation, the filter is split into two, one with even-indexed coefficients of the original filter and the other with odd-indexed coefficients. The common input to these filters comes at half the sampling rate of the upsampled input in the case when the upsampler is placed before the filter. The outputs of these filters are merged to form the interpolated output. In the direct form, the upsampler produces $X(z^2)$ from the input $X(z)$. The output is $Y(z) = X(z^2)H(z)$, where $H(z)$ is the z-transform transfer function of the filter. The even powers in $H(z)$ produces the even powers in $Y(z)$. The odd powers in $H(z)$ produces the odd powers in $Y(z)$. For this example,

$$X(z^2) = 4 + 2z^{-2} + z^{-4} + 3z^{-6} \quad \text{and} \quad H(z) = 1 + 2z^{-1} + 3z^{-2} + 4z^{-3}$$

The output $Y(z) = X(z^2)H(z)$ is given by

$$Y(z) = (4)\underline{(1)} + (4)\overline{(2z^{-1})} + ((2z^{-2})(\underline{1}) + (4)(\underline{3z^{-2}})) + ((4)\overline{(4z^{-3})} + (2z^{-2})\overline{(2z^{-1})}) + \cdots$$

The outputs of convolving $x(n)$ with the even- and odd-indexed coefficients of $h(n)$ are

$$y_e(n) = x(n) * h_e(n) = \{4, 2, 1, 3\} * \{1, 3\} = \{4, 14, 7, 6, 9\}$$
$$y_o(n+1) = x(n) * h_o(n) = \{4, 2, 1, 3\} * \{2, 4\} = \{8, 20, 10, 10, 12\}$$

The output $y(n)$ is obtained by interleaving the two partial outputs.

$$y(n) = 4 \searrow 8 \nearrow 14 \searrow 20 \nearrow 7 \searrow 10 \nearrow 6 \searrow 10 \nearrow 9 \searrow 12$$

Using the z-transform, we get

$$Y_e(z) = (1 + 3z^{-1})(4 + 2z^{-1} + z^{-2} + 3z^{-3}) = (4 + 14z^{-1} + 7z^{-2} + 6z^{-3} + 9z^{-4})$$
$$zY_o(z) = (2 + 4z^{-1})(4 + 2z^{-1} + z^{-2} + 3z^{-3}) = (8 + 20z^{-1} + 10z^{-2} + 10z^{-3} + 12z^{-4})$$

These are the outputs of the polyphase filters. After passing through the upsamplers, they become

$$Y_e(z^2) = (4 + 14z^{-2} + 7z^{-4} + 6z^{-6} + 9z^{-8})$$
$$zY_o(z^2) = (8 + 20z^{-2} + 10z^{-4} + 10z^{-6} + 12z^{-8})$$

The delayed odd-phase output is

$$Y_o(z^2) = (8z^{-1} + 20z^{-3} + 10z^{-5} + 10z^{-7} + 12z^{-9})$$

The output is the sum of the upsampled even-phase component and the upsampled and delayed odd-phase component

$$Y(z) = (4 + 8z^{-1} + 14z^{-2} + 20z^{-3} + 7z^{-4} + 10z^{-5} + 6z^{-6} + 10z^{-7} + 9z^{-8} + 12z^{-9})$$ ■

At each sample instant at the output side, only one of the two output components is nonzero. Two output samples in the two phases are produced for each input sample by the two polyphase filters, which work at half the sampling rate of the output. Overall, the polyphase implementation is a parallel realization of the polyphase filters.

The polyphase implementation of an FIR interpolation filter

$$H(z) = h(0) + h(1)z^{-1} + h(2)z^{-2} + h(3)z^{-3}$$

with four coefficients is shown in Figure 7.10. The input is common to both the even- and odd-indexed coefficient filters. The first filter produces even-indexed output, and the second produces odd-indexed output. Both the outputs are upsampled by a factor of 2. The odd-indexed output is delayed and then added to the first partial output to get the interpolated output.

Noting that the odd-indexed samples are delayed by a sampling interval and taking the z-transform, we get

$$Y_e(z) = X(z)H_e(z)$$
$$zY_o(z) = X(z)H_o(z)$$

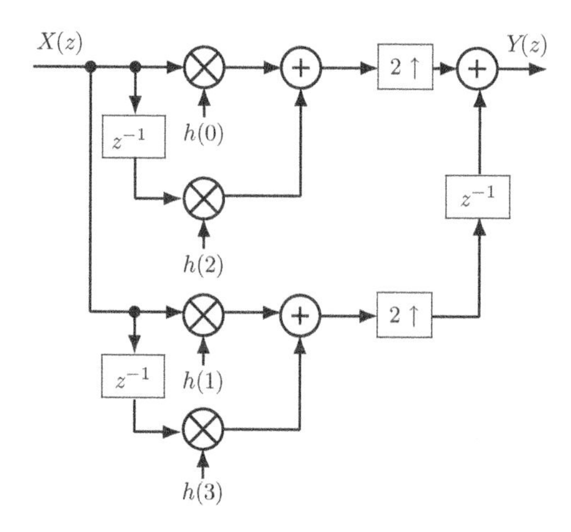

Figure 7.10 Polyphase form of an FIR interpolation filter

The z-transform of the synthesis filter bank output is the sum of the lower and upper channels, which are

$$X_e(z) = X_\phi(z)\widetilde{L}_e(z) + X_\psi(z)\widetilde{H}_e(z)$$
$$zX_o(z) = X_\phi(z)\widetilde{L}_o(z) + X_\psi(z)\widetilde{H}_o(z)$$

In matrix form,

$$\begin{bmatrix} X_e(z) \\ zX_o(z) \end{bmatrix} = \begin{bmatrix} \widetilde{L}_e(z) & \widetilde{H}_e(z) \\ \widetilde{L}_o(z) & \widetilde{H}_o(z) \end{bmatrix} \begin{bmatrix} X_\phi(z) \\ X_\psi(z) \end{bmatrix} = \widetilde{P}(z) \begin{bmatrix} X_\phi(z) \\ X_\psi(z) \end{bmatrix} \tag{7.19}$$

where $\widetilde{P}(z)$ is the polyphase matrix of the synthesis filter bank.

Example 7.8 Let the input signal be

$$x(n) = \{2, -1, 3, 0, 4, -2, 3, 2\}$$

Let the polyphase matrices be

$$P(z) = \widetilde{P}(z) = \begin{bmatrix} 1 & 1 \\ 1 & -1 \end{bmatrix}$$

Find the output of the analysis filter bank. Using this output, reconstruct the input signal in the synthesis filter bank.

Solution

From Equation (7.16), we get

$$\begin{bmatrix} X_\phi(2,n) \\ X_\psi(2,n) \end{bmatrix} = \begin{bmatrix} 1 & 1 \\ 1 & -1 \end{bmatrix} \begin{bmatrix} 2 & 3 & 4 & 3 \\ -1 & 0 & -2 & 2 \end{bmatrix} = \begin{bmatrix} 1 & 3 & 2 & 5 \\ 3 & 3 & 6 & 1 \end{bmatrix}$$

From Equation (7.19) and scaling by 2, we get

$$\begin{bmatrix} x_e(n) \\ x_o(n) \end{bmatrix} = \frac{1}{2}\begin{bmatrix} 1 & 1 \\ 1 & -1 \end{bmatrix}\begin{bmatrix} 1 & 3 & 2 & 5 \\ 3 & 3 & 6 & 1 \end{bmatrix} = \begin{bmatrix} 2 & 3 & 4 & 3 \\ -1 & 0 & -2 & 2 \end{bmatrix}$$

The scaling by 2 is required as

$$P(z)\widetilde{P}(z) = 2I$$

■

7.4.3 *Polyphase Form of the Filter Bank*

Figure 7.11 shows a two-channel filter bank with no filtering or with an identity polyphase matrix.

$$\begin{bmatrix} X_\phi(z) \\ X_\psi(z) \end{bmatrix} = \begin{bmatrix} 1 & 0 \\ 0 & 1 \end{bmatrix}\begin{bmatrix} X_e(z) \\ z^{-1}X_o(z) \end{bmatrix} = \begin{bmatrix} X_e(z) \\ z^{-1}X_o(z) \end{bmatrix} \tag{7.20}$$

In a practical filter bank, the polyphase matrix is not an identity matrix. Still, it is important to study the filter bank with no filtering for understanding the flow of the data sequence through the filter bank. With no filtering, the input data remain the same. With filters, the data are modified in a desired way, but the flow of the data sequence still remains the same. In the upper channel, the downsampling and upsampling of $x(n)$ replace all the odd-indexed values by zeros ($xe_u(n)$), and this sequence is delayed by one sample interval ($xe_u(n-1)$). In the lower channel, input $x(n)$ is first delayed, and the downsampling and upsampling of $x(n-1)$ replace all the even-indexed values by zeros ($xo_u(n-1)$). The sum of these two sequences is $x(n-1)$, the delayed input. Let the signal be

$$x(n) = \ldots, x(-1), x(0), x(1), \ldots$$

The value of the independent variable n is shown in the first row of Table 7.1, and $x(n)$ is shown in the second row. Then, the downsampled signal $xe(n)$ in the upper channel

$$\ldots, x(-2), x(0), x(2), \ldots$$

is shown in the third row. In order to get the odd-indexed values of the input $x(n)$, we have to delay and then downsample. The downsampled signal $xo(n-1)$ in the lower channel

$$\ldots, x(-3), x(-1), x(1), \ldots$$

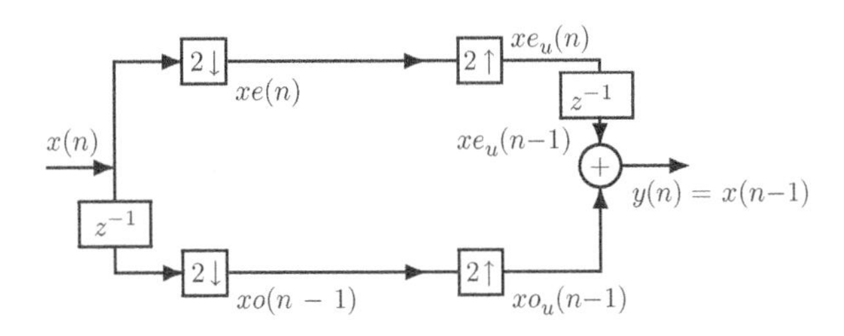

Figure 7.11 A two-channel filter bank with no filtering

Table 7.1 The transformation of a sequence $x(n)$ through a two-channel filter bank with no filtering

n	...	-3	-2	-1	0	1	2	3	4	5	...
$x(n)$	...	$x(-3)$	$x(-2)$	$x(-1)$	$x(0)$	$x(1)$	$x(2)$	$x(3)$	$x(4)$	$x(5)$	...
$xe(n)$	...	$x(-6)$	$x(-4)$	$x(-2)$	$x(0)$	$x(2)$	$x(4)$	$x(6)$	$x(8)$	$x(10)$	...
$xo(n-1)$	...	$x(-7)$	$x(-5)$	$x(-3)$	$x(-1)$	$x(1)$	$x(3)$	$x(5)$	$x(7)$	$x(9)$	...
$xe_u(n)$	...	0	$x(-2)$	0	$x(0)$	0	$x(2)$	0	$x(4)$	0	...
$xo_u(n-1)$	...	0	$x(-3)$	0	$x(-1)$	0	$x(1)$	0	$x(3)$	0	...
$xe_u(n-1)$	...	$x(4)$	0	$x(-2)$	0	$x(0)$	0	$x(2)$	0	$x(4)$	...
$y(n) = x(n-1)$	...	$x(-4)$	$x(-3)$	$x(-2)$	$x(-1)$	$x(0)$	$x(1)$	$x(2)$	$x(3)$	$x(4)$	...

is shown in the fourth row. The corresponding upsampled signals $xe_u(n)$ and $xo_u(n-1)$

$$\ldots, x(-2), 0, x(0), 0, x(2), 0, \ldots$$

$$\ldots, x(-3), 0, x(-1), 0, x(1), 0 \ldots$$

are shown, respectively, in the fifth and sixth rows. The signal in the upper channel $xe_u(n)$ is delayed by one sample interval to get $xe_u(n-1)$

$$\ldots, 0, x(-2), 0, x(0), 0, x(2), 0, \ldots$$

which is shown in the seventh row. This delay is required to compensate the delay introduced for the odd part. Now, the sum of the signals in the sixth and seventh rows is the reconstructed signal, which is the same as the input but delayed by one sample interval, as shown in the last row.

$$\ldots, x(-3), 0, x(-1), 0, x(1), 0 \ldots + \ldots, 0, x(-2), 0, x(0), 0, x(2), 0, \ldots = x(n-1)$$

Summarizing, the input sequence is split into even- and odd-indexed sequences, upsampled, realigned using a delay in the upper channel, and added to get the delayed version of the input. Whether the signal is filtered or not in the filter bank, the input signal must be reconstructed perfectly, except for a delay, from the unaltered decomposed components in a perfect reconstruction filter bank. The delay of the reconstructed output will be proportional to the length of the filters.

Polyphase approach of implementing a single filter can be easily extended to the implementation of the filter banks. The downsamplers and upsamplers are placed inside in the direct implementation of the filter bank shown in Figure 7.8, which is inefficient. In contrast, the downsamplers and upsamplers are placed outside, and the polyphase filters are placed side by side in the polyphase implementation of the filter bank, which is efficient. Polyphase form of a two-channel filter bank is shown in Figure 7.12 and, in detail, in Figure 7.13. The polyphase filters are formed using the even- and odd-indexed coefficients of the analysis and synthesis filters, and they are time invariant. The polyphase matrices can be inverses of each other, and their elements are polynomials. With no processing of the analysis coefficients, the synthesis bank exactly reconstructs the input to the analysis bank with a delay. That is,

$$P(z)\widetilde{P}(z) = z^{-K}\mathbf{I} \qquad (7.21)$$

While the input–output relationship of both forms of filter bank is the same, the polyphase form is more efficient and easy to analyze using the matrix Equations (7.16) and (7.19). In the

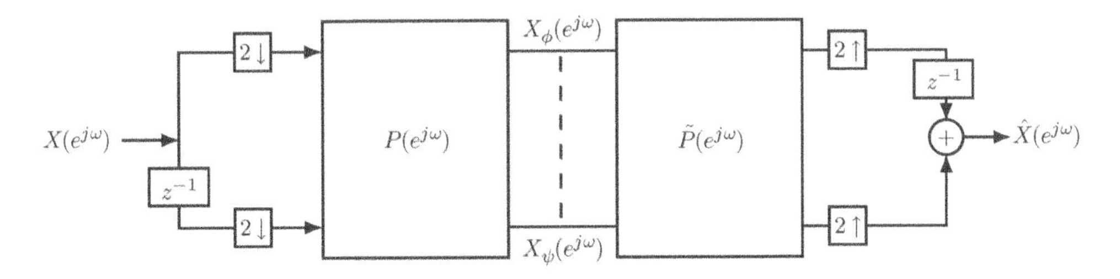

Figure 7.12 Polyphase form of a two-channel analysis and synthesis filter bank

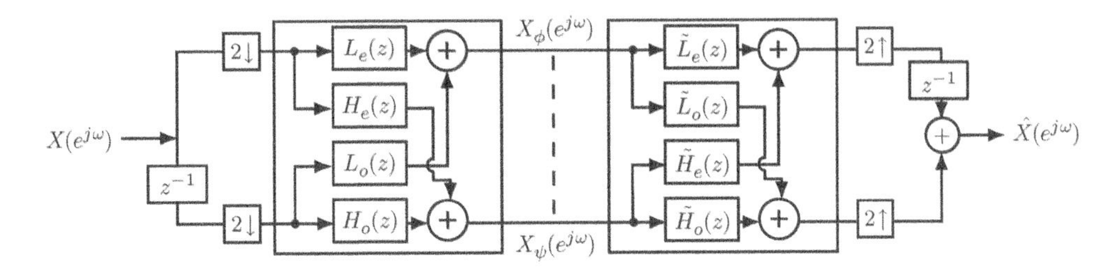

Figure 7.13 Polyphase form of a two-channel analysis and synthesis filter bank, in detail

direct form of the filter bank, the input is processed separately by the two channels, and their outputs are added to get the reconstructed signal. In the polyphase form, the input is split into even- and odd-phase components and processed to produce the two analysis bank outputs. The inputs to the synthesis part are at half the rate, while the output is at full rate (the same as that of the input to the filter bank). The synthesis part produces the even- and odd-phase components of the output and merges them to produce the reconstructed signal. At each sampling instant at the output side, only one output component is nonzero.

Example 7.9 Assume that the polyphase matrices are

$$P(z) = \widetilde{P}(z) = \begin{bmatrix} 1 & 1 \\ 1 & -1 \end{bmatrix}$$

The filter bank is shown in Figure 7.14. Make a table to show the processing of the input $x(n)$ by the filter bank.

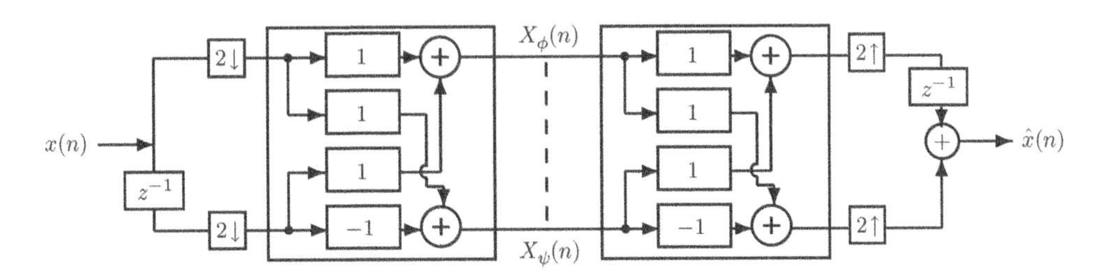

Figure 7.14 Example of the polyphase form of a two-channel analysis and synthesis filter bank

Table 7.2 The transformation of a sequence $x(n)$ through a two-channel filter bank with filtering

n	-1	0	1	2
$x(n)$	$x(-1)$	$x(0)$	$x(1)$	$x(2)$
$xe(n)$	$x(-2)$	$x(0)$	$x(2)$	$x(4)$
$xo(n-1)$	$x(-3)$	$x(-1)$	$x(1)$	$x(3)$
$X_\phi(n) = xe(n) + xo(n-1)$	$x(-2)+x(-3)$	$x(0)+x(-1)$	$x(2)+x(1)$	$x(4)+x(3)$
$X_\psi(n) = xe(n) - xo(n-1)$	$x(-2)-x(-3)$	$x(0)-x(-1)$	$x(2)-x(1)$	$x(4)-x(3)$
$xa(n) = X_\phi(n) + X_\psi(n)$	$2x(-2)$	$2x(0)$	$2x(2)$	$2x(4)$
$xb(n) = X_\phi(n) - X_\psi(n)$	$2x(-3)$	$2x(-1)$	$2x(1)$	$2x(3)$
$xa_u(n-1)$	$2x(-2)$	0	$2x(0)$	0
$xb_u(n)$	0	$2x(-1)$	0	$2x(1)$
$y(n) = 2x(n-1)$	$2x(-2)$	$2x(-1)$	$2x(0)$	$2x(1)$

Solution

The processing of the input $x(n)$, shown in Table 7.2, is similar to that in the filter bank with no filtering, except that the output of the analysis filter bank is a linear combination of the samples of the input. Synthesis bank filters and samplers reconstruct the input from the output of the analysis filter bank. The factor 2 in the output is due to the fact that

$$P(z)\widetilde{P}(z) = 2I$$

■

7.5 Summary

- Processing a signal at more than one sampling rate is called multirate digital signal processing.
- Multirate digital signal processing provides higher efficiency in signal processing applications.
- Downsampling and upsampling are the two basic operations used, respectively, to reduce and increase the sampling rate in multirate digital signal processing.
- Filtering followed by downsampling is called the decimation operation. Upsampling followed by filtering is called the interpolation operation.
- Sampling rate conversion is realized using decimation and interpolation operations.
- Filter banks are the basic implementation structures of the DWT. A filter bank is a set of filters, samplers, and delay units.
- A two-channel analysis filter bank consists of a lowpass filter followed by a downsampler in one channel and a highpass filter followed by a downsampler in the other channel.
- A two-channel synthesis filter bank consists of a lowpass filter preceded by an upsampler in one channel and a highpass filter preceded by an upsampler in the other channel.
- Polyphase filter structures provide an efficient implementation of sampling rate conversion systems.

Exercises

7.1 Prove that downsamplers are linear and time-varying systems. Assume that the downsamping factor is 2.

7.2 Given a signal $x(n)$, find an expression for its DTFT. Using the DTFT, find an expression for the DTFT of $x(2n)$. Verify the answer by computing the DTFT of $x(2n)$ directly.

7.2.1 $x(-1) = 1, x(0) = 1, x(1) = 1$ and $x(n) = 0$ otherwise.
7.2.2 $x(-1) = 1, x(0) = 2, x(1) = 3$ and $x(n) = 0$ otherwise.
***7.2.3** $x(-2) = 5, x(-1) = 1, x(0) = 1, x(1) = 1, x(2) = 5$ and $x(n) = 0$ otherwise.
7.2.4 $x(n) = a^n u(n), |a| < 1$.

7.3 Given a signal $x(n)$, find an expression for its DTFT. Using the DTFT, find an expression for the DTFT of $x(n/2)$. Verify the answer by computing the DTFT of $x(n/2)$ directly.

7.3.1 $x(-1) = 1, x(0) = 1, x(1) = 1$ and $x(n) = 0$ otherwise.
7.3.2 $x(-1) = 1, x(0) = 2, x(1) = 3$ and $x(n) = 0$ otherwise.
7.3.3 $x(-2) = 1, x(-1) = 1, x(0) = 1, x(1) = 1, x(2) = 1$ and $x(n) = 0$ otherwise.
***7.3.4** $x(n) = (0.8)^n u(n)$.

7.4 Figure 7.15 shows a two-channel filter bank. Derive the input–output relationship $y(n) = x(n-1)$ in the time domain and the frequency domain.

$$x(n) = \dots, x(-1), x(0), x(1), \dots$$

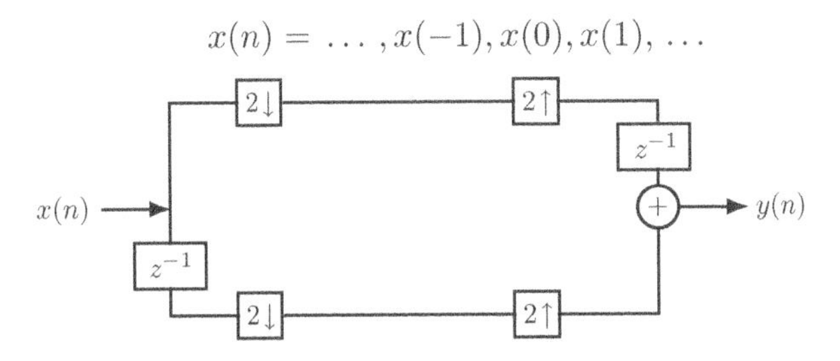

Figure 7.15 A two-channel filter bank

7.5 Verify perfect reconstruction conditions for the filters given as follows in the frequency domain.

$$L(e^{j\omega}) = \frac{1}{4\sqrt{2}}((1-\sqrt{3}) + (3-\sqrt{3})e^{-j\omega} + (3+\sqrt{3})e^{-j2\omega} + (1+\sqrt{3})e^{-j3\omega})$$

$$H(e^{j\omega}) = \frac{1}{4\sqrt{2}}(-(1+\sqrt{3}) + (3+\sqrt{3})e^{-j\omega} - (3-\sqrt{3})e^{-j2\omega} + (1-\sqrt{3})e^{-j3\omega})$$

$$\widetilde{L}(e^{j\omega}) = \frac{1}{4\sqrt{2}}((1+\sqrt{3}) + (3+\sqrt{3})e^{-j\omega} + (3-\sqrt{3})e^{-j2\omega} + (1-\sqrt{3})e^{-j3\omega})$$

$$\widetilde{H}(e^{j\omega}) = \frac{1}{4\sqrt{2}}((1-\sqrt{3}) - (3-\sqrt{3})e^{-j\omega} + (3+\sqrt{3})e^{-j2\omega} - (1+\sqrt{3})e^{-j3\omega})$$

7.6 Given the input $x(n)$ and the impulse response $h(n)$, find the decimated output of a two-phase polyphase filter: (a) in the time domain and (b) in the z-domain. Verify the output by directly convolving $x(n)$ and $h(n)$ and then downsampling the result.

7.6.1

$$x(n) = \{-2, -1, 3, 4, -4, 3, -2, 1\}$$
$$h(n) = \{1, -1, 3, 2\}$$

7.6.2

$$x(n) = \{5, -3, 1, 2, 1, -1, 3, 2\}$$
$$h(n) = \{4, 1, -3, 2\}$$

***7.6.3**

$$x(n) = \{1, 4, 3, -1, 3, 2, 2, -4\}$$
$$h(n) = \{3, 1, -3, 2\}$$

7.7 Given the input $x(n)$ and the impulse response $h(n)$, find the interpolated output of a two-phase polyphase filter: (a) in the time domain and (b) in the z-domain. Verify the output by upsampling the input $x(n)$ and then convolving it with $h(n)$.

7.7.1

$$x(n) = \{-2, -1, 3, 4\}$$
$$h(n) = \{1, -1, 3, 2\}$$

***7.7.2**

$$x(n) = \{1, 4, 3, -1\}$$
$$h(n) = \{3, 1, -3, 2\}$$

7.7.3

$$x(n) = \{5, -3, 1, 2\}$$
$$h(n) = \{4, 1, -3, 2\}$$

7.8 Let the input signal to a two-channel filter bank be $x(n)$. Let the polyphase matrices be

$$P(z) = \widetilde{P}(z) = \begin{bmatrix} 1 & 1 \\ 1 & -1 \end{bmatrix}$$

Verify that the output of the filter bank is the same as the input multiplied by 2.

7.8.1

$$x(n) = \{1, 4, 3, -1, 3, 2, 2, -4\}$$

7.8.2

$$x(n) = \{5, -3, 1, 2, 1, -1, 3, 2\}$$

7.8.3

$$x(n) = \{-2, -1, 3, 4, -4, 3, -2, 1\}$$

8

The Haar Discrete Wavelet Transform

8.1 Introduction

8.1.1 Signal Representation

In signal processing, appropriate representation of practical signals, which usually have arbitrary amplitude profile, is the first step. The most important representations of signals are in terms of impulse and sinusoidal signals, respectively, in the time and frequency domains. The representation in terms of impulses provides the ideal time resolution, and the representation in terms of sinusoids provides the ideal frequency resolution. The spectrum of an impulse is uniform and of infinite bandwidth. The impulse is defined as a rectangular pulse of unit area with its width approaching zero. Instead of allowing the width to approach zero, let us use rectangular pulses of finite width. Then, the spectrum of the signal becomes finite approximately. Now, we get neither the ideal time resolution nor the ideal frequency resolution. However, we get a compromised time and frequency resolutions. Therefore, an arbitrary signal may be considered as a linear combination of appropriately selected rectangular pulses. The pulses may be positive or negative and of suitable widths and heights. In this chapter, we use rectangular pulses to represent signals. However, pulses of arbitrary shapes are also used, as presented later. This is the essence of the wavelet transform. What is more is that a lot of details have to be taken into account so that this representation becomes practical.

Now, we consider the wavelet transform from sinusoidal waveform point of view. Instead of considering individual frequency components, we consider frequency components over subbands of the spectrum of a signal. The corresponding time-domain signals are approximately of finite duration, resulting in the time-frequency representation of signals. Using the linearity property, a signal can be considered as composed of two components. One is the low-frequency component, and the other is the high-frequency component. Both the components are obtained from the signal using ideal lowpass and highpass filters with the same cutoff frequency. The high-frequency component is the difference between the original signal and the low-frequency component. The high-frequency component represents the rapidly varying component of the signal that is complementary to the low-frequency component.

Discrete Wavelet Transform: A Signal Processing Approach, First Edition. D. Sundararajan.

Companion Website: www.wiley.com/go/sundararajan/wavelet

Modifying the Fourier transform to obtain the time-frequency representation of signals has been in practice for a long time. The wavelet transform approach has also been in existence for a long time. In recent times, the details of the theory of time-frequency representation have been rigorously developed. The result is that this representation, called the wavelet transform representation of signals, is inherently suitable to nonstationary signal analysis and is widely used in applications.

8.1.2 The Wavelet Transform Concept

The key idea of the wavelet transform is the multiresolution decomposition of signals and images. This is advantageous because small objects require high resolution, and low resolution is enough for large objects. As typical images contain objects of varying sizes, multiresolution study is required. The general approach in multiresolution decomposition is to create an approximation component using a scaling function (a lowpass filter) and detail components using wavelet functions (highpass filters). A series of approximations of a signal or image is created, each differing in resolution by a factor of 2. The detail components contain the difference between adjacent approximations. Similar to Fourier analysis, there are different versions of wavelet analysis, the continuous wavelet series and transform and the DWT. The basis functions for continuous wavelet transform are the time-shifted and scaled versions of the scaling and wavelet functions. The basis functions for the DWT are the filter coefficients. Each of these versions differs in that signals and their transforms are continuous or discrete. The DWT is similar to the DFT of the Fourier analysis and is widely used in practical applications. The principle remains the same as that of the other transforms. The signal is transformed into a different form using basis functions that are more suitable for the required processing, and the transformation can be carried out efficiently using fast algorithms. Basis functions are linearly independent functions that can express any admissible function as a linear combination of them. One distinct feature of the wavelet transform basis functions is that all of them are derived from two transient functions, a scaling function and a wavelet function, by time-shifting and scaling. The wavelet function itself is defined as a linear combination of scaled and shifted scaling functions.

8.1.3 Fourier and Wavelet Transform Analyses

In Fourier analysis, a signal is decomposed in terms of sinusoidal components spanning the whole spectrum. The wavelet transform analysis is based on the fact that the combination of a continuous group of frequency components of the spectrum yields a transient time-domain signal. The basis signals for DWT are related to this type of signals and derived from that relation. Therefore, the DWT is local to the extent possible in both time and frequency. That is, the time interval of the occurrence of a component composed of a continuous group of frequencies can be determined. However, the time instance of the location of a single frequency component cannot be precisely found. Therefore, it is inherently suitable to represent functions that are local in time and frequency. The Fourier transform is perfectly local in frequency, but is global in time. In Fourier analysis, the basis functions are periodic waves, while the wavelets (short waves) are the basis functions in the wavelet transform. Typical basis waveforms for Fourier and wavelet transforms are shown, respectively, in Figures 8.1(a) and (b).

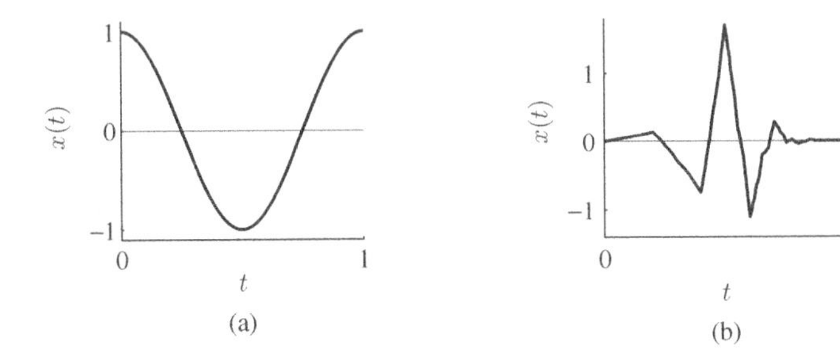

Figure 8.1 Typical basis functions for Fourier and wavelet transforms; (a) one cycle of a sinusoid; (b) a wavelet

The sinusoidal waveforms oscillate from minus infinity to plus infinity, whereas the wavelets oscillate for a short duration. A major advantage of the sinusoidal basis function is its derivative property. That is, the shape remains the same, irrespective of the number of derivatives taken. This property leads to the convolution operation in the time domain reducing to much simpler multiplication operation in the frequency domain. A major advantage of the wavelet basis function is that it is local. Because of this, a small change in the time-domain waveform affects only a few coefficients. As the Fourier basis functions are periodic, a small change in the input affects all the coefficients. It is possible to determine the location of the change in the input in the case of the wavelet transform. A short pulse is reconstructed using a large number of coefficients in the Fourier transform. Most of them are used for canceling various components to get the long zero portions of the waveform. The wavelet transform, due to its local nature, requires fewer coefficients. In contrast to Fourier analysis, there exist an infinite number of DWT basis signals. However, only a small number of them are used in practical applications. While most of these type of signals cannot be defined by analytical expressions, the Haar basis signals are exceptions. It is the oldest and the simplest of all the DWT basis signals. In addition to its usefulness in practical applications, due to its simplicity, it serves as the starting point in the study of the DWT. The other DWT basis signals can be considered as a generalization of the Haar basis signals.

8.1.4 Time-Frequency Domain

Fourier analysis is a tool to convert a time-domain signal into a frequency-domain signal and back. Wavelet transform analysis is a tool to convert a time-domain signal into a time-frequency-domain signal and back. As the basis signals are of transient nature and correspond to a part of the spectrum, we are able to locate the occurrence of an event in the time domain. That is, the correlation between a given signal and a basis signal can be determined at various instants of time by shifting it. The second observation is that the signal spectrum is split into unequal parts. This suits signal analysis as longer bandwidth and shorter time intervals are appropriate to detect a high-frequency component and vice versa. These are the two salient features of the wavelet transform analysis, and the advantages of the DWT in signal analysis accrue from these features. The DWT is usually presented in three forms: (i) matrix formulation, (ii) lowpass and highpass filtering (filter bank), and (iii) polyphase matrix factorization. The matrix formulation of the DWT is similar to that of the other discrete transforms with some differences. The output vector is obtained by multiplying the input vector with the transform matrix. However, there are several transform matrices, and each one is more suitable for some applications. The output vector has two independent variables corresponding to time and frequency. By its nature, the DWT is suitable to the analysis of nonstationary signals. If the spectrum of a signal varies at intervals in its duration, then the signal is called a nonstationary signal. If a time-frequency representation, rather than a pure frequency representation, is more suitable for a particular application, then the DWT is the choice.

In this chapter, the Haar DWT is presented primarily in terms of transform matrices. Formulas and examples of transform matrices for various data lengths and levels of decomposition are given. The difference between the Haar DWT and the 2-point DFT is pointed out. The decomposition of a signal into various resolutions is described. The continuous form of the wavelet transform basis functions is derived from the DWT filter coefficients by a limit process. The two-scale relation is presented. This relation establishes the connection between the scaling and wavelet functions, and the DWT filter coefficients. The row-column method of computing the 2-D Haar DWT is explored. Application of the DWT for the detection of discontinuities in signals is discussed.

8.2 The Haar Discrete Wavelet Transform

The 2-point 1-level DWT of the time-domain signal $\{x(0), x(1)\}$ is defined as

$$\begin{bmatrix} X_\phi(0,0) \\ X_\psi(0,0) \end{bmatrix} = \frac{1}{\sqrt{2}} \begin{bmatrix} 1 & 1 \\ 1 & -1 \end{bmatrix} \begin{bmatrix} x(0) \\ x(1) \end{bmatrix} \quad \text{or} \quad \boldsymbol{X} = \boldsymbol{W}_{2,0}\boldsymbol{x} \tag{8.1}$$

where $\boldsymbol{X}$, $\boldsymbol{W}_{2,0}$, and $\boldsymbol{x}$ represent, respectively, the coefficient, transform, and input matrices. The subscript 2 indicates the input data length, and 0 indicates the scale of decomposition. For $N = 2$, there is only one scale. It is assumed that the length N of a column sequence $x(n)$

$$\{x(0), x(1), x(2), \ldots, x(N-1)\}$$

is a power of 2. That is, $N = 2^J$, where J is a positive integer. In general, the coefficients are designated as $X_\phi(j_0, k)$ and $X_\psi(j, k)$, where j_0 is a given starting scale in the range $0, 1, \ldots, J-1$. The range of index j, which specifies the scale of decomposition, is

$$j = j_0, j_0 + 1, \ldots, J - 1$$

and it is an indicator of the subband of the spectrum of the corresponding basis function. The range of index k, which indicates the time span of the basis function, is

$$k = 0, 1, \ldots, 2^j - 1$$

Together, indices j and k indicate specified group of frequency components at specified times and lead to time-frequency representation. With $N = 2$, $J = 1$, $j = 0$, and $k = 0$. The letter ϕ in $X_\phi(j_0, k)$ indicates that it is an approximation coefficient obtained by some averaging of the input signal. The letter ψ in $X_\psi(j, k)$ indicates that it is a detail coefficient obtained by some differencing of the input signal. Haar DWT basis functions, with $N = 2$, are shown in Figures 8.2(a) and (b). Basically, in computing the Haar DWT, scaled sums and differences

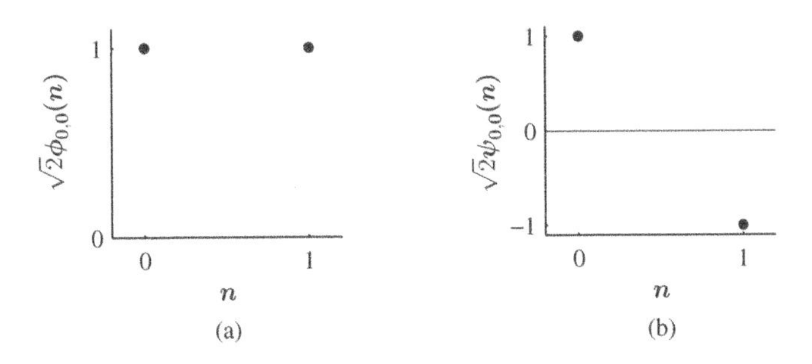

Figure 8.2 Haar DWT basis functions with $N = 2$. (a) $\sqrt{2}\phi_{0,0}(n)$; (b) $\sqrt{2}\psi_{0,0}(n)$

of adjacent pairs of input values are formed; the sums form the top half of the output, and the differences form the bottom half.

The 2-point 1-level inverse DWT (IDWT) of the coefficients $\{X_\phi(0,0), X_\psi(0,0)\}$ is defined as

$$\begin{bmatrix} x(0) \\ x(1) \end{bmatrix} = \frac{1}{\sqrt{2}} \begin{bmatrix} 1 & 1 \\ 1 & -1 \end{bmatrix} \begin{bmatrix} X_\phi(0,0) \\ X_\psi(0,0) \end{bmatrix} \quad \text{or} \quad \boldsymbol{x} = \boldsymbol{W}_{2,0}^{-1}\boldsymbol{X} \tag{8.2}$$

In computing the Haar IDWT, scaled sums and differences, of the values taken, respectively, from the top half and the bottom half of the input, are formed and stored consecutively as adjacent pairs in the output.

The inverse transform matrix $\boldsymbol{W}_{2,0}^{-1} = \boldsymbol{W}_{2,0}^{T}$ is the inverse as well as the transpose of the transform matrix $\boldsymbol{W}_{2,0}$. The forward and inverse transform matrices are real and orthogonal. That is, their product is an identity matrix. The product of a 2×2 transform matrix, its transpose and the input, for arbitrary real values, is

$$\begin{bmatrix} x \\ y \end{bmatrix} = \begin{bmatrix} a & c \\ b & d \end{bmatrix} \begin{bmatrix} a & b \\ c & d \end{bmatrix} \begin{bmatrix} x \\ y \end{bmatrix}$$

$$= \begin{bmatrix} (a^2 + c^2)\,x + (ab + cd)y \\ (ba + dc)x + (b^2 + d^2)y \end{bmatrix} = \begin{bmatrix} x \\ y \end{bmatrix}$$

With the orthogonality conditions,

$$a^2 + b^2 = 1 = c^2 + d^2 \quad \text{and} \quad ac + bd = 0$$

and

$$a^2 + c^2 = 1 = b^2 + d^2 \quad \text{and} \quad ab + cd = 0$$

the equation is satisfied. If the rows of a matrix are mutually orthogonal unit vectors, then the columns automatically satisfy a similar condition. An orthogonal matrix of order N is a set of N mutually orthogonal unit vectors.

Example 8.1 Using the Haar transform matrix, find the DWT of $x(n)$. Verify that $x(n)$ is reconstructed by computing the IDWT. Verify Parseval's theorem.

$$\{x(0) = 1, x(1) = -2\}$$

Solution

$$\begin{bmatrix} X_\phi(0,0) \\ X_\psi(0,0) \end{bmatrix} = \frac{1}{\sqrt{2}} \begin{bmatrix} 1 & 1 \\ 1 & -1 \end{bmatrix} \begin{bmatrix} 1 \\ -2 \end{bmatrix} = \begin{bmatrix} -\frac{1}{\sqrt{2}} \\ \frac{3}{\sqrt{2}} \end{bmatrix}$$

As in common with other transforms, the DWT output for an input is a set of coefficients that express the input in terms of the basis functions. Each basis function, during its existence, contributes to the value of the time-domain signal at each sample point. Therefore, multiplying each basis function by the corresponding DWT coefficient and summing the products get back the time-domain signal.

$$\begin{pmatrix} \frac{1}{\sqrt{2}} & \frac{1}{\sqrt{2}} \end{pmatrix} \left(-\frac{1}{\sqrt{2}}\right) +$$

$$\begin{pmatrix} \frac{1}{\sqrt{2}} & -\frac{1}{\sqrt{2}} \end{pmatrix} \left(\frac{3}{\sqrt{2}}\right) =$$

$$\begin{matrix} ---- & ---- \\ 1 & -2 \end{matrix}$$

Formally, the IDWT gets back the original input samples.

$$\begin{bmatrix} x(0) \\ x(1) \end{bmatrix} = \frac{1}{\sqrt{2}} \begin{bmatrix} 1 & 1 \\ 1 & -1 \end{bmatrix} \begin{bmatrix} -\frac{1}{\sqrt{2}} \\ \frac{3}{\sqrt{2}} \end{bmatrix} = \begin{bmatrix} 1 \\ -2 \end{bmatrix}$$

Parseval's theorem states that the sum of the squared-magnitude of a time-domain sequence (energy) equals the sum of the squared-magnitude of the corresponding transform coefficients. For the example sequence,

$$1^2 + (-2)^2 = 5 = \left(-\frac{1}{\sqrt{2}}\right)^2 + \left(\frac{3}{\sqrt{2}}\right)^2$$

■

8.2.1 *The Haar DWT and the 2-Point DFT*

The 2-point DFT of the time-domain signal $\{x(0) = 1, x(1) = -2\}$ is given by

$$\begin{bmatrix} X(0) \\ X(1) \end{bmatrix} = \begin{bmatrix} 1 & 1 \\ 1 & -1 \end{bmatrix} \begin{bmatrix} 1 \\ -2 \end{bmatrix} = \begin{bmatrix} -1 \\ 3 \end{bmatrix} \tag{8.3}$$

which is the same as that of the DWT except for a constant factor. But, there is a difference in the interpretation of the DFT and the DWT. The signal is composed of two components, one with zero frequency and another with frequency π radians. In terms of its components,

$$\{1, -2\} = \left(\frac{-1\{1,1\} = \{-1,-1\}}{2}\right) + \left(\frac{3\{1,-1\} = \{3,-3\}}{2}\right)$$

In the DFT interpretation, the pair of values $\{-1, 3\}$ is considered as the scaled coefficients of the two sinusoids, $\{1, 1\}$ and $\{1, -1\}$, constituting the signal $\{1, -2\}$. In 1-level DWT, the signal is expressed in terms of its components corresponding to the lower half and upper half of the spectrum. As the spectrum is divided into two equal parts, one sample is enough to represent each part. The other sample can be interpolated in reconstructing the signal. The coefficient $X_{\phi}(0,0) = -1/\sqrt{2}$ represents the component of the signal corresponding to the lower half of the spectrum, and $X_{\psi}(0,0) = 3/\sqrt{2}$ represents the component of the signal corresponding to the upper half of the spectrum. For a sequence of length 2, the decomposition of the two components of the signal is perfect. With longer sequence lengths, the signal contains more frequency components. Correlation of the signal with shifted versions of $\{1/\sqrt{2}, 1/\sqrt{2}\}$ yields a smaller value with increasing frequencies. At frequency π radians, the response is zero. The magnitude of the frequency response is one-quarter of a cosine wave, which is a lowpass filter (although not a good one). Correlation with the other basis function $\{1/\sqrt{2}, -1/\sqrt{2}\}$ yields a response that resembles a highpass filter.

8.2.2 The Haar Transform Matrix

Kronecker products are used in generating higher-order matrices from lower-order matrices. Let $\boldsymbol{A}$ be an $M \times N$ matrix and $\boldsymbol{B}$ be a $P \times Q$ matrix. Their Kronecker product is given by

$$\boldsymbol{A} \otimes \boldsymbol{B} = a_{ij}\boldsymbol{B} = \begin{bmatrix} a_{11}\boldsymbol{B} & \cdots & a_{1N}\boldsymbol{B} \\ \cdots & \cdots & \cdots \\ a_{M1}\boldsymbol{B} & \cdots & a_{MN}\boldsymbol{B} \end{bmatrix}$$

This is an $M \times N$ block matrix with each of its elements a $P \times Q$ matrix. The size of the block matrix is $MP \times NQ$. A matrix is a block matrix if its elements themselves are matrices. In a diagonal block matrix, the blocks off the diagonal consist of zero values. One advantage of block matrices is that matrix operations often reduce to the operations of the individual blocks.

The Haar transform matrix for 1-level (scale $\log_2(N) - 1$) DWT of a sequence of length N (N is a power of 2) is defined as

$$\boldsymbol{W}_{N,(\log_2(N)-1)} = \begin{bmatrix} \boldsymbol{I}_{\frac{N}{2}} \otimes [\frac{1}{\sqrt{2}} \quad \frac{1}{\sqrt{2}}] \\ \boldsymbol{I}_{\frac{N}{2}} \otimes [\frac{1}{\sqrt{2}} \quad -\frac{1}{\sqrt{2}}] \end{bmatrix} \tag{8.4}$$

where $\boldsymbol{I}_N$ is the $N \times N$ identity matrix. The 1-level (scale 1) 4-point Haar DWT is

$$\begin{bmatrix} X_{\phi}(1,0) \\ X_{\phi}(1,1) \\ X_{\psi}(1,0) \\ X_{\psi}(1,1) \end{bmatrix} = \frac{1}{\sqrt{2}} \begin{bmatrix} 1 & 1 & 0 & 0 \\ 0 & 0 & 1 & 1 \\ 1 & -1 & 0 & 0 \\ 0 & 0 & 1 & -1 \end{bmatrix} \begin{bmatrix} x(0) \\ x(1) \\ x(2) \\ x(3) \end{bmatrix} \quad \text{or} \quad \boldsymbol{X} = \boldsymbol{W}_{4,1}\boldsymbol{x}$$

Each DFT coefficient is computed using all the input values, whereas the DWT coefficients are computed using values taken over subintervals of the input. The IDWT, with $\boldsymbol{W}_{4,1}^{-1} = \boldsymbol{W}_{4,1}^{T}$, is

$$\begin{bmatrix} x(0) \\ x(1) \\ x(2) \\ x(3) \end{bmatrix} = \frac{1}{\sqrt{2}} \begin{bmatrix} 1 & 0 & 1 & 0 \\ 1 & 0 & -1 & 0 \\ 0 & 1 & 0 & 1 \\ 0 & 1 & 0 & -1 \end{bmatrix} \begin{bmatrix} X_{\phi}(1,0) \\ X_{\phi}(1,1) \\ X_{\psi}(1,0) \\ X_{\psi}(1,1) \end{bmatrix} \quad \text{or} \quad \boldsymbol{x} = \boldsymbol{W}_{4,1}^{-1}\boldsymbol{X}$$

The 1-level (scale 2) 8-point Haar DWT is

$$\begin{bmatrix} X_\phi(2,0) \\ X_\phi(2,1) \\ X_\phi(2,2) \\ X_\phi(2,3) \\ X_\psi(2,0) \\ X_\psi(2,1) \\ X_\psi(2,2) \\ X_\psi(2,3) \end{bmatrix} = \frac{1}{\sqrt{2}} \begin{bmatrix} 1 & 1 & 0 & 0 & 0 & 0 & 0 & 0 \\ 0 & 0 & 1 & 1 & 0 & 0 & 0 & 0 \\ 0 & 0 & 0 & 0 & 1 & 1 & 0 & 0 \\ 0 & 0 & 0 & 0 & 0 & 0 & 1 & 1 \\ 1 & -1 & 0 & 0 & 0 & 0 & 0 & 0 \\ 0 & 0 & 1 & -1 & 0 & 0 & 0 & 0 \\ 0 & 0 & 0 & 0 & 1 & -1 & 0 & 0 \\ 0 & 0 & 0 & 0 & 0 & 0 & 1 & -1 \end{bmatrix} \begin{bmatrix} x(0) \\ x(1) \\ x(2) \\ x(3) \\ x(4) \\ x(5) \\ x(6) \\ x(7) \end{bmatrix} \tag{8.5}$$

The IDWT is

$$\begin{bmatrix} x(0) \\ x(1) \\ x(2) \\ x(3) \\ x(4) \\ x(5) \\ x(6) \\ x(7) \end{bmatrix} = \frac{1}{\sqrt{2}} \begin{bmatrix} 1 & 0 & 0 & 0 & 1 & 0 & 0 & 0 \\ 1 & 0 & 0 & 0 & -1 & 0 & 0 & 0 \\ 0 & 1 & 0 & 0 & 0 & 1 & 0 & 0 \\ 0 & 1 & 0 & 0 & 0 & -1 & 0 & 0 \\ 0 & 0 & 1 & 0 & 0 & 0 & 1 & 0 \\ 0 & 0 & 1 & 0 & 0 & 0 & -1 & 0 \\ 0 & 0 & 0 & 1 & 0 & 0 & 0 & 1 \\ 0 & 0 & 0 & 1 & 0 & 0 & 0 & -1 \end{bmatrix} \begin{bmatrix} X_\phi(2,0) \\ X_\phi(2,1) \\ X_\phi(2,2) \\ X_\phi(2,3) \\ X_\psi(2,0) \\ X_\psi(2,1) \\ X_\psi(2,2) \\ X_\psi(2,3) \end{bmatrix} \tag{8.6}$$

There are $J = \log_2(N)$ levels of decomposition, $\{j = 0, 1, 2, \ldots, J-1\}$. The scale $j = J-1$ indicates that the signal is decomposed into

$$J - j + 1 = J - (J-1) + 1 = 2$$

components. The scale $j = 0$ indicates that the signal is decomposed into a maximum of

$$J - j + 1 = J - 0 + 1 = J + 1 = \log_2(N) + 1$$

components. The transform matrix for the scale $J-1$ is defined by Equation (8.4). For other scales, the transform matrix is developed recursively as follows. The 2-level N-point DWT is obtained by taking the 1-level DWT of the first half of the N 1-level DWT (approximation part) coefficients, leaving the second half (detail part) unchanged. This process continues until we take the 1-level DWT of the last pair of the DWT coefficients. For example,

$$\boldsymbol{W}_{4,0} = \begin{bmatrix} \boldsymbol{W}_{2,0} & \boldsymbol{0}_2 \\ \boldsymbol{0}_2 & \boldsymbol{I}_2 \end{bmatrix} \boldsymbol{W}_{4,1} = \begin{bmatrix} \frac{1}{\sqrt{2}} & \frac{1}{\sqrt{2}} & 0 & 0 \\ \frac{1}{\sqrt{2}} & -\frac{1}{\sqrt{2}} & 0 & 0 \\ 0 & 0 & 1 & 0 \\ 0 & 0 & 0 & 1 \end{bmatrix} \frac{1}{\sqrt{2}} \begin{bmatrix} 1 & 1 & 0 & 0 \\ 0 & 0 & 1 & 1 \\ 1 & -1 & 0 & 0 \\ 0 & 0 & 1 & -1 \end{bmatrix} \tag{8.7}$$

$$= \begin{bmatrix} \frac{1}{2} & \frac{1}{2} & \frac{1}{2} & \frac{1}{2} \\ \frac{1}{2} & \frac{1}{2} & -\frac{1}{2} & -\frac{1}{2} \\ \frac{1}{\sqrt{2}} & -\frac{1}{\sqrt{2}} & 0 & 0 \\ 0 & 0 & \frac{1}{\sqrt{2}} & -\frac{1}{\sqrt{2}} \end{bmatrix} = \begin{bmatrix} \phi_{0,0}(n) \\ \psi_{0,0}(n) \\ \psi_{1,0}(n) \\ \psi_{1,1}(n) \end{bmatrix} = \begin{bmatrix} \boldsymbol{W}_{2,0} \otimes [\frac{1}{\sqrt{2}} & \frac{1}{\sqrt{2}}] \\ \boldsymbol{I}_2 \otimes [\frac{1}{\sqrt{2}} & -\frac{1}{\sqrt{2}}] \end{bmatrix} \tag{8.8}$$

The aforementioned example for higher levels of the DWT transform matrix is also the efficient algorithm for the computation of the DWT. 1-level DWT is computed in each stage with the sequence length reduced by 2. The use of a single transform matrix to compute a higher-level DWT is not efficient. However, a single formula is available for higher-level transform matrices. The transform matrix for the j-scale DWT of a sequence of length N (N is a power of 2) is defined by

$$\boldsymbol{W}_{N,j} = \begin{bmatrix} \boldsymbol{W}_{\frac{N}{2},j} \otimes [\frac{1}{\sqrt{2}} \quad \frac{1}{\sqrt{2}}] \\ \boldsymbol{I}_{\frac{N}{2}} \otimes [\frac{1}{\sqrt{2}} \quad -\frac{1}{\sqrt{2}}] \end{bmatrix}, \quad j = 0, 1, \ldots, (\log_2(N) - 2) \tag{8.9}$$

The pattern of computation is the same for all the other DWT matrices to be studied later with a sequence length that is a power of 2. In the implementation of the DWT, the convolution operation is used to compute the DWT coefficients.

Example 8.2 Using the Haar transform matrix, find the 2-level (scale 0) DWT of $x(n)$. Verify that $x(n)$ is reconstructed by computing the IDWT. Verify Parseval's theorem.

$$\{x(0) = 3, x(1) = 4, x(2) = 3, x(3) = -2\}$$

Solution

Using $\boldsymbol{W}_{4,0}$ defined in Equation (8.8) (or computing recursively),

$$\begin{bmatrix} X_\phi(0,0) \\ X_\psi(0,0) \\ X_\psi(1,0) \\ X_\psi(1,1) \end{bmatrix} = \begin{bmatrix} \frac{1}{2} & \frac{1}{2} & \frac{1}{2} & \frac{1}{2} \\ \frac{1}{2} & \frac{1}{2} & -\frac{1}{2} & -\frac{1}{2} \\ \frac{1}{\sqrt{2}} & -\frac{1}{\sqrt{2}} & 0 & 0 \\ 0 & 0 & \frac{1}{\sqrt{2}} & -\frac{1}{\sqrt{2}} \end{bmatrix} \begin{bmatrix} 3 \\ 4 \\ 3 \\ -2 \end{bmatrix} = \begin{bmatrix} 4 \\ 3 \\ -\frac{1}{\sqrt{2}} \\ \frac{5}{\sqrt{2}} \end{bmatrix}$$

Multiplying each basis function, during its existence, by the corresponding DWT coefficient and summing the products give the time-domain signal.

$$\begin{array}{llllr} (\frac{1}{2} & \frac{1}{2} & \frac{1}{2} & \frac{1}{2}) & (4)+ \\ (\frac{1}{2} & \frac{1}{2} & -\frac{1}{2} & -\frac{1}{2}) & (3)+ \\ (\frac{1}{\sqrt{2}} & -\frac{1}{\sqrt{2}} & 0 & 0) & \left(-\frac{1}{\sqrt{2}}\right)+ \\ (0 & 0 & \frac{1}{\sqrt{2}} & -\frac{1}{\sqrt{2}}) & \left(\frac{5}{\sqrt{2}}\right) = \\ \text{----} & \text{----} & \text{----} & \text{----} & \\ 3 & 4 & 3 & -2 & \end{array}$$

Formally, the IDWT gets back the original input samples.

$$\begin{bmatrix} x(0) \\ x(1) \\ x(2) \\ x(3) \end{bmatrix} = \begin{bmatrix} \frac{1}{2} & \frac{1}{2} & \frac{1}{\sqrt{2}} & 0 \\ \frac{1}{2} & \frac{1}{2} & -\frac{1}{\sqrt{2}} & 0 \\ \frac{1}{2} & -\frac{1}{2} & 0 & \frac{1}{\sqrt{2}} \\ \frac{1}{2} & -\frac{1}{2} & 0 & -\frac{1}{\sqrt{2}} \end{bmatrix} \begin{bmatrix} 4 \\ 3 \\ -\frac{1}{\sqrt{2}} \\ \frac{5}{\sqrt{2}} \end{bmatrix} = \begin{bmatrix} 3 \\ 4 \\ 3 \\ -2 \end{bmatrix}$$

Applying Parseval's theorem, we get

$$3^2 + 4^2 + 3^2 + (-2)^2 = 38 = 4^2 + 3^2 + \left(-\frac{1}{\sqrt{2}}\right)^2 + \left(\frac{5}{\sqrt{2}}\right)^2$$

■

The 2-level (scale 1) Haar DWT transform matrix with $N = 8$ is

$$\mathbf{W}_{8,1} = \begin{bmatrix} \phi_{1,0}(n) \\ \phi_{1,1}(n) \\ \psi_{1,0}(n) \\ \psi_{1,1}(n) \\ \psi_{2,0}(n) \\ \psi_{2,1}(n) \\ \psi_{2,2}(n) \\ \psi_{2,3}(n) \end{bmatrix} = \begin{bmatrix} \frac{1}{\sqrt{4}} & \frac{1}{\sqrt{4}} & \frac{1}{\sqrt{4}} & \frac{1}{\sqrt{4}} & 0 & 0 & 0 & 0 \\ 0 & 0 & 0 & 0 & \frac{1}{\sqrt{4}} & \frac{1}{\sqrt{4}} & \frac{1}{\sqrt{4}} & \frac{1}{\sqrt{4}} \\ \frac{1}{\sqrt{4}} & \frac{1}{\sqrt{4}} & -\frac{1}{\sqrt{4}} & -\frac{1}{\sqrt{4}} & 0 & 0 & 0 & 0 \\ 0 & 0 & 0 & 0 & \frac{1}{\sqrt{4}} & \frac{1}{\sqrt{4}} & -\frac{1}{\sqrt{4}} & -\frac{1}{\sqrt{4}} \\ \frac{1}{\sqrt{2}} & -\frac{1}{\sqrt{2}} & 0 & 0 & 0 & 0 & 0 & 0 \\ 0 & 0 & \frac{1}{\sqrt{2}} & -\frac{1}{\sqrt{2}} & 0 & 0 & 0 & 0 \\ 0 & 0 & 0 & 0 & \frac{1}{\sqrt{2}} & -\frac{1}{\sqrt{2}} & 0 & 0 \\ 0 & 0 & 0 & 0 & 0 & 0 & \frac{1}{\sqrt{2}} & -\frac{1}{\sqrt{2}} \end{bmatrix} \tag{8.10}$$

The 3-level (scale 0) Haar DWT transform matrix with $N = 8$ is

$$\mathbf{W}_{8,0} = \begin{bmatrix} \phi_{0,0}(n) \\ \psi_{0,0}(n) \\ \psi_{1,0}(n) \\ \psi_{1,1}(n) \\ \psi_{2,0}(n) \\ \psi_{2,1}(n) \\ \psi_{2,2}(n) \\ \psi_{2,3}(n) \end{bmatrix} = \begin{bmatrix} \frac{1}{\sqrt{8}} & \frac{1}{\sqrt{8}} & \frac{1}{\sqrt{8}} & \frac{1}{\sqrt{8}} & \frac{1}{\sqrt{8}} & \frac{1}{\sqrt{8}} & \frac{1}{\sqrt{8}} & \frac{1}{\sqrt{8}} \\ \frac{1}{\sqrt{8}} & \frac{1}{\sqrt{8}} & \frac{1}{\sqrt{8}} & \frac{1}{\sqrt{8}} & -\frac{1}{\sqrt{8}} & -\frac{1}{\sqrt{8}} & -\frac{1}{\sqrt{8}} & -\frac{1}{\sqrt{8}} \\ \frac{1}{\sqrt{4}} & \frac{1}{\sqrt{4}} & -\frac{1}{\sqrt{4}} & -\frac{1}{\sqrt{4}} & 0 & 0 & 0 & 0 \\ 0 & 0 & 0 & 0 & \frac{1}{\sqrt{4}} & \frac{1}{\sqrt{4}} & -\frac{1}{\sqrt{4}} & -\frac{1}{\sqrt{4}} \\ \frac{1}{\sqrt{2}} & -\frac{1}{\sqrt{2}} & 0 & 0 & 0 & 0 & 0 & 0 \\ 0 & 0 & \frac{1}{\sqrt{2}} & -\frac{1}{\sqrt{2}} & 0 & 0 & 0 & 0 \\ 0 & 0 & 0 & 0 & \frac{1}{\sqrt{2}} & -\frac{1}{\sqrt{2}} & 0 & 0 \\ 0 & 0 & 0 & 0 & 0 & 0 & \frac{1}{\sqrt{2}} & -\frac{1}{\sqrt{2}} \end{bmatrix} \tag{8.11}$$

The norm (the square root of the sum of the squares of the values) of each row of all the basis functions is 1. The first row of the transform matrix $\phi_{0,0}(n)$ in $\mathbf{W}_{8,0}$ is used to find the approximation coefficient. It has the same nonzero value for all its samples. The rest of the rows $\psi_{j,k}(n)$ are used to find the detail coefficients. The second row is a square pulse with mean zero. The third row is a compressed (by a factor of 2) and scaled (by $\sqrt{2}$) version the second row. The fourth row is a shifted version of the third row. The last four rows are the compressed, shifted, and scaled versions of the third row. The shifting of a finite-extent basis function is one distinct feature of the DWT. This feature makes it possible for time-frequency analysis. For example, if the DWT coefficient $X_{\psi}(1,1)$ of a signal corresponding to the fourth row of the basis function $\psi_{1,1}(n)$ is zero, it indicates that there is no component in the signal corresponding to the frequency band of the basis function during the last four samples. Another distinct feature is that the duration of the nonzero samples of the basis functions varies, giving the multiresolution characteristic.

8.3 The Time-Frequency Plane

The time-frequency resolutions of the DWT, with $N = 8$, are shown in Figure 8.3. The time-frequency plane is a plot to show how the signal strength is distributed with respect to time and frequency variables. The horizontal axis represents the sample index (time), and the vertical axis represents frequency in radians. A better frequency resolution (a shorter frequency band) at low frequencies and vice versa is appropriate for signal analysis in several applications. The Fourier transform provides no time resolution, as the time variable disappears in the transformation process. Each group of the DWT basis functions corresponds to a distinct band of the spectrum of the signal with frequency varying from 0 to π radians. The combination of all the subbands spans the complete spectrum of the signal. The decomposition of a 1-D signal by k levels produces $k + 1$ components of the signal. With eight samples, the DWT can decompose a signal into a maximum of four components (three scales, $j = 0, 1, 2$) corresponding to spectral bands

$$\left(0 - \frac{\pi}{8}\right), \left(\frac{\pi}{8} - \frac{\pi}{4}\right), \left(\frac{\pi}{4} - \frac{\pi}{2}\right) \quad \text{and} \quad \left(\frac{\pi}{2} - \pi\right) \quad \text{radians}$$

This decomposition corresponds to letting the starting scale $j_0 = 0$. The DWT coefficients, shown in groups corresponding to the frequency bands, are

$$\{X_\phi(0,0)\}, \{X_\psi(0,0)\}, \{X_\psi(1,0), X_\psi(1,1)\}, \{X_\psi(2,0), X_\psi(2,1), X_\psi(2,2), X_\psi(2,3)\}$$

With $j_0 = 1$ (two scales, $j = 1, 2$), the spectral range is decomposed into

$$\left(0 - \frac{\pi}{4}\right), \left(\frac{\pi}{4} - \frac{\pi}{2}\right) \quad \text{and} \quad \left(\frac{\pi}{2} - \pi\right) \quad \text{radians}$$

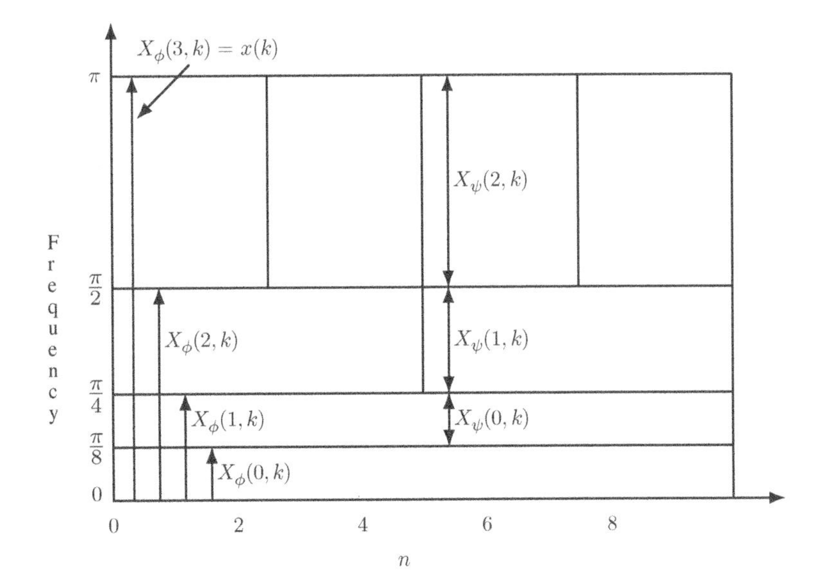

Figure 8.3 Time-frequency plane of the DWT, with $N = 8$

and the signal is decomposed into three components. The corresponding DWT coefficients are

$$\{X_\phi(1,0), X_\phi(1,1)\}, \{X_\psi(1,0), X_\psi(1,1)\}, \{X_\psi(2,0), X_\psi(2,1), X_\psi(2,2), X_\psi(2,3)\}$$

With $j_0 = 2$, the spectral range is decomposed into

$$\left(0 - \frac{\pi}{2}\right) \quad \text{and} \quad \left(\frac{\pi}{2} - \pi\right) \quad \text{radians}$$

and the signal is decomposed into two components. The corresponding DWT coefficients are

$$\{X_\phi(2,0), X_\phi(2,1), X_\phi(2,2), X_\phi(2,3)\}, \{X_\psi(2,0), X_\psi(2,1), X_\psi(2,2), X_\psi(2,3)\}$$

With $j_0 = 3$, the spectral range is 0 to π radians. That is, the signal samples themselves are the approximation coefficients, and no detail coefficients are required. Scale $j_0 = 3$ is the finest, and $j_0 = 0$ is the coarsest. All of the decompositions are shown in Figure 8.3. However, the decompositions, in practice, are not perfect, as the frequency response of practical filters is not ideal. Fortunately, despite some errors in the decomposition, the original signal can be perfectly reconstructed, as the 2-point DFT is a reversible operation. The spectrum is divided equally on a logarithmic frequency scale, which resembles human perception. In effect, the DWT is a set of bandpass filters.

The 2-level decomposition of a 512-point signal $x(n)$ by the DWT with ideal filters is shown in Figure 8.4. The input signal $x(n)$, with spectral components in the range 0 to π radians, is also considered as the approximation coefficients $X_\phi(9, 0 : 511)$. As the filters are ideal, filtering $x(n)$ using these filters with passbands 0 to $\pi/2$ and $\pi/2$ to π radians decomposes the signal into low- and high-frequency components exactly. As the components are of half the bandwidth of $x(n)$, they can be downsampled by a factor of 2 without losing any information. The results of the first stage of decomposition are two sets of coefficients, the approximation coefficients $X_\phi(8, 0 : 255)$ and the detail coefficients $X_\psi(8, 256 : 511)$ with frequency components in the range 0 to $\pi/2$ and $\pi/2$ to π radians, respectively. These coefficients multiplied by the corresponding basis functions and summed get back the original input. The approximation coefficients $X_\phi(8, 0 : 255)$ are again decomposed using the same set of filters, in the second stage, to get $X_\phi(7, 0 : 127)$ and $X_\psi(7, 128 : 255)$ with frequency components in the range 0 to $\pi/4$ and $\pi/4$ to $\pi/2$ radians, respectively. With $N = 512$, the decomposition can continue until coefficients $X_\phi(0, 0)$ and $X_\psi(0, 0)$ are obtained. The final decomposition

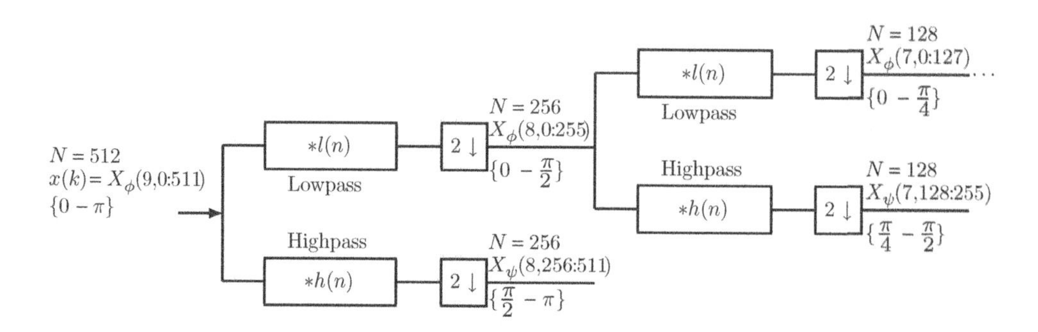

Figure 8.4 The 2-level decomposition of a 512-point signal $x(n)$ by the DWT

(9th level) produces just one coefficient in each set of coefficients. Typically, three or four levels of decompositions are used in applications. The whole process is similar to the decomposition of a signal by the DFT with the following differences: (i) the basis functions are single frequencies in the DFT, and they are composed of a set of frequencies in the DWT; (ii) there is only one level of decomposition in the DFT, whereas multilevel decompositions are carried out in the DWT; (iii) there is only one set of basis functions used by the DFT in contrast to several sets used in the DWT.

Example 8.3 Draw the graphs of all the components of

$$\begin{aligned} x(n) &= (0.8)^n, \quad n = 0, 1, \dots, 7 \\ &= \{1, 0.8, 0.64, 0.512, 0.4096, 0.3277, 0.2621, 0.2097\} \end{aligned}$$

obtained by computing the 3-level Haar DWT.

Solution
The 1-level DWT of $x(n)$, shown in Figure 8.5(a), using $\boldsymbol{W}_{8,2}$ defined in Equation (8.5) is given by

$$\begin{aligned} &\{X_\phi(2,0), X_\phi(2,1), X_\phi(2,2), X_\phi(2,3), X_\psi(2,0), X_\psi(2,1), X_\psi(2,2), X_\psi(2,3)\} \\ &= \{1.2728, 0.8146, 0.5213, 0.3337, \mathbf{0.1414}, \mathbf{0.0905}, \mathbf{0.0579}, \mathbf{0.0371}\} \end{aligned}$$

The detail coefficients are shown in boldface. The magnitude of the detail coefficients is much smaller than that of the approximation coefficients. The approximation component, which can be obtained from the first four coefficients multiplied by their respective basis functions, is shown in Figure 8.5(b).

$$xa_2(n) = \{0.9, 0.9, 0.576, 0.576, 0.3686, 0.3686, 0.2359, 0.2359\}$$

The detail component

$$xd_2(n) = \{0.1, -0.1, 0.064, -0.064, 0.041, -0.041, 0.0262, -0.0262\}$$

is shown in Figure 8.5(c).

$$x(n) = xa_3(n) = xa_2(n) + xd_2(n)$$

That is, the sum of the approximation and detail components of a signal at a scale is equal to its approximation at the next higher scale.

Using $\boldsymbol{W}_{8,1}$ defined in Equation (8.10) (or computing recursively), the 2-level DWT of $x(n)$ is given by

$$\begin{aligned} &\{\phi_{1,0}(n), \phi_{1,1}(n), \psi_{1,0}(n), \psi_{1,1}(n), \psi_{2,0}(n), \psi_{2,1}(n), \psi_{2,2}(n), \psi_{2,3}(n)\} \\ &= \{1.476, 0.6046, \mathbf{0.324}, \mathbf{0.1327}, \mathbf{0.1414}, \mathbf{0.0905}, \mathbf{0.0579}, \mathbf{0.0371}\} \end{aligned}$$

The approximation component

$$xa_1(n) = \{0.738, 0.738, 0.738, 0.738, 0.3023, 0.3023, 0.3023, 0.3023\}$$

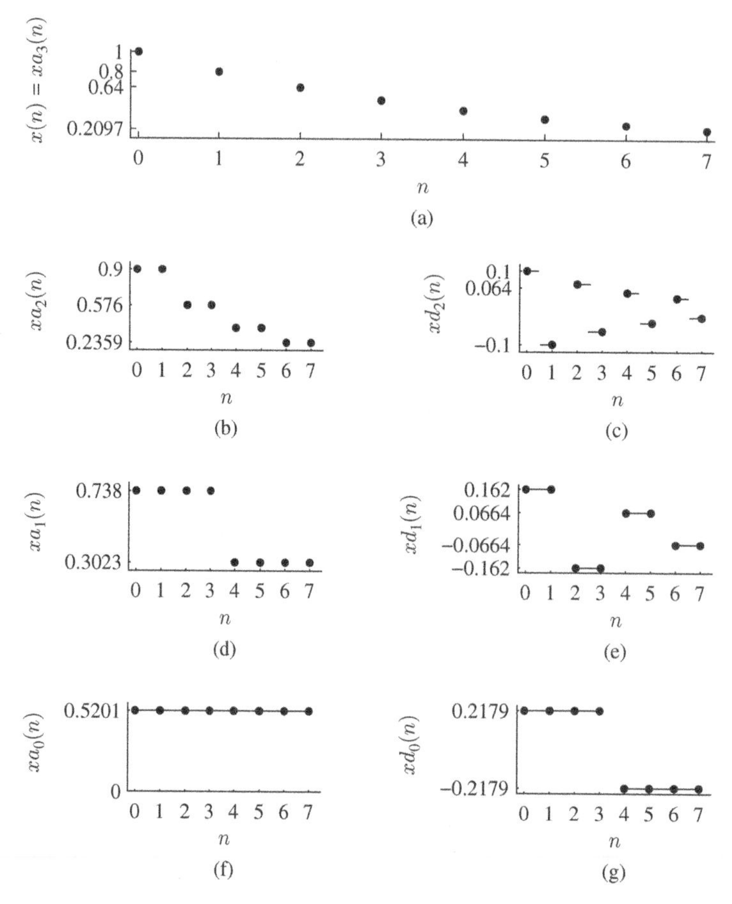

Figure 8.5 (a) A signal $x(n) = xa_3(n)$ with $N = 8$ samples; (b) signal in (a) decomposed into an approximation component $xa_2(n)$ and (c) a detail component $xd_2(n)$. $x(n) = xa_3(n) = xa_2(n) + xd_2(n)$; (d) signal in (b) decomposed into an approximation component $xa_1(n)$ and (e) a detail component $xd_1(n)$. $xa_2(n) = xa_1(n) + xd_1(n)$ and $x(n) = xa_3(n) = xa_1(n) + xd_1(n) + xd_2(n)$; (f) signal in (d) decomposed into an approximation component $xa_0(n)$ and (g) a detail component $xd_0(n)$. $xa_1(n) = xa_0(n) + xd_0(n)$ and $x(n) = xa_3(n) = xa_0(n) + xd_0(n) + xd_1(n) + xd_2(n)$

is shown in Figure 8.5(d). The detail component

$$xd_1(n) = \{0.162, 0.162, -0.162, -0.162, 0.0664, 0.0664, -0.0664, -0.0664\}$$

is shown in Figure 8.5(e).

$$xa_2(n) = xa_1(n) + xd_1(n) \quad \text{and} \quad x(n) = xa_1(n) + xd_1(n) + xd_2(n)$$

Using $\mathbf{W}_{8,0}$ defined in Equation (8.11) (or computing recursively), the 3-level DWT of $x(n)$ is given by

$$\{\phi_{0,0}(n), \psi_{0,0}(n), \psi_{1,0}(n), \psi_{1,1}(n), \psi_{2,0}(n), \psi_{2,1}(n), \psi_{2,2}(n), \psi_{2,3}(n)\}$$
$$= \{1.4712, \mathbf{0.6162}, \mathbf{0.324}, \mathbf{0.1327}, \mathbf{0.1414}, \mathbf{0.0905}, \mathbf{0.0579}, \mathbf{0.0371}\}$$

The approximation component

$$xa_0(n) = \{0.5201, 0.5201, 0.5201, 0.5201, 0.5201, 0.5201, 0.5201, 0.5201\}$$

is shown in Figure 8.5(f). The detail component

$$xd_0(n) = \{0.2179, 0.2179, 0.2179, 0.2179, -0.2179, -0.2179, -0.2179, -0.2179\}$$

is shown in Figure 8.5(g).

$$xa_1(n) = xa_0(n) + xd_0(n) \quad \text{and} \quad x(n) = xa_0(n) + xd_0(n) + xd_1(n) + xd_2(n)$$

Figures 8.5(c), (e), (g) and (f) shows the components of the input waveform with decreasing frequencies. The samples are connected by continuous lines, clearly showing that the basis functions are a set of rectangular pulses. As more and more details are added to the approximation, the resultant signal resembles more like the original signal, as can be seen from Figures 8.5(f), (d), (b), and (a). It is very similar to the approximation of waveforms by Fourier analysis. The difference is that the basis functions are different, and each transform is more suitable for different applications. ■

8.4 Wavelets from the Filter Coefficients

In the case of Fourier analysis, the basis functions (the sinusoids or complex exponentials) are well defined. For discrete Fourier analysis, the basis functions are obtained by sampling the continuous basis functions. In the case of the wavelet transform, most of the basis functions are not well defined. Therefore, the continuous wavelet basis functions are obtained from the corresponding filter coefficients by a limit process. While the DWT analysis can be carried out using the filter coefficients without any knowledge of the corresponding continuous basis functions, it is good to look at them for a better understanding of the transform. The process involves finding the inverse transform. The inverse is obtained by the repeated use of the upsampling and convolution operations. For ease of understanding, let us consider the IDFT first. The N-point IDFT of the frequency coefficients $X(k)$ is defined as

$$x(n) = \frac{1}{N}\sum_{k=0}^{N-1} X(k)W_N^{-nk}, \quad n = 0, 1, \ldots, N-1 \tag{8.12}$$

Let all the values of the transform $X(k)$ be zero, except that the value of $X(k_0) = N$. Then,

$$x(n) = W_N^{-k_0 n} = e^{j\frac{2\pi}{N}k_0 n}, \quad n = 0, 1, \ldots, N-1$$

which is the basis function with index k_0.

While there are N basis functions in the DFT, there are just two basis functions in wavelet transform, a scaling function and a wavelet function corresponding to each set of DWT lowpass and highpass filter coefficients. The task is to construct these functions from the filter coefficients. The scaling function is obtained by applying a unit-impulse to the first lowpass

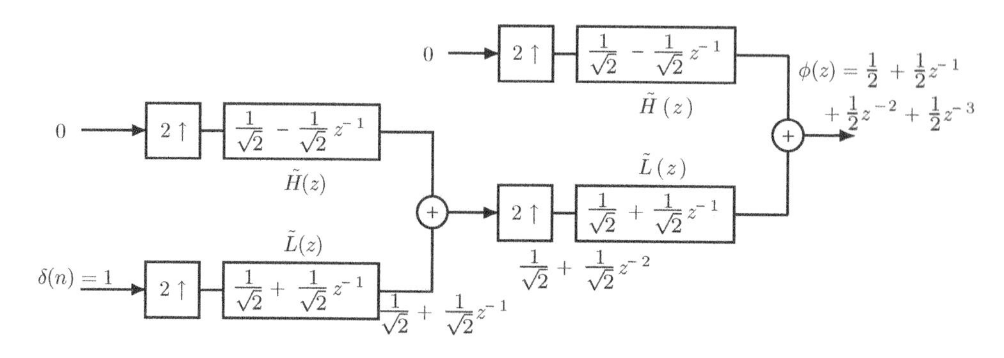

Figure 8.6 Haar scaling function from lowpass synthesis filter coefficients. $\phi(z) = \tilde{L}(z^2)\tilde{L}(z)$

channel of an infinite cascade of reconstruction filter banks. Figure 8.6 shows a two-stage cascade. The upsampled impulse is $\{1, 0\}$. When multiplied by the transfer function of the Haar lowpass filter,

$$\frac{1}{\sqrt{2}} + \frac{1}{\sqrt{2}}z^{-1}$$

we get the same. Now, the transfer function, after upsampling in the second stage, becomes

$$\frac{1}{\sqrt{2}} + \frac{1}{\sqrt{2}}z^{-2}$$

This function multiplied by the transfer function in the second stage results in

$$\left(\frac{1}{\sqrt{2}} + \frac{1}{\sqrt{2}}z^{-2}\right)\left(\frac{1}{\sqrt{2}} + \frac{1}{\sqrt{2}}z^{-1}\right) = \frac{1}{2} + \frac{1}{2}z^{-1} + \frac{1}{2}z^{-2} + \frac{1}{2}z^{-3}$$

Scaling the amplitude by $\sqrt{2}$ after each stage, we get

$$1 + z^{-1} + z^{-2} + z^{-3}$$

The inverse z-transform gives the interpolated samples of the scaling function as

$$\{1, 1, 1, 1\}$$

The time axis has to be rescaled by a factor of 2 after each stage to get a limit. With increasing number of iterations, the waveform becomes denser with more interpolated values. After infinite iterations with rescaling, we get the continuous Haar scaling function.

The wavelet function is obtained by applying a unit-impulse to the first highpass channel of a cascade of reconstruction filter banks. Figure 8.7 shows a two-stage cascade. The upsampled impulse is $\{1, 0\}$. When multiplied by the transfer function of the Haar highpass filter,

$$\frac{1}{\sqrt{2}} - \frac{1}{\sqrt{2}}z^{-1}$$

we get the same. Now, the transfer function, after upsampling in the second stage, becomes

$$\frac{1}{\sqrt{2}} - \frac{1}{\sqrt{2}}z^{-2}$$

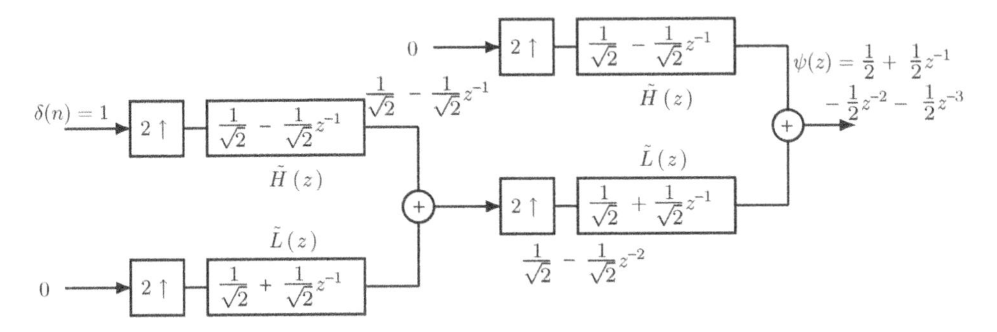

Figure 8.7 Haar wavelet function from highpass synthesis filter coefficients. $\psi(z) = \tilde{H}(z^2)\tilde{L}(z)$

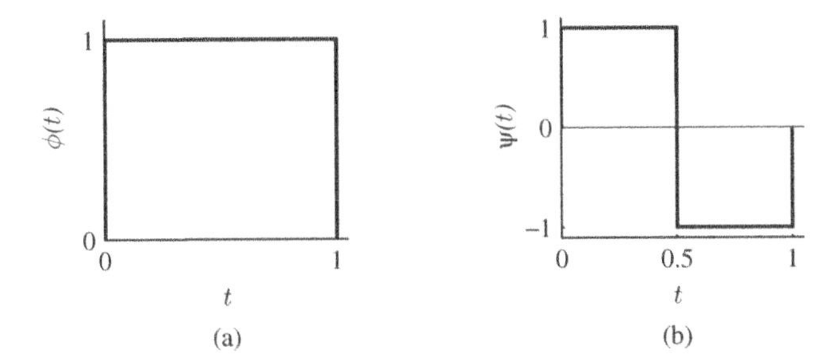

Figure 8.8 (a) Haar scaling function; (b) Haar wavelet function

This function multiplied by the lowpass transfer function in the second stage results in

$$\left(\frac{1}{\sqrt{2}} - \frac{1}{\sqrt{2}}z^{-2}\right)\left(\frac{1}{\sqrt{2}} + \frac{1}{\sqrt{2}}z^{-1}\right) = \frac{1}{2} + \frac{1}{2}z^{-1} - \frac{1}{2}z^{-2} - \frac{1}{2}z^{-3}$$

Scaling the amplitude by $\sqrt{2}$ after each stage , we get

$$1 + z^{-1} - z^{-2} - z^{-3}$$

The inverse z-transform gives the interpolated samples of the wavelet function as

$$\{1, 1, -1, -1\}$$

For the Haar transform, the scaling function is a square, and the wavelet function is made up of two rectangles of half the width of the scaling function, one with a positive peak and the other with a negative peak, as shown in Figures 8.8(a) and (b).

The presentation of the interpolation process to get the continuous scaling and wavelet functions is clearer than exact. The rescaling of the time axis after each stage is not taken into account. Now, we present the exact limit process. Let the scaling function and its FT be

$$\phi(t) \leftrightarrow \phi(j\omega)$$

The FT $\phi(j\omega)$ is derived from the filter coefficients. This transform can be inverted only by numerical approximation, as the basis functions of the wavelet transform are usually irregular. That is, only samples can be given by the IDFT. Then, a staircase approximation of the samples is taken as the desired function. This procedure is common in approximating continuous functions or spectra by the IDFT/DFT. The DTFT of a sequence $\phi(nT_s)$ and the FT of the corresponding sampled signal, $\sum_{n=-\infty}^{\infty}\phi(nT_s)\delta(t-nT_s)$, are the same when the DTFT version includes the sampling interval, T_s. A scaling of the frequency axis is also adequate to obtain the DTFT spectrum for any sampling interval, after computing it with $T_s = 1$ second. Therefore, we can use the inverse DTFT equation to compute the samples of the scaling function.

The FT transfer function of the Haar lowpass filter (sampled data), which is the same as the DTFT transfer function, is

$$\phi_s(j\omega) = \phi(e^{j\omega}) = \frac{1}{\sqrt{2}} + \frac{1}{\sqrt{2}}e^{-j\omega}$$

If we sample $\phi(t)$ in the time domain with $T_s = 1$ seconds, then the transfer function gets multiplied by the factor $1/T_s = 1/1 = 1$ and becomes periodic with period 2π and the corresponding inverse DTFT is a set of impulses with the sampling interval $T_s = 1$ second. In order to reduce T_s by a factor of 2, using the DTFT scaling theorem (just replace ω by $\omega/2$) and multiplying by $1/T_s = 1/(1/2) = 2$ for sampling, the transfer function is given by

$$2\left(\frac{1}{\sqrt{2}} + \frac{1}{\sqrt{2}}e^{-j\frac{\omega}{2}}\right) = 2\tilde{L}\left(e^{j\frac{\omega}{2}}\right)$$

The value of $\tilde{L}\left(e^{j\frac{\omega}{2}}\right)$, at $\omega = 0$, is $\sqrt{2}$. With the inclusion of the normalization factor $(1/\sqrt{2})$ so that $\tilde{L}\left(e^{j\frac{\omega}{2}}\right) = 1$ at $\omega = 0$, we get

$$2\frac{1}{\sqrt{2}}\left(\frac{1}{\sqrt{2}} + \frac{1}{\sqrt{2}}e^{-j\frac{\omega}{2}}\right) = 2\frac{1}{\sqrt{2}}\tilde{L}\left(e^{j\frac{\omega}{2}}\right)$$

This transfer function is periodic with period 4π, and the corresponding inverse DTFT is a set of impulses with the sampling interval $T_s = 1/2$ second. The transfer function of the first two stages becomes

$$4\left(\frac{1}{\sqrt{2}}\tilde{L}\left(e^{j\frac{\omega}{2}}\right)\frac{1}{\sqrt{2}}\tilde{L}\left(e^{j\frac{\omega}{4}}\right)\right)$$

and so on. This transfer function is periodic with period 8π, and the corresponding inverse DTFT is a set of impulses with the sampling interval $T_s = 1/4$ second. For each additional stage, the sampling interval gets reduced by a factor of 2. As the stages are connected in cascade, the DTFT transfer function of the scaling function is

$$\phi(e^{j\omega}) = \frac{1}{T_s}\left(\frac{1}{\sqrt{2}}\tilde{L}\left(e^{j\frac{\omega}{2}}\right)\frac{1}{\sqrt{2}}\tilde{L}\left(e^{j\frac{\omega}{4}}\right)\frac{1}{\sqrt{2}}\tilde{L}\left(e^{j\frac{\omega}{8}}\right)\cdots\right) \tag{8.13}$$

The transfer function, when it becomes an infinite product, is aperiodic, and the corresponding inverse FT

$$\phi(t) = \frac{1}{2\pi}\int_{-\infty}^{\infty}\phi(j\omega)e^{j\omega t}\,d\omega \tag{8.14}$$

is a continuous function (the scaling function).

As the Haar scaling function is well defined, it can be derived (not approximated as in most cases) analytically. As always, the simplicity of the Haar wavelet transform helps our understanding. Let us find a closed-form expression for the first N terms of the infinite product defining the scaling function.

$$\phi_N(j\omega) = \frac{1}{2}\left(1 + e^{-j\frac{\omega}{2^1}}\right)\frac{1}{2}\left(1 + e^{-j\frac{\omega}{2^2}}\right)\cdots\frac{1}{2}\left(1 + e^{-j\frac{\omega}{2^N}}\right)$$

$$= \frac{1}{2^N}\sum_{k=0}^{2^N-1} e^{-j\frac{\omega}{2^N}k} = \frac{1}{2^N}\frac{1 - e^{-j\omega}}{1 - e^{-j\frac{\omega}{2^N}}}$$

We used the closed-form expression for the sum of a geometric progression

$$\sum_{k=0}^{N-1} e^{-j\frac{\omega}{N}k} = \frac{1 - e^{-j\omega}}{1 - e^{-j\frac{\omega}{N}}}$$

Now, we take the limit of the partial product as the number of terms, N, tends to infinity.

$$\phi(j\omega) = \frac{1}{2^N}\lim_{N\to\infty}\frac{1 - e^{-j\omega}}{1 - e^{-j\frac{\omega}{2^N}}} = \frac{1 - e^{-j\omega}}{j\omega}$$

Note that

$$2^N \lim_{N\to\infty}\left(1 - e^{-j\frac{\omega}{2^N}}\right) = 2^N\left(1 - \left(1 - j\frac{\omega}{2^N} + \cdots\right)\right) = j\omega$$

To put it another way,

$$2^N \lim_{N\to\infty}\left(1 - e^{-j\frac{\omega}{2^N}}\right) = 2^N \lim_{N\to\infty}\left(1 - \left(\cos\left(\frac{\omega}{2^N}\right) - j\sin\left(\frac{\omega}{2^N}\right)\right)\right) = j\omega$$

As $N \to \infty$, $\frac{\omega}{2^N} \to 0$ and $\cos\left(\frac{\omega}{2^N}\right)$ tends to 1 and $\sin\left(\frac{\omega}{2^N}\right)$ tends to $\frac{\omega}{2^N}$. The inverse FT of $\phi(j\omega)$ is

$$\phi(t) = u(t) - u(t-1)$$

which is a square (the Haar scaling function) located at the origin with width 1 and height 1. The inverse FT can be verified using the definition.

$$\phi(j\omega) = \int_0^1 e^{-j\omega t}\,dt = \frac{1 - e^{-j\omega}}{j\omega}$$

In most cases, the task is to evaluate the inverse DTFT by numerical methods. The defining equation of the inverse DTFT equation is

$$x(n) = \frac{1}{2\pi}\int_{-\pi}^{\pi} X(e^{j\omega})e^{j\omega n}\,d\omega, \quad n = 0, \pm 1, \pm 2, \ldots$$

In this equation, it is assumed that the sampling interval of the time-domain signal, T_s, is 1 second. For other values of T_s, only scaling of the frequency axis is required. The IDFT of the samples of the spectrum gives the output. Let k be the number of stages in the cascade. Then, the spectral range becomes $(2^k)(2\pi)$ radians, and the sampling interval is $1/(2^k)$ seconds. If there is one stage, spectral range is $(2^1)(2\pi) = 4\pi$ radians and the sampling interval is

$1/(2^1) = 1/2$ seconds. For two stages, we get 8π radians and $1/4$ seconds and so on. For the wavelet function, the DTFT transfer function is

$$\psi(e^{j\omega}) = \frac{1}{T_s}\left(\frac{1}{\sqrt{2}}\tilde{H}\left(e^{j\frac{\omega}{2}}\right)\frac{1}{\sqrt{2}}\tilde{L}\left(e^{j\frac{\omega}{4}}\right)\frac{1}{\sqrt{2}}\tilde{L}\left(e^{j\frac{\omega}{8}}\right)\cdots\right) \tag{8.15}$$

Example 8.4 Approximate the Haar scaling and wavelet functions using one-stage IDWT with the sampling interval $T_s = 0.5$ seconds.

Solution

With $T_s = 0.5$,

$$\phi(e^{j\omega}) = 2\frac{1}{\sqrt{2}}\frac{1}{\sqrt{2}}\left(1 + e^{-j\frac{\omega}{2}}\right) = \left(1 + e^{-j\frac{\omega}{2}}\right)$$

The bandwidth is $2\pi/0.5 = 4\pi$ radians. The samples of the spectrum at $\omega = 0, \pi, 2\pi, 3\pi$ are

$$\phi(k) = \{2, 1 - j1, 0, 1 + j1\}$$

The IDFT of $\phi(k)$ is the samples of the scaling function with a sampling interval 0.5 second.

$$\{\phi(0) = 1, \phi(1/2) = 1, \phi(1) = 0, \phi(3/2) = 0\}$$

Of course, we could have solved this problem with two samples, as the support width of the Haar basis function is 1 second. To ensure that it is so, we computed four samples. For the wavelet function,

$$\psi(e^{j\omega}) = \left(1 - e^{-j\frac{\omega}{2}}\right)$$

the samples of the spectrum at $\omega = 0, 2\pi$ are

$$\phi(k) = \{0, 2\}$$

The IDFT of $\phi(k)$, $\{0 + 2 = 2, 0 - 2 = -2\}/2 = \{1, -1\}$, is the samples of the wavelet function.

$$\{\phi(0) = 1, \phi(1/2) = -1\}$$

The convergence of Haar basis function is immediate. ■

8.4.1 Two Scale Relations

The two scale relations establish the relation between the DWT filter coefficients and the corresponding wavelet basis functions. There is a continuous scaling function $\phi(t)$ corresponding to the lowpass filter coefficients $\tilde{l}(n)$ of the filter bank. Similarly, there is a continuous wavelet function $\psi(t)$ corresponding to the highpass filter $\tilde{h}(n)$. The FT transfer function of the scaling function $\phi(t)$, by definition, is

$$\phi(j\omega) = \frac{1}{\sqrt{2}}\tilde{L}\left(e^{j\frac{\omega}{2}}\right)\frac{1}{\sqrt{2}}\tilde{L}\left(e^{j\frac{\omega}{4}}\right)\frac{1}{\sqrt{2}}\tilde{L}\left(e^{j\frac{\omega}{8}}\right)\frac{1}{\sqrt{2}}\tilde{L}\left(e^{j\frac{\omega}{16}}\right)\cdots$$

Replacing ω by $\omega/2$, we get

$$\phi\left(j\frac{\omega}{2}\right) = \frac{1}{\sqrt{2}}\tilde{L}\left(e^{j\frac{\omega}{4}}\right)\frac{1}{\sqrt{2}}\tilde{L}\left(e^{j\frac{\omega}{8}}\right)\frac{1}{\sqrt{2}}\tilde{L}\left(e^{j\frac{\omega}{16}}\right)\cdots$$

Therefore,

$$\phi(j\omega) = \frac{1}{\sqrt{2}}\tilde{L}\left(e^{j\frac{\omega}{2}}\right)\phi\left(j\frac{\omega}{2}\right)$$

This equation implies that the transform $\phi(j\omega)$ of the scaling function $\phi(t)$ is the expanded (by a factor of 2) version of itself multiplied by the polynomial $\frac{1}{\sqrt{2}}\tilde{L}\left(e^{j\frac{\omega}{2}}\right)$. The inverse FT $\phi(t)$ of $\phi(j\omega)$ is defined as

$$\phi(t) = \frac{1}{2\pi}\int_{-\infty}^{\infty}\phi(j\omega)e^{j\omega t}\,d\omega \tag{8.16}$$

The term $\tilde{L}\left(e^{j\frac{\omega}{2}}\right)$ is defined by

$$\tilde{L}\left(e^{j\frac{\omega}{2}}\right) = \sum_{k=-\infty}^{\infty}\tilde{l}(k)e^{-j\frac{\omega}{2}k}$$

Now, Equation (8.16) can be written as

$$\begin{aligned}\phi(t) &= \frac{1}{\sqrt{2}}\frac{1}{2\pi}\int_{-\infty}^{\infty}\sum_{k=-\infty}^{\infty}\tilde{l}(k)e^{-j\frac{\omega}{2}k}\phi\left(j\frac{\omega}{2}\right)e^{j\omega t}\,d\omega \\ &= \frac{1}{\sqrt{2}}\sum_{k=-\infty}^{\infty}\tilde{l}(k)\left(\frac{1}{2\pi}\int_{-\infty}^{\infty}\phi\left(j\frac{\omega}{2}\right)e^{j\omega\left(t-\frac{k}{2}\right)}\,d\omega\right)\end{aligned}$$

Let $\omega = 2x$, Then, $d\omega = 2dx$. Using the variable x, we get

$$\begin{aligned}\phi(t) &= \sqrt{2}\sum_{k=-\infty}^{\infty}\tilde{l}(k)\left(\frac{1}{2\pi}\int_{-\infty}^{\infty}\phi\left(jx\right)e^{jx(2t-k)}\,dx\right) \\ &= \sqrt{2}\sum_{k=-\infty}^{\infty}\tilde{l}(k)\phi(2t-k)\end{aligned}$$

The two scale relations are as follows:

$$\phi(t) = \sqrt{2}\sum_{n=-\infty}^{\infty}\tilde{l}(n)\phi(2t-n)$$

$$\psi(t) = \sqrt{2}\sum_{n=-\infty}^{\infty}\tilde{h}(n)\phi(2t-n)$$

The two-scale (t and $2t$) relation (a difference equation with two scales) expresses a scaling function $\phi(t)$ as a linear combination of its scaled and shifted versions. A wavelet function $\psi(t)$ is a linear combination of scaled and shifted versions of the corresponding scaling function $\phi(t)$. The coefficient vectors $\tilde{l}(n)$ and $\tilde{h}(n)$ form the transform matrix of the DWT. For Haar scaling and wavelet functions, the two scale relations are simple.

$$\phi(t) = \phi(2t) + \phi(2t-1)$$
$$\psi(t) = \phi(2t) - \phi(2t-1)$$

where

$$\phi(t) = \begin{cases} 1, & 0 \le t < 1 \\ 0, & \text{otherwise} \end{cases}$$

The Haar scaling function is a square with width 1. When compressed by a factor of 2, the width becomes 0.5. Summing the compressed waveform with its shifted version is the uncompressed scaling function. The wavelet function is the sum of the compressed waveform and its shifted version with the sign of the latter negated. The scaling function has to be obtained from a limit process. Similarly, the wavelet function can also be obtained. However, the wavelet function is given in just one step by the two-scale relation. All of the properties of the scaling and wavelet functions, such as their support interval, orthogonality, and smoothness, can be determined by the corresponding DWT filter coefficients. For example, the fact that Haar filter coefficients are orthogonal implies that the corresponding scaling and wavelet functions are orthogonal. Therefore, we thoroughly study the properties of the filter coefficients of some of the members of the commonly used wavelets in the following two chapters, in both time and frequency domains. It was noted in Chapter 7 that the frequency response of the DWT lowpass filter is zero at $\omega = 0$. This requirement is essential for a well-defined solution to the two scale relations.

8.5 The 2-D Haar Discrete Wavelet Transform

The computation of the 2-D DWT requires four 2-D filters, one scaling filter $\phi(n_1, n_2)$ and three wavelet filters $H(n_1, n_2)$, $V(n_1, n_2)$, and $D(n_1, n_2)$. For separable filters, the 2-D filters are the product of 1-D filters. The four filters are given by

$$\phi(n_1, n_2) = \phi(n_1)\phi(n_2)$$
$$V(n_1, n_2) = \psi(n_1)\phi(n_2)$$
$$H(n_1, n_2) = \phi(n_1)\psi(n_2)$$
$$D(n_1, n_2) = \psi(n_1)\psi(n_2)$$

and the scaled Haar 2-D filters are shown in Figure 8.9. These filters are straightforward extensions of the 1-D filters in the other coordinate. The 1-D filters are lines, whereas the 2-D filters are surfaces. A stack of N 1-D lowpass filters is the 2-D lowpass filter $\phi(n_1)\phi(n_2)$, shown in Figure 8.9(a). Similarly, a stack of 1-D highpass filters along the columns is the 2-D highpass filter $V(n_1, n_2)$, shown in Figure 8.9(b). A stack of 1-D highpass filters along the rows is the 2-D highpass filter $H(n_1, n_2)$, shown in Figure 8.9(c). The diagonal highpass filter $D(n_1, n_2)$, shown in Figure 8.9(d), is the result of interaction of two 1-D highpass filters in the two directions, in contrast to the interaction of lowpass and highpass filters in the earlier two cases. The 2×2 scaled Haar 2-D filters are shown in Figure 8.10.

$$\phi(n_1, n_2) = \begin{bmatrix} 1 \\ 1 \end{bmatrix} \begin{bmatrix} 1 & 1 \end{bmatrix} = \begin{bmatrix} 1 & 1 \\ 1 & 1 \end{bmatrix}$$

$$V(n_1, n_2) = \begin{bmatrix} 1 \\ -1 \end{bmatrix} \begin{bmatrix} 1 & 1 \end{bmatrix} = \begin{bmatrix} 1 & 1 \\ -1 & -1 \end{bmatrix}$$

$$H(n_1, n_2) = \begin{bmatrix} 1 \\ 1 \end{bmatrix} \begin{bmatrix} 1 & -1 \end{bmatrix} = \begin{bmatrix} 1 & -1 \\ 1 & -1 \end{bmatrix}$$

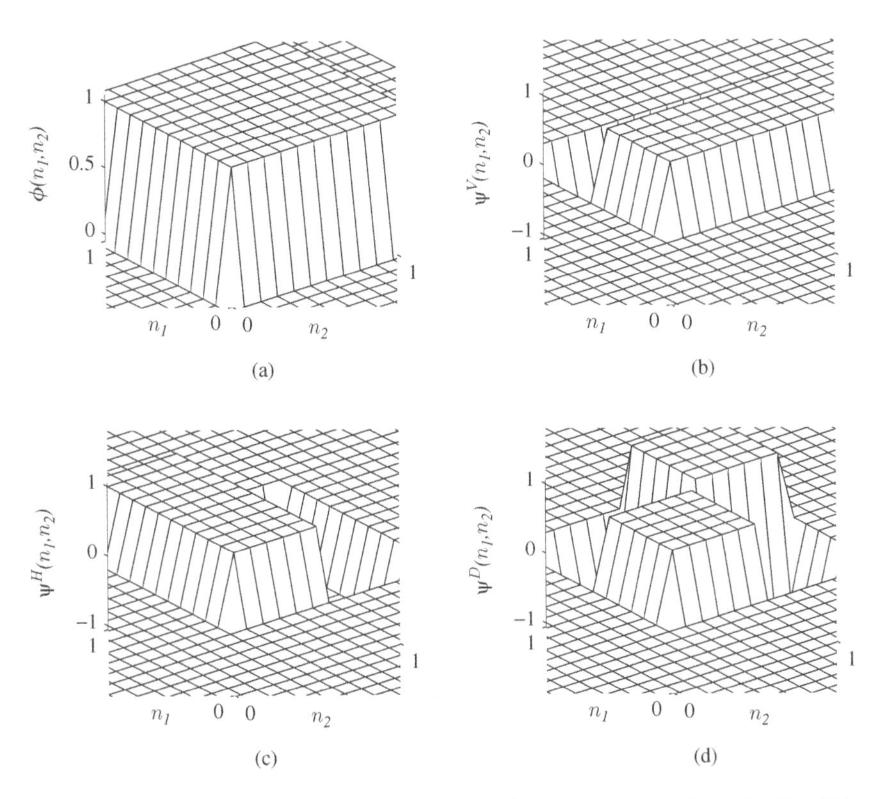

Figure 8.9 (a) The Haar 2-D scaled scaling filter $\phi(n_1, n_2)$; (b) the 2-D scaled wavelet filter $V(n_1, n_2)$; (c) the 2-D scaled wavelet filter $H(n_1, n_2)$; (d) the 2-D scaled wavelet filter $D(n_1, n_2)$

$\phi\ (n_1, n_2)$

1	1
1	1

$V\ (n_1, n_2)$

1	1
-1	-1

$H\ (n_1, n_2)$

1	-1
1	-1

$D\ (n_1, n_2)$

1	-1
-1	1

Figure 8.10 2×2 scaled Haar 2-D DWT filters $\phi(n_1, n_2)$, $V(n_1, n_2)$, $H(n_1, n_2)$, and $D(n_1, n_2)$

$$D(n_1, n_2) = \begin{bmatrix} 1 \\ -1 \end{bmatrix} \begin{bmatrix} 1 & -1 \end{bmatrix} = \begin{bmatrix} 1 & -1 \\ -1 & 1 \end{bmatrix}$$

For separable filters, the 2-D DWT can be computed by the row-column method used for computing the 2-D DFT. The 1-D DWT algorithm can be used to compute the 1-D DWT of each column of the input data followed by computing the 1-D DWT of each row of the resulting data. Of course, the order of the computation can also be reversed. In contrast to the computation of the 2-D DFT, 1-level DWT has to be computed in a recursive manner to compute the 2-D DWT for different scales. Each decomposition of an image produces four quarter-size subimages, one approximation component $X^j_\phi(n_1, n_2)$ and three sets of detail components

$X^j_H(n_1,n_2)$, $X^j_V(n_1,n_2)$, and $X^j_D(n_1,n_2)$. Component $X^j_\phi(n_1,n_2)$ is a smoothed version of the input. It almost resembles the original image, as it is well known that the low-frequency part of the spectrum approximates an image well. Components $X^j_H(n_1,n_2)$, $X^j_V(n_1,n_2)$, and $X^j_D(n_1,n_2)$ are measures of horizontal (variation along rows), vertical, and diagonal differences, respectively. They are effective in detecting vertical, horizontal, and diagonal edges, respectively.

The 2-D DWT of an image $\boldsymbol{x}$ and its inverse is defined, respectively, as

$$\boldsymbol{X} = \boldsymbol{W}\,\boldsymbol{x}\,\boldsymbol{W}^T \quad \text{and} \quad \boldsymbol{x} = \boldsymbol{W}^T\boldsymbol{X}\,\boldsymbol{W} \tag{8.17}$$

where $\boldsymbol{W}$ is the transform matrix defined for 1-D DWT. Postmultiplying the image matrix by the right transform matrix is computing the 1-D DWT of the rows. Premultiplying the image matrix by the left transform matrix is computing the 1-D DWT of the columns. Note that

$$\boldsymbol{x} = \boldsymbol{W}^T\boldsymbol{X}\,\boldsymbol{W} = \boldsymbol{W}^T\boldsymbol{W}\,\boldsymbol{x}\,\boldsymbol{W}^T\boldsymbol{W} = \boldsymbol{x}$$

With the 1-level transform matrix $\boldsymbol{W}$ partitioned as

$$\boldsymbol{W} = \begin{bmatrix} \boldsymbol{L} \\ \boldsymbol{H} \end{bmatrix}, \boldsymbol{W}\,\boldsymbol{x}\,\boldsymbol{W}^T = \begin{bmatrix} \boldsymbol{L} \\ \boldsymbol{H} \end{bmatrix} \boldsymbol{x} \begin{bmatrix} \boldsymbol{L} \\ \boldsymbol{H} \end{bmatrix}^T = \begin{bmatrix} \boldsymbol{Lx} \\ \boldsymbol{Hx} \end{bmatrix} \begin{bmatrix} \boldsymbol{L}^T & \boldsymbol{H}^T \end{bmatrix}$$
$$= \begin{bmatrix} \boldsymbol{LxL}^T & \boldsymbol{LxH}^T \\ \boldsymbol{HxL}^T & \boldsymbol{HxH}^T \end{bmatrix} = \begin{bmatrix} \boldsymbol{X}^j_\phi & \boldsymbol{X}^j_H \\ \boldsymbol{X}^j_V & \boldsymbol{X}^j_D \end{bmatrix} \tag{8.18}$$

Example 8.5 Compute the 2-D DWT of

$$\begin{bmatrix} 2 & 5 \\ 1 & 7 \end{bmatrix}$$

in different ways.

Solution

Computing the 1-D DWT of the columns followed by computing the 1-D DWT of the rows of the resulting data, we get

$$\begin{bmatrix} \frac{1}{\sqrt{2}} & \frac{1}{\sqrt{2}} \\ \frac{1}{\sqrt{2}} & -\frac{1}{\sqrt{2}} \end{bmatrix} \begin{bmatrix} 2 & 5 \\ 1 & 7 \end{bmatrix} \rightarrow \begin{bmatrix} \frac{3}{\sqrt{2}} & \frac{12}{\sqrt{2}} \\ \frac{1}{\sqrt{2}} & -\frac{2}{\sqrt{2}} \end{bmatrix} \begin{bmatrix} \frac{1}{\sqrt{2}} & \frac{1}{\sqrt{2}} \\ \frac{1}{\sqrt{2}} & -\frac{1}{\sqrt{2}} \end{bmatrix} = \begin{bmatrix} \frac{15}{2} & -\frac{9}{2} \\ -\frac{1}{2} & \frac{3}{2} \end{bmatrix}$$

Now, the same 2-D DWT is computed by first computing the 1-D DWT of the rows followed by computing the 1-D DWT of the columns of the resulting data.

$$\begin{bmatrix} 2 & 5 \\ 1 & 7 \end{bmatrix} \begin{bmatrix} \frac{1}{\sqrt{2}} & \frac{1}{\sqrt{2}} \\ \frac{1}{\sqrt{2}} & -\frac{1}{\sqrt{2}} \end{bmatrix} \rightarrow \begin{bmatrix} \frac{1}{\sqrt{2}} & \frac{1}{\sqrt{2}} \\ \frac{1}{\sqrt{2}} & -\frac{1}{\sqrt{2}} \end{bmatrix} \begin{bmatrix} \frac{7}{\sqrt{2}} & -\frac{3}{\sqrt{2}} \\ \frac{8}{\sqrt{2}} & -\frac{6}{\sqrt{2}} \end{bmatrix} = \begin{bmatrix} \frac{15}{2} & -\frac{9}{2} \\ -\frac{1}{2} & \frac{3}{2} \end{bmatrix}$$

The same computation is made using Equation (8.18) yielding the same result.

$$\begin{bmatrix} \boldsymbol{LxL}^T & \boldsymbol{LxH}^T \\ \boldsymbol{HxL}^T & \boldsymbol{HxH}^T \end{bmatrix} = \begin{bmatrix} \begin{bmatrix} \frac{1}{\sqrt{2}} & \frac{1}{\sqrt{2}} \end{bmatrix} \begin{bmatrix} 2 & 5 \\ 1 & 7 \end{bmatrix} \begin{bmatrix} \frac{1}{\sqrt{2}} \\ \frac{1}{\sqrt{2}} \end{bmatrix} & \begin{bmatrix} \frac{1}{\sqrt{2}} & \frac{1}{\sqrt{2}} \end{bmatrix} \begin{bmatrix} 2 & 5 \\ 1 & 7 \end{bmatrix} \begin{bmatrix} \frac{1}{\sqrt{2}} \\ -\frac{1}{\sqrt{2}} \end{bmatrix} \\ \begin{bmatrix} \frac{1}{\sqrt{2}} & -\frac{1}{\sqrt{2}} \end{bmatrix} \begin{bmatrix} 2 & 5 \\ 1 & 7 \end{bmatrix} \begin{bmatrix} \frac{1}{\sqrt{2}} \\ \frac{1}{\sqrt{2}} \end{bmatrix} & \begin{bmatrix} \frac{1}{\sqrt{2}} & -\frac{1}{\sqrt{2}} \end{bmatrix} \begin{bmatrix} 2 & 5 \\ 1 & 7 \end{bmatrix} \begin{bmatrix} \frac{1}{\sqrt{2}} \\ -\frac{1}{\sqrt{2}} \end{bmatrix} \end{bmatrix}$$

$$= \begin{bmatrix} \frac{15}{2} & -\frac{9}{2} \\ -\frac{1}{2} & \frac{3}{2} \end{bmatrix} = \begin{bmatrix} \boldsymbol{X}_\phi^0 & \boldsymbol{X}_H^0 \\ \boldsymbol{X}_V^0 & \boldsymbol{X}_D^0 \end{bmatrix}$$

■

Example 8.6 Compute the 1-level 2-D DWT of the following 4×4 2-D data. Verify that the input is reconstructed by computing the IDWT of the DWT coefficients. Verify Parseval's theorem.

$$x(n_1, n_2) = \begin{matrix} n_1 \\ \downarrow \end{matrix} \overset{n_2 \rightarrow}{\begin{bmatrix} 1 & 4 & -1 & 2 \\ 2 & 3 & -1 & 1 \\ 2 & 1 & 1 & -3 \\ 1 & 2 & -1 & 3 \end{bmatrix}}$$

Solution

Computing the 1-D DWT of the rows of $x(n_1, n_2)$, we get

$$\begin{bmatrix} 1 & 4 & -1 & 2 \\ 2 & 3 & -1 & 1 \\ 2 & 1 & 1 & -3 \\ 1 & 2 & -1 & 3 \end{bmatrix} \frac{1}{\sqrt{2}} \begin{bmatrix} 1 & 0 & 1 & 0 \\ 1 & 0 & -1 & 0 \\ 0 & 1 & 0 & 1 \\ 0 & 1 & 0 & -1 \end{bmatrix} = \frac{1}{\sqrt{2}} \begin{bmatrix} 5 & 1 & -3 & -3 \\ 5 & 0 & -1 & -2 \\ 3 & -2 & 1 & 4 \\ 3 & 2 & -1 & -4 \end{bmatrix}$$

Computing the 1-D DWT of the columns of the partially transformed matrix, we get

$$\frac{1}{\sqrt{2}} \begin{bmatrix} 1 & 1 & 0 & 0 \\ 0 & 0 & 1 & 1 \\ 1 & -1 & 0 & 0 \\ 0 & 0 & 1 & -1 \end{bmatrix} \frac{1}{\sqrt{2}} \begin{bmatrix} 5 & 1 & -3 & -3 \\ 5 & 0 & -1 & -2 \\ 3 & -2 & 1 & 4 \\ 3 & 2 & -1 & -4 \end{bmatrix} = \frac{1}{2} \begin{bmatrix} 10 & 1 & -4 & -5 \\ 6 & 0 & 0 & 0 \\ 0 & 1 & -2 & -1 \\ 0 & -4 & 2 & 8 \end{bmatrix}$$

the 1-level 2-D DWT of the input as

$$\begin{bmatrix} 5.0 & 0.5 & -2.0 & -2.5 \\ 3.0 & 0 & 0 & 0 \\ 0 & 0.5 & -1.0 & -0.5 \\ 0 & -2.0 & 1.0 & 4.0 \end{bmatrix}$$

Both the sum of the squared-magnitude of the input and that of its DWT are 67.

Computing the 1-D IDWT of the columns of the 2-D DWT, we get

$$\frac{1}{\sqrt{2}}\begin{bmatrix} 1 & 0 & 1 & 0 \\ 1 & 0 & -1 & 0 \\ 0 & 1 & 0 & 1 \\ 0 & 1 & 0 & -1 \end{bmatrix}\begin{bmatrix} 5.0 & 0.5 & -2.0 & -2.5 \\ 3.0 & 0 & 0 & 0 \\ 0 & 0.5 & -1.0 & -0.5 \\ 0 & -2.0 & 1.0 & 4.0 \end{bmatrix} = \frac{1}{\sqrt{2}}\begin{bmatrix} 5 & 1 & -3 & -3 \\ 5 & 0 & -1 & -2 \\ 3 & -2 & 1 & 4 \\ 3 & 2 & -1 & -4 \end{bmatrix}$$

Computing the 1-D IDWT of the rows of partial result, we get the input as

$$\frac{1}{\sqrt{2}}\begin{bmatrix} 5 & 1 & -3 & -3 \\ 5 & 0 & -1 & -2 \\ 3 & -2 & 1 & 4 \\ 3 & 2 & -1 & -4 \end{bmatrix}\frac{1}{\sqrt{2}}\begin{bmatrix} 1 & 1 & 0 & 0 \\ 0 & 0 & 1 & 1 \\ 1 & -1 & 0 & 0 \\ 0 & 0 & 1 & -1 \end{bmatrix} = \begin{bmatrix} 1 & 4 & -1 & 2 \\ 2 & 3 & -1 & 1 \\ 2 & 1 & 1 & -3 \\ 1 & 2 & -1 & 3 \end{bmatrix}$$

■

Example 8.7 Compute the 2-level 2-D DWT of the following 4×4 2-D data. Verify that the input is reconstructed by computing the IDWT of the DWT coefficients. Verify Parseval's theorem.

$$x(n_1, n_2) = \begin{matrix} & n_2 \rightarrow \\ n_1 \downarrow & \begin{bmatrix} 1 & 2 & -1 & 3 \\ 2 & 1 & 4 & 3 \\ 1 & 1 & 2 & 2 \\ 4 & 2 & 1 & 3 \end{bmatrix} \end{matrix}$$

Solution

Computing the 1-D DWT of the rows of $x(n_1, n_2)$, we get

$$\frac{1}{\sqrt{2}}\begin{bmatrix} 3 & 2 & -1 & -4 \\ 3 & 7 & 1 & 1 \\ 2 & 4 & 0 & 0 \\ 6 & 4 & 2 & -2 \end{bmatrix}$$

Computing the 1-D DWT of the columns of the partially transformed matrix, we get the 1-level 2-D DWT of $x(n_1, n_2)$, shown in Figure 8.11(a). Both the sum of the squared-magnitude of

(a)

$X_\phi^1(n_1, n_2)$ (top-left), $X_H^1(n_1, n_2)$ (top-right), $X_V^1(n_1, n_2)$ (bottom-left), $X_D^1(n_1, n_2)$ (bottom-right):

$$\left[\begin{array}{cc|cc} 3 & 4.5 & 0 & -1.5 \\ 4 & 4 & 1 & -1 \\ \hline 0 & -2.5 & -1 & -2.5 \\ -2 & 0 & -1 & 1 \end{array}\right]$$

(b)

$X_\phi^0(0,0)$, $X_H^0(0,1)$, $X_V^0(1,0)$, $X_D^0(1,1)$:

$$\begin{bmatrix} 7.75 & -0.75 \\ -0.25 & -0.75 \end{bmatrix}$$

Figure 8.11 (a) The 1-level and (b) 2-level 2-D Haar DWT of $x(n_1, n_2)$. The top-left corner coefficients alone are shown. The other coefficients are the same as in (a)

the input and that of its DWT are 85. 1-level decomposition produces four components of a 2-D signal. The decomposition of a 2-D signal by k levels produces $3k + 1$ components of the signal.

Computing the 1-D IDWT of the rows of the DWT coefficients shown in Figure 8.11(a), we get

$$\frac{1}{\sqrt{2}}\begin{bmatrix} 3 & 3 & 3 & 6 \\ 5 & 3 & 3 & 5 \\ -1 & 1 & -5 & 0 \\ -3 & -1 & 1 & -1 \end{bmatrix}$$

Note that the inverse transform matrix is the transpose of that of the forward transform matrix. Computing the 1-D IDWT of the columns of the partially transformed matrix, we get back $x(n_1, n_2)$.

Computing the 1-D row DWT of the 2×2 DWT coefficients shown at the top-left corner in Figure 8.11(a), we get

$$\frac{1}{\sqrt{2}}\begin{bmatrix} 7.5 & -1.5 \\ 8 & 0 \end{bmatrix}$$

Computing the 1-D column DWT of the partially transformed matrix, we get the 2-level 2-D DWT of $x(n_1, n_2)$, shown in Figure 8.11(b). The top-left corner coefficients alone are shown in Figure 8.11(b). The other coefficients are the same as in (a). Computing the 1-D row IDWT of the 2×2 DWT coefficients shown in Figure 8.11(b), we get

$$\frac{1}{\sqrt{2}}\begin{bmatrix} 7 & 8.5 \\ -1 & 0.5 \end{bmatrix}$$

Computing the 1-D column IDWT of the of the partially transformed matrix, we get the 1-level DWT of $x(n_1, n_2)$ shown in Figure 8.11(a). A 1-level 2-D IDWT will get back $x(n_1, n_2)$. ■

Figure 8.12 shows a 512×512 image with 256 gray levels. The 1-level Haar DWT decomposition of the image, shown in Figure 8.13, has the approximation component in the top left-hand corner, and the other three are the vertical (bottom left-hand corner), horizontal, and diagonal (bottom right-hand corner) detail components, each of size 256×256. As the magnitudes of the detail components are small, these components are scaled to make them more visible in Figure 8.13. Each of the subbands represents a specific group of frequencies of the image spectrum. The decompositions are carried out using analysis filter bank with Haar filters. Of course, a variety of DWT filters with different characteristics are available to suit the needs of the application. In DWT processing of signals, the filters and the number of levels of the decomposition have to be carefully selected. After the components are suitably processed, the output image can be reconstructed using the synthesis filter bank. The histogram of the image and that of its 1-level DWT coefficients are shown, respectively, in Figures 8.14(a) and (b). The histograms of the four components are shown in Figure 8.14(c–f). Using these components appropriately, the input image can be approximated with multiresolutions. The histograms of the image and its approximation part are widespread. The other three histograms are very narrow. Most of the values of the other three histograms of the detail components are close to zero. They can be coded with less number of bits or can be discarded with acceptable image quality.

Figure 8.12 A 512×512 image with 256 gray levels

Figure 8.13 The 1-level Haar DWT of the image in Figure 8.12

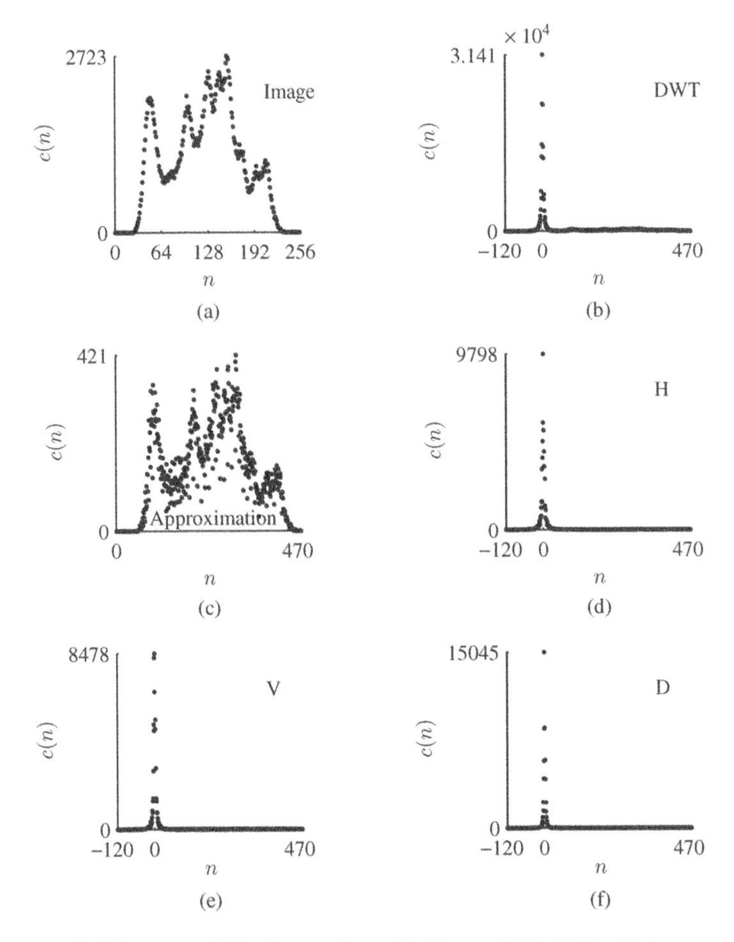

Figure 8.14 (a) The histogram of the image shown in Figure 8.12; (b) the histogram of the 1-level DWT of the image shown in Figure 8.13; (c) the histogram of the approximation part of the DWT; (d) the histogram of the horizontal component of the detail part of the DWT; (e) the histogram of the vertical component; (f) the histogram of the diagonal component

This characteristic of typical images enables good compression with acceptable fidelity. Applications such as image compression depends on the fact that the magnitudes of high-frequency components of practical signals are much smaller than those of the low-frequency components. That is, the spectral values decrease with increasing frequencies. The purpose of a transform is to transform the data into a more suitable form for processing. The separation of the components of a signal corresponding to various subbands by the DWT makes it easy and efficient to carry out operations such as compression, denoising, edge detection, and so on. The 2-level Haar DWT decomposition of the image, shown in Figure 8.15, has approximation component and vertical, horizontal, and diagonal detail components, each of size 128×128 in the top left quarter of the image. The total number of components is $3(2) + 1 = 7$. With more levels of decomposition, we get the image components with more number of resolutions.

Figure 8.15 The 2-level Haar DWT of the image in Figure 8.12

8.6 Discontinuity Detection

In the DWT, the representation and analysis of signals are carried out at more than one resolution. Important features of a signal are likely be detected at least with one of the representations. DWT is very useful in detecting short discontinuities of the signal or its difference, which may be difficult to detect in the time-domain or frequency-domain representation. Consider a sinusoidal signal with a discontinuity, shown in Figure 8.16(a). The 1-level Haar DWT of the

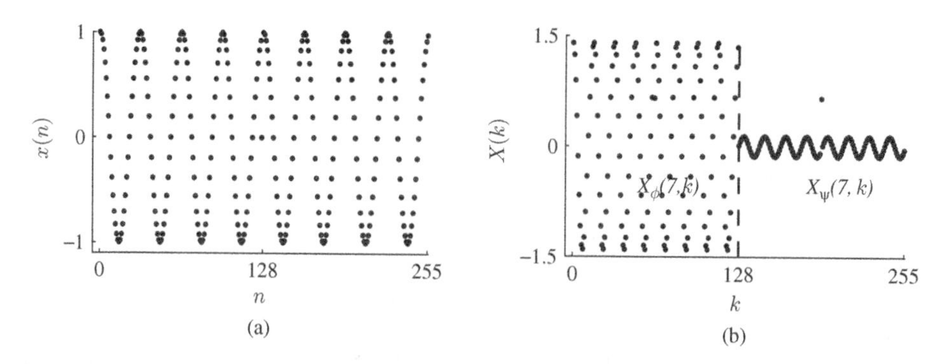

Figure 8.16 (a) A signal with a discontinuity; (b) the 1-level Haar DWT of the signal showing the occurrence of the discontinuity in the detail part

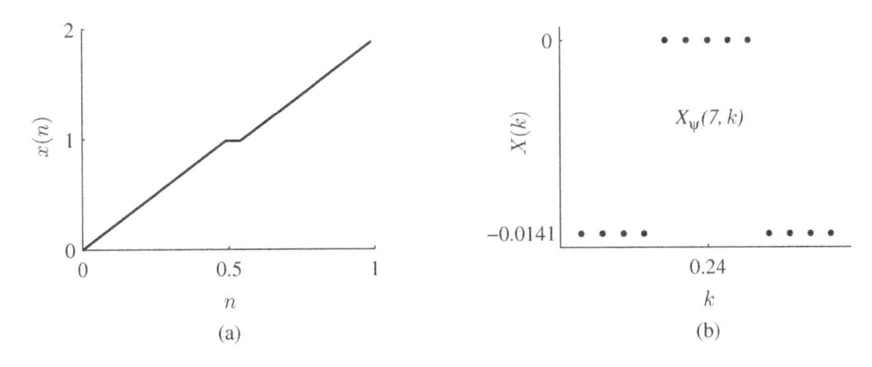

Figure 8.17 (a) A signal with two discontinuities of its difference; (b) the 1-level Haar DWT of the signal showing the occurrence of its discontinuities in the detail part

signal, shown in Figure 8.16(b), clearly shows the occurrence of the discontinuity in the detail part, as a discontinuity is characterized by high-frequency components.

A signal with two discontinuities, at an interval of 10 samples, of its difference is shown in Figure 8.17(a). The signal is discrete with the sampling interval of 0.01 seconds. Due to the density of the 100 samples, the graph of the signal looks like that of a continuous signal. The value of the detail coefficients between the two discontinuities is zero, as it is a horizontal line. At other places, it is $-0.02/\sqrt{2} = -0.0141421$. The 1-level Haar DWT detail coefficients in the neighbourhood of the discontinuities of the signal show their occurrence, as shown in Figure 8.17(b).

8.7 Summary

- The DWT decomposes a signal in terms of its components corresponding to different spectral bands, called subbands of the spectrum.
- The DWT is essentially a set of bandpass filters.
- The basis functions of the DWT are localized in both time and frequency.
- The DWT provides a time-frequency analysis of a signal.
- The decomposition of a signal by the DWT and its reconstruction are carried out using a set of lowpass and highpass filters in a recursive and efficient manner.
- Haar filters are the oldest of all the DWT filters. In addition to its usefulness in applications, it is the starting point in the study of the DWT, as it is also the shortest and simplest.
- Haar lowpass filter computes a simple average. Haar highpass filter computes a simple difference.
- The multiresolution capability of the DWT makes it possible to detect features of a signal at a resolution that may go undetected at another.
- The 2-D DWT is computed using the row-column method for separable filters.
- The DWT is inherently suitable for processing nonstationary signals. The DWT is widely used in applications such as signal analysis, compression, and denoising.

Exercises

8.1 Using the Haar transform matrix, find the DWT of $x(n)$. Verify that $x(n)$ is reconstructed by computing the IDWT. Verify Parseval's theorem.

***8.1.1**

$$x(n) = \{1, -6\}$$

8.1.2

$$x(n) = \{-5, 3\}$$

8.2 Using the Haar transform matrix, find the 1-level and 2-level DWT of $x(n)$. Verify that $x(n)$ is reconstructed by computing the IDWT. Verify Parseval's theorem.

8.2.1

$$x(n) = \{4, 1, 2, -4\}$$

***8.2.2**

$$x(n) = \{3, 1, -2, 4\}$$

8.3 Using the Haar transform matrix, find the 1-level, 2-level, and 3-level DWT of $x(n)$. Verify that $x(n)$ is reconstructed by computing the IDWT. Verify Parseval's theorem.

***8.3.1**

$$x(n) = \{3, 1, 2, 5, 3, -2, -1, 2\}$$

8.3.2

$$x(n) = \{4, -2, 3, 1, 2, 2, -3, 1\}$$

8.4 Draw the time-frequency plane of the DWT, with $N = 16$.

8.5 Draw the graphs of all the components of

$$x(n) = \begin{cases} 2 & \text{for} \quad n = 0, 1, 2, 3, 4 \\ -1 & \text{for} \quad n = 5, 6, 7 \end{cases}$$

obtained by computing the 3-level Haar DWT.

***8.6** Draw the graphs of all the components of

$$x(n) = \{\cos(2\pi n/8), \quad n = 0, 1, \ldots, 7\}$$

obtained by computing the 3-level Haar DWT.

8.7 Draw the graphs of all the components of

$$x(n) = \{n^2/49, \quad n = 0, 1, \ldots, 7\}$$

obtained by computing the 3-level Haar DWT.

8.8 Draw the graphs of all the components of

$$x(n) = \{n/7, \ \ n = 0, 1, \ldots, 7\}$$

obtained by computing the 3-level Haar DWT.

8.9 Draw the graphs of all the components of

$$x(n) = \{0, 1, 2, 3, 4, 3, 2, 1\}$$

obtained by computing the 3-level Haar DWT.

8.10 Compute the 1-level 2-D Haar DWT of

$$\begin{bmatrix} 1 & 1 & 1 & 1 \\ 1 & 1 & 1 & 1 \\ 1 & 1 & 1 & 1 \\ 0 & 0 & 0 & 0 \end{bmatrix}$$

***8.11** Compute the 1-level 2-D Haar DWT of

$$\begin{bmatrix} 1 & 1 & 1 & 0 \\ 1 & 1 & 1 & 0 \\ 1 & 1 & 1 & 0 \\ 1 & 1 & 1 & 0 \end{bmatrix}$$

8.12 Compute the 1-level 2-D Haar DWT of

$$\begin{bmatrix} 1 & 0 & 0 & 0 \\ 0 & 1 & 0 & 0 \\ 0 & 0 & 1 & 0 \\ 0 & 0 & 0 & 1 \end{bmatrix}$$

8.13 Compute the 1-level 2-D Haar DWT of

$$\begin{bmatrix} 1 & 0 & 0 & 0 \\ 1 & 1 & 0 & 0 \\ 1 & 1 & 1 & 0 \\ 1 & 1 & 1 & 1 \end{bmatrix}$$

8.14 Compute the 2-level 2-D Haar DWT of

$$\begin{bmatrix} 4 & 2 & -1 & -3 \\ 3 & -1 & 4 & 2 \\ 1 & 3 & 2 & -2 \\ 2 & -2 & 4 & 3 \end{bmatrix}$$

Verify that the input is reconstructed by computing the IDWT. Verify Parseval's theorem.

***8.15** Compute the 2-level 2-D Haar DWT of

$$\begin{bmatrix} 1 & -2 & 1 & 3 \\ 2 & 1 & -3 & 3 \\ 1 & -4 & 2 & 3 \\ -2 & 2 & 1 & -3 \end{bmatrix}$$

Verify that the input is reconstructed by computing the IDWT. Verify Parseval's theorem.

8.16 Compute the 1-level 2-D Haar DWT of

$$\begin{bmatrix} 80 & 100 & 111 & 108 \\ 79 & 78 & 72 & 82 \\ 71 & 69 & 68 & 80 \\ 60 & 78 & 83 & 84 \end{bmatrix}$$

Verify that the detail components are much smaller than the approximation component.

8.17 Compute the 1-level 2-D Haar DWT of

$$\begin{bmatrix} 127 & 124 & 130 & 132 \\ 126 & 125 & 131 & 128 \\ 118 & 132 & 127 & 128 \\ 123 & 125 & 121 & 130 \end{bmatrix}$$

Verify that the detail components are much smaller than the approximation component.

8.18 Compute the 1-level 2-D Haar DWT of

$$\begin{bmatrix} 131 & 127 & 127 & 130 \\ 132 & 132 & 129 & 126 \\ 132 & 132 & 134 & 133 \\ 135 & 131 & 135 & 133 \end{bmatrix}$$

Verify that the detail components are much smaller than the approximation component.

8.19 Find the eight samples of the Haar scaling function after three stages of reconstruction. Use both the z-transform and IDFT methods.

***8.20** Find the eight samples of the Haar wavelet function after three stages of reconstruction. Use both the z-transform and IDFT methods.

9

Orthogonal Filter Banks

The basis functions of the DWT are a linear combination of sinusoids. The transform, which decomposes a signal into a set of components, of transient nature, corresponding to a set of continuous groups of frequencies of the spectrum, is, in effect, a set of bandpass filters. The effect of bandpass filters is achieved using a set of lowpass and highpass filters recursively. These filters require certain properties that are not emphasized in the conventional filter design methods used in digital signal processing. In this chapter, the design is illustrated by deriving the impulse response of some commonly used filters. A filter bank is referred as orthogonal if the corresponding transform matrix $\boldsymbol{W}_N$ satisfies the property

$$\boldsymbol{W}_N \boldsymbol{W}_N^{-1} = \boldsymbol{W}_N \boldsymbol{W}_N^T = \boldsymbol{I}_N$$

where $\boldsymbol{I}_N$ is the $N \times N$ identity matrix. The transpose is the inverse.

The design of filters for DWT is different from that of the conventional filter design methods. One difference is that the effect of a bandpass filter is achieved using a set of lowpass and highpass filters recursively. The frequency response $L(e^{j\omega})$ of the lowpass filters must have a number of zeros at $\omega = \pi$. Another difference is that the original signal should be reconstructed perfectly from its unaltered components. The set of equations formulated for the design includes quadratic equations. Furthermore, different filters are used to suit different applications. A pair of lowpass and highpass filters, along with samplers and delays, constitutes a two-channel filter bank. There is an analysis filter bank and a synthesis filter bank. If the transform matrices of the two filter banks are transposes as well as inverses of each other, then the filter bank, for example, those shown in Figures 9.1 and 9.3, is called orthogonal. The problem reduces to the design of a set of lowpass and highpass filters for analysis and another set for reconstruction. The filters are typically FIR type. The difference in the design procedure is that the constraints imposed on the frequency response of the filters are different to suit the requirements of the DWT.

In this chapter, the design of filters for orthogonal filter banks is presented. The design is illustrated for the commonly used filters: Haar filter, Daubechies filter of length four (D4), and Coiflet filter of length six (C6). These filters are designed using the othogonality of the inverse and forward transform matrices and certain constraints on the frequency response of the filter. Different filter design methods are available, and the method selected is due to its simplicity. The coefficients and characteristics of other DWT filters often used in practice can be found in the current edition of MATLAB® manuals.

Discrete Wavelet Transform: A Signal Processing Approach, First Edition. D. Sundararajan.

Companion Website: www.wiley.com/go/sundararajan/wavelet

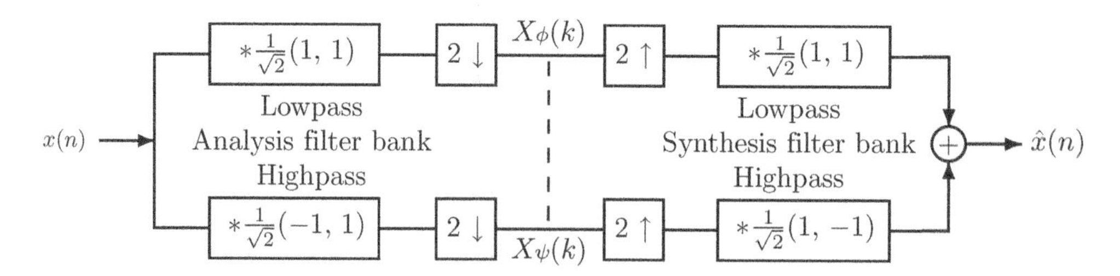

Figure 9.1 A two-channel analysis and synthesis filter bank with Haar filters

9.1 Haar Filter

The 1-level 2-point Haar DWT of the time-domain signal $\{x(0), x(1)\}$ is given as

$$\begin{bmatrix} X_\phi(0,0) \\ X_\psi(0,0) \end{bmatrix} = \frac{1}{\sqrt{2}} \begin{bmatrix} 1 & 1 \\ 1 & -1 \end{bmatrix} \begin{bmatrix} x(0) \\ x(1) \end{bmatrix} \tag{9.1}$$

The first row of the transform matrix represents the basis function that separates the low-frequency component of the signal, while the second row represents the basis function that separates the high-frequency component. As each of the components is composed of half of the spectral components of the signal, they are represented using half the number of samples in the leftmost coefficient vector. As the basis functions are nonideal, the separation of the signal into two components corresponding to the lower (frequency 0 to $\pi/2$ radians) and upper (frequency $\pi/2$ to π radians) parts of the spectrum is imperfect. The objective is that it should be possible to reconstruct the original signal from its unaltered components. This objective is achieved if the inverse transform matrix is orthogonal to the transform matrix. As the matrices are real valued, orthogonal forward and inverse transform matrices are transposes as well as inverses of each other. That is, for this example,

$$\begin{bmatrix} x(0) \\ x(1) \end{bmatrix} = \frac{1}{\sqrt{2}} \begin{bmatrix} 1 & 1 \\ 1 & -1 \end{bmatrix} \begin{bmatrix} X_\phi(0,0) \\ X_\psi(0,0) \end{bmatrix} \tag{9.2}$$

and

$$\frac{1}{\sqrt{2}} \begin{bmatrix} 1 & 1 \\ 1 & -1 \end{bmatrix} \frac{1}{\sqrt{2}} \begin{bmatrix} 1 & 1 \\ 1 & -1 \end{bmatrix} = \begin{bmatrix} 1 & 0 \\ 0 & 1 \end{bmatrix} \tag{9.3}$$

The derivation of the Haar transform lowpass filter impulse response

$$l(n) = \{l(0), l(1)\} = \frac{1}{\sqrt{2}}\{1, 1\}$$

is as follows. The Haar basis function is well defined, and the impulse response can be derived from it. However, the design is presented as it is the simplest and serves as the starting point for the design of other filters. The 1-level 2-point DWT, using Haar basis functions, is defined as

$$\begin{bmatrix} X_\phi(0,0) \\ X_\psi(0,0) \end{bmatrix} = \begin{bmatrix} l(1) & l(0) \\ h(1) & h(0) \end{bmatrix} \begin{bmatrix} x(0) \\ x(1) \end{bmatrix} \tag{9.4}$$

The 2×2 transform matrix can be expressed, using block matrix representation for the low-pass and highpass portions, as

$$\mathbf{W}_2 = \begin{bmatrix} l(1) & l(0) \\ h(1) & h(0) \end{bmatrix} = \begin{bmatrix} \mathbf{L} \\ \mathbf{H} \end{bmatrix} \tag{9.5}$$

where

$$\mathbf{L} = \begin{bmatrix} l(1) & l(0) \end{bmatrix} \quad \text{and} \quad \mathbf{H} = \begin{bmatrix} h(1) & h(0) \end{bmatrix} \tag{9.6}$$

As the inverse of the transform matrix is its transpose due to orthogonality, we get

$$\mathbf{W}_2 \mathbf{W}_2^T = \begin{bmatrix} \mathbf{L} \\ \mathbf{H} \end{bmatrix} \begin{bmatrix} \mathbf{L}^T & \mathbf{H}^T \end{bmatrix} = \begin{bmatrix} \mathbf{L}\mathbf{L}^T & \mathbf{L}\mathbf{H}^T \\ \mathbf{H}\mathbf{L}^T & \mathbf{H}\mathbf{H}^T \end{bmatrix} = \begin{bmatrix} 1 & 0 \\ 0 & 1 \end{bmatrix} = \mathbf{I}_2 \tag{9.7}$$

In orthogonal filter banks, the analysis lowpass filter is usually designed, and the other three filters are derived from it. From Equation (9.7),

$$\mathbf{L}\mathbf{L}^T = \begin{bmatrix} l(1) & l(0) \end{bmatrix} \begin{bmatrix} l(1) \\ l(0) \end{bmatrix} = l(1)^2 + l(0)^2 = 1$$

Therefore, we get the orthogonality condition

$$l(1)^2 + l(0)^2 = 1 \tag{9.8}$$

For DWT lowpass filters, the frequency response $L(e^{j\omega})$ at $\omega = \pi$ is defined as

$$L(e^{j\pi}) = 0$$

For the Haar filter with impulse response $l(n) = \{l(0), l(1)\}$, the frequency response is given by

$$L(e^{j\omega}) = l(0) + l(1)e^{-j\omega}$$

At $\omega = \pi$, we get

$$L(e^{j\pi}) = l(0) - l(1) = 0$$

and $l(0) = l(1)$. From the orthogonality condition (Equation 9.8),

$$l(1)^2 + l(0)^2 = l(0)^2 + l(0)^2 = 2l(0)^2 = 1$$

Taking the positive root, we get

$$l(n) = \{l(0), l(1)\} = \frac{1}{\sqrt{2}}\{1, 1\}$$

Now, the frequency response at $\omega = 0$ is given by

$$L(e^{j0}) = l(0) + l(1) = \frac{1}{\sqrt{2}}(1 + 1) = \sqrt{2}$$

The orthogonality condition for highpass filter is

$$h(1)^2 + h(0)^2 = 1 \tag{9.9}$$

For DWT highpass filters, the frequency response $H(e^{j\omega})$ at $\omega = 0$ is defined as

$$H(e^{j0}) = 0$$

and we get $h(1) = -h(0)$. Using the orthogonality condition, we get one of the two possible sets for highpass filter coefficients as

$$h(n) = \{h(0), h(1)\} = \frac{1}{\sqrt{2}}\{-1, 1\}$$

We list the impulse responses of all the four Haar filters.
Lowpass analysis filter

$$l(n) = \{l(0), l(1)\} = \frac{1}{\sqrt{2}}\{1, 1\}$$

Highpass analysis filter

$$h(n) = \{h(0), h(1)\} = \frac{1}{\sqrt{2}}\{-1, 1\}$$

The synthesis filters are the time-reversed and shifted versions (as derived later) of their respective analysis filters.
Lowpass synthesis filter

$$\widetilde{l}(n) = \{\widetilde{l}(0), \widetilde{l}(1)\} = \frac{1}{\sqrt{2}}\{1, 1\}$$

Highpass synthesis filter

$$\widetilde{h}(n) = \{\widetilde{h}_0, \widetilde{h}_1\} = \frac{1}{\sqrt{2}}\{1, -1\}$$

Examples of forward and inverse transform matrices are given in Chapter 8. A two-channel filter bank with Haar filters is shown in Figure 9.1. The asterisk symbol near the impulse responses of the filters indicates convolution. The frequency responses of the Haar filters are

$$L(e^{j\omega}) = \frac{1}{\sqrt{2}}(1 + e^{-j\omega}) \quad H(e^{j\omega}) = \frac{1}{\sqrt{2}}(-1 + e^{-j\omega})$$
$$\widetilde{L}(e^{j\omega}) = \frac{1}{\sqrt{2}}(1 + e^{-j\omega}) \quad \widetilde{H}(e^{j\omega}) = \frac{1}{\sqrt{2}}(1 - e^{-j\omega})$$

The magnitude of the frequency responses of the Haar analysis filters

$$|L(e^{j\omega})| = \left|\frac{1}{\sqrt{2}}\left(1 + e^{-j\omega}\right)\right| = \left|\frac{1}{\sqrt{2}}e^{-j\frac{\omega}{2}}\left(e^{j\frac{\omega}{2}} + e^{-j\frac{\omega}{2}}\right)\right| = \sqrt{2}\cos\left(\frac{\omega}{2}\right)$$
$$|H(e^{j\omega})| = \left|\frac{1}{\sqrt{2}}\left(-1 + e^{-j\omega}\right)\right| = \left|-je^{-j\omega/2}\frac{1}{j\sqrt{2}}\left(e^{j\frac{\omega}{2}} - e^{-j\frac{\omega}{2}}\right)\right| = \sqrt{2}\sin\left(\frac{\omega}{2}\right)$$

is shown in Figure 9.2(a), and the phase response of the lowpass filter

$$\angle L(e^{j\omega}) = -\frac{\omega}{2}$$

is shown in Figure 9.2(b). The phase response is linear as the filter is symmetric. At $\omega = \pi$, the phase is $-\pi/2$ radians.
Synthesis filters: The conditions for perfect reconstruction are

$$\widetilde{L}(e^{j\omega})L(e^{j\omega}) + \widetilde{H}(e^{j\omega})H(e^{j\omega}) = 2e^{-jK\omega}$$
$$\widetilde{L}(e^{j\omega})L(e^{j(\omega+\pi)}) + \widetilde{H}(e^{j\omega})H(e^{j(\omega+\pi)}) = 0$$

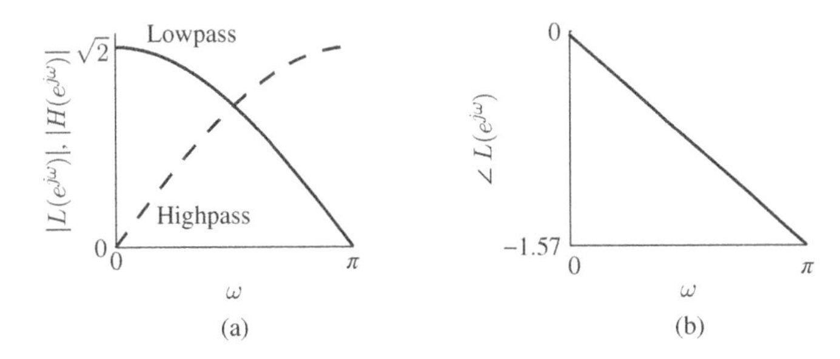

Figure 9.2 (a) The magnitude of the frequency responses of the Haar analysis filters; (b) the phase response of the lowpass filter

From the second condition, one of the two possible solutions is to set

$$\widetilde{L}(e^{j\omega}) = -H(e^{j(\omega+\pi)}) \quad \text{and} \quad \widetilde{H}(e^{j\omega}) = L(e^{j(\omega+\pi)})$$

This implies, in the time domain,

$$\widetilde{l}(n) = (-1)^{n+1}h(n) \quad \text{and} \quad \widetilde{h}(n) = (-1)^{n}l(n)$$

As shown later,

$$h(n) = -(-1)^{n}l(N-1-n) \quad \text{and} \quad l(n) = (-1)^{n}h(N-1-n)$$

and we get from the two expressions

$$\widetilde{l}(n) = (-1)^{n+1}h(n) = (-1)^{n+1}(-1)^{n+1}l(N-1-n) = l(N-1-n)$$
$$\widetilde{h}(n) = (-1)^{n}l(n) = (-1)^{n}(-1)^{n}h(N-1-n) = h(N-1-n)$$

When the transform matrix is transposed, the order of the filter coefficients is reversed. Note that in order to form the DFT transform matrix, we traverse the unit circle in the clockwise direction, and in the other direction to form the IDFT transform matrix.

Let $l(n) = \{a, b, c, d\}$. Then

$$h(n) = \{-(-1)^{n}l(3-n),\ n = 0, 1, 2, 3\} = \{-d, c, -b, a\}$$

In the frequency domain,

$$\widetilde{L}(e^{j\omega}) = -H(e^{j(\omega+\pi)}) = d + ce^{-j\omega} + be^{-j2\omega} + ae^{-j3\omega}$$
$$\widetilde{H}(e^{j\omega}) = L(e^{j(\omega+\pi)}) = a - be^{-j\omega} + ce^{-j2\omega} - de^{-j3\omega}$$

9.2 Daubechies Filter

Let the impulse response of the analysis lowpass D4 filter be

$$l(n) = \{l(0), l(1), l(2), l(3)\}$$

Writing the filter coefficients in the first row followed by the overlapping double-shifted versions, we get

$$\begin{array}{cccccc} l(0) & l(1) & l(2) & l(3) & & \\ l(0) & l(1) & l(2) & l(3) & & \\ & & l(0) & l(1) & l(2) & l(3) \end{array}$$

Now, the orthogonality conditions are

$$l^2(0) + l^2(1) + l^2(2) + l^2(3) = 1 \tag{9.10}$$

$$l(0)l(2) + l(1)l(3) = 0 \tag{9.11}$$

The double-shift orthogonality condition does not allow odd-length filters. The lowpass condition ($L(e^{j\pi}) = 0$) gives

$$l(0) - l(1) + l(2) - l(3) = 0 \tag{9.12}$$

These conditions imply that

$$l(0) + l(1) + l(2) + l(3) = \sqrt{2} \tag{9.13}$$

Using Equations (9.10) and (9.12), we get

$$(l(0) + l(1) + l(2) + l(3))^2 = 1 + 2(l(0)l(1) + l(0)l(3) + l(1)l(2) + l(2)l(3))$$

$$(l(0) - l(1) + l(2) - l(3))^2 = 0 = 1 - 2(l(0)l(1) + l(0)l(3) + l(1)l(2) + l(2)l(3))$$

The result in Equation (9.13) is obtained from the last two expressions.

An ideal lowpass filter frequency response is zero from some $0 < \omega < \pi$ to $\omega = \pi$. To approximate this response as closely as possible, we set an additional lowpass condition, using the derivative of the frequency response,

$$L\prime(e^{j\pi}) = 0$$

This condition makes the response to approach $\omega = \pi$ tangentially from the left. Differentiating the lowpass filter frequency response

$$L(e^{j\omega}) = l(0) + l(1)e^{-j\omega} + l(2)e^{-j2\omega} + l(3)e^{-j3\omega}$$

we get

$$L'(e^{j\omega}) = -jl(1)e^{-j\omega} - j2l(2)e^{-j2\omega} - j3l(3)e^{-j3\omega} \tag{9.14}$$

The additional lowpass condition is obtained by letting $\omega = \pi$ in the aforementioned equation.

$$-jl(1)e^{-j\pi} - j2l(2)e^{-j2\pi} - j3l(3)e^{-j3\pi} = 0$$

$$l(1) - 2l(2) + 3l(3) = 0 \tag{9.15}$$

Now, the set of linear and quadratic equations to be solved for this filter is

$$l(0) + l(1) + l(2) + l(3) = \sqrt{2} \tag{9.16}$$

$$l(0) - l(1) + l(2) - l(3) = 0 \tag{9.17}$$

$$l(1) - 2l(2) + 3l(3) = 0 \tag{9.18}$$

$$l(0)l(2) + l(1)l(3) = 0 \tag{9.19}$$

Adding and subtracting Equations (9.16) and (9.17), we get

$$l(0) + l(2) = \frac{1}{\sqrt{2}} \quad \text{and} \quad l(1) + l(3) = \frac{1}{\sqrt{2}}$$

Using these results, Equations (9.16)–(9.18) can be replaced by

$$l(1) = \frac{1}{\sqrt{2}} - l(3)$$

$$\left(\frac{1}{\sqrt{2}} - l(3)\right) - 2\left(\frac{1}{\sqrt{2}} - l(0)\right) + 3l(3) = 0, \quad l(0) = \frac{1}{2\sqrt{2}} - l(3)$$

$$\left(\frac{1}{2\sqrt{2}} - l(3)\right) + l(2) - \left(\frac{1}{\sqrt{2}}\right) = 0, \quad l(2) = \frac{1}{2\sqrt{2}} + l(3)$$

Eliminating $l(0), l(1)$, and $l(2)$ in Equation (9.19), we get

$$\left(\frac{1}{2\sqrt{2}} - l(3)\right)\left(\frac{1}{2\sqrt{2}} + l(3)\right) + \left(\frac{1}{\sqrt{2}} - l(3)\right) l(3) = -2\left(l^2(3) - \frac{1}{2\sqrt{2}} l(3) - \frac{1}{16}\right) = 0$$

Solving for $l(3)$, we find that one of the two values is

$$l(3) = \frac{1 + \sqrt{3}}{4\sqrt{2}}$$

The values of $l(0), l(1)$, and $l(2)$ are found as follows:

$$l(0) = \frac{1}{2\sqrt{2}} - l(3) = \frac{1}{2\sqrt{2}} - \frac{1 + \sqrt{3}}{4\sqrt{2}} = \frac{1 - \sqrt{3}}{4\sqrt{2}}$$

$$l(2) = \frac{1}{2\sqrt{2}} + l(3) = \frac{1}{2\sqrt{2}} + \frac{1 + \sqrt{3}}{4\sqrt{2}} = \frac{3 + \sqrt{3}}{4\sqrt{2}}$$

$$l(1) = \frac{1}{\sqrt{2}} - l(3) = \frac{1}{\sqrt{2}} - \frac{1 + \sqrt{3}}{4\sqrt{2}} = \frac{3 - \sqrt{3}}{4\sqrt{2}}$$

We list the impulse responses of all the four D4 filters.
Lowpass analysis filter

$$l(n) = \{l(0), l(1), l(2), l(3)\} = \frac{1}{4\sqrt{2}}\{(1 - \sqrt{3}), (3 - \sqrt{3}), (3 + \sqrt{3}), (1 + \sqrt{3})\}$$

One choice for highpass filter coefficients is

$$h(n) = \{-l(3), l(2), -l(1), l(0)\}$$

This is the time-reversed and sign-alternated version of $l(n)$. If the double-shifted versions of $l(n)$ are orthogonal, then those of $h(n)$ are also orthogonal.

$$\begin{array}{cccccc} -l(3) & l(2) & -l(1) & l(0) & & \\ -l(3) & l(2) & -l(1) & l(0) & & \\ & & -l(3) & l(2) & -l(1) & l(0) \end{array}$$

Writing the lowpass filter coefficients in the first row followed by the double-shifted versions of the highpass filter coefficients, we get

$$\begin{array}{cccccc} l(0) & l(1) & l(2) & l(3) & & \\ -l(3) & l(2) & -l(1) & l(0) & & \\ & & -l(3) & l(2) & -l(1) & l(0) \end{array}$$

Now, the orthogonality conditions are

$$-l(0)l(3) + l(1)l(2) - l(2)l(1) + l(3)l(0) = 0 \tag{9.20}$$

$$-l(2)l(3) + l(3)l(2) = 0 \tag{9.21}$$

Therefore, highpass analysis filter is obtained using the rule

$$h(n) = -(-1)^n l(N-1-n), \quad N = 4$$

where N, the filter length, is equal to 4.

$$h(n) = \{h(0), h(1), h(2), h(3)\} = \frac{1}{4\sqrt{2}}\left\{-\left(1+\sqrt{3}\right), (3+\sqrt{3}), -(3-\sqrt{3}), (1-\sqrt{3})\right\}$$

The synthesis filters are the time-reversed and shifted versions of their respective analysis filters.

Lowpass synthesis filter

$$\widetilde{l}(n) = \left\{\widetilde{l}(0), \widetilde{l}(1), \widetilde{l}(2), \widetilde{l}(3)\right\} = \frac{1}{4\sqrt{2}}\left\{\left(1+\sqrt{3}\right), (3+\sqrt{3}), (3-\sqrt{3}), (1-\sqrt{3})\right\}$$

Highpass synthesis filter

$$\widetilde{h}(n) = \left\{\widetilde{h}(0), \widetilde{h}(1), \widetilde{h}(2), \widetilde{h}(3)\right\} = \frac{1}{4\sqrt{2}}\left\{\left(1-\sqrt{3}\right), -(3-\sqrt{3}), (3+\sqrt{3}), -(1+\sqrt{3})\right\}$$

A two-channel filter bank with D4 filters is shown in Figure 9.3. Note that if $\{l(0), l(1), l(2), l(3)\}$ is a filter, then the reflection $\{l(3), l(2), l(1), l(0)\}$ and the negations $\{-l(0), -l(1), -l(2), -l(3)\}$ and $\{-l(3), -l(2), -l(1), -l(0)\}$ are also filters with the same magnitude of the frequency response. If $X(e^{j\omega})$ is the DTFT of $x(n)$, then $X(e^{-j\omega})$ is the DTFT of $x(-n)$. Furthermore, the DTFT of real-valued signals is conjugate-symmetric. The frequency responses of the D4 filters are

$$L(e^{j\omega}) = \frac{1}{4\sqrt{2}}((1-\sqrt{3}) + (3-\sqrt{3})e^{-j\omega} + (3+\sqrt{3})e^{-j2\omega} + (1+\sqrt{3})e^{-j3\omega})$$

$$H(e^{j\omega}) = \frac{1}{4\sqrt{2}}(-(1+\sqrt{3}) + (3+\sqrt{3})e^{-j\omega} - (3-\sqrt{3})e^{-j2\omega} + (1-\sqrt{3})e^{-j3\omega})$$

$$\widetilde{L}(e^{j\omega}) = \frac{1}{4\sqrt{2}}((1+\sqrt{3}) + (3+\sqrt{3})e^{-j\omega} + (3-\sqrt{3})e^{-j2\omega} + (1-\sqrt{3})e^{-j3\omega})$$

$$\widetilde{H}(e^{j\omega}) = \frac{1}{4\sqrt{2}}((1-\sqrt{3}) - (3-\sqrt{3})e^{-j\omega} + (3+\sqrt{3})e^{-j2\omega} - (1+\sqrt{3})e^{-j3\omega})$$

The magnitude of the frequency responses of the D4 analysis filters is shown in Figure 9.4(a). Verify that the magnitude response is $\sqrt{2}$ at $\omega = 0$ and 0 at $\omega = \pi$ for lowpass filters and $\sqrt{2}$ at

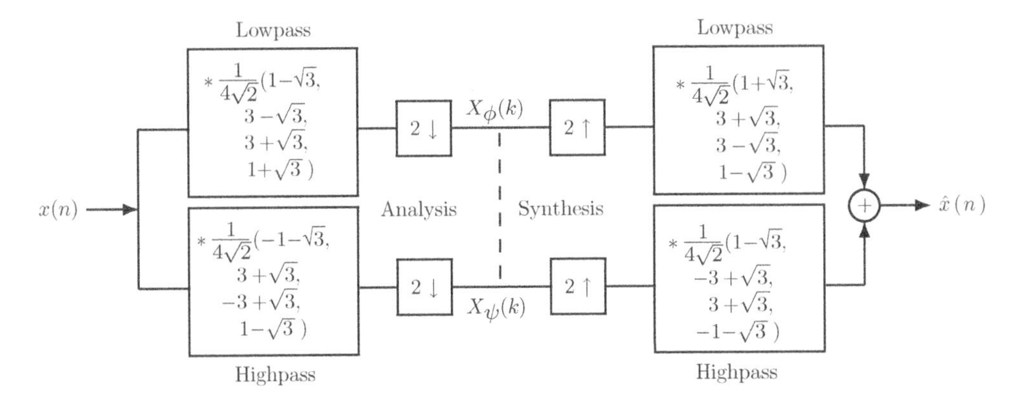

Figure 9.3 A two-channel analysis and synthesis filter bank with D4 filters

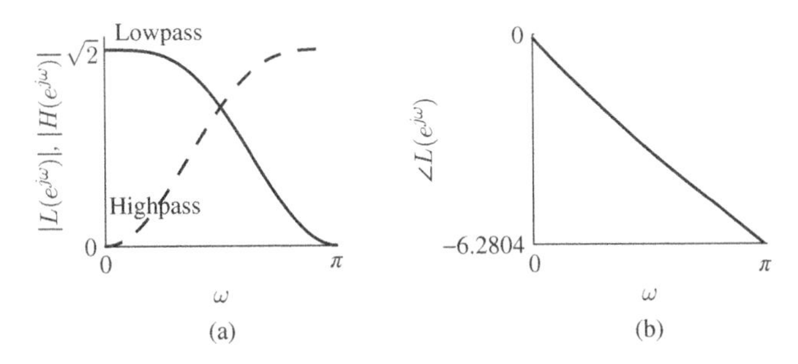

Figure 9.4 The magnitude of the frequency responses of the D4 analysis filters; (b) the phase response of the lowpass filter

$\omega = \pi$ and 0 at $\omega = 0$ for highpass filters. The phase response of the lowpass filter is shown in Figure 9.4(b), which is nonlinear.

Approximations of the D4 scaling function are shown in Figure 9.5. This function is derived using the Equation (8.13). Approximations of the wavelet function are shown in Figure 9.6. This function is derived using Equation (8.15).

The 8×8 1-level DWT transform matrix is

$$W_8 = \begin{bmatrix} 0.4830 & 0.8365 & 0.2241 & -0.1294 & 0 & 0 & 0 & 0 \\ 0 & 0 & 0.4830 & 0.8365 & 0.2241 & -0.1294 & 0 & 0 \\ 0 & 0 & 0 & 0 & 0.4830 & 0.8365 & 0.2241 & -0.1294 \\ 0.2241 & -0.1294 & 0 & 0 & 0 & 0 & 0.4830 & 0.8365 \\ -0.1294 & -0.2241 & 0.8365 & -0.4830 & 0 & 0 & 0 & 0 \\ 0 & 0 & -0.1294 & -0.2241 & 0.8365 & -0.4830 & 0 & 0 \\ 0 & 0 & 0 & 0 & -0.1294 & -0.2241 & 0.8365 & -0.4830 \\ 0.8365 & -0.4830 & 0 & 0 & 0 & 0 & -0.1294 & -0.2241 \end{bmatrix}$$

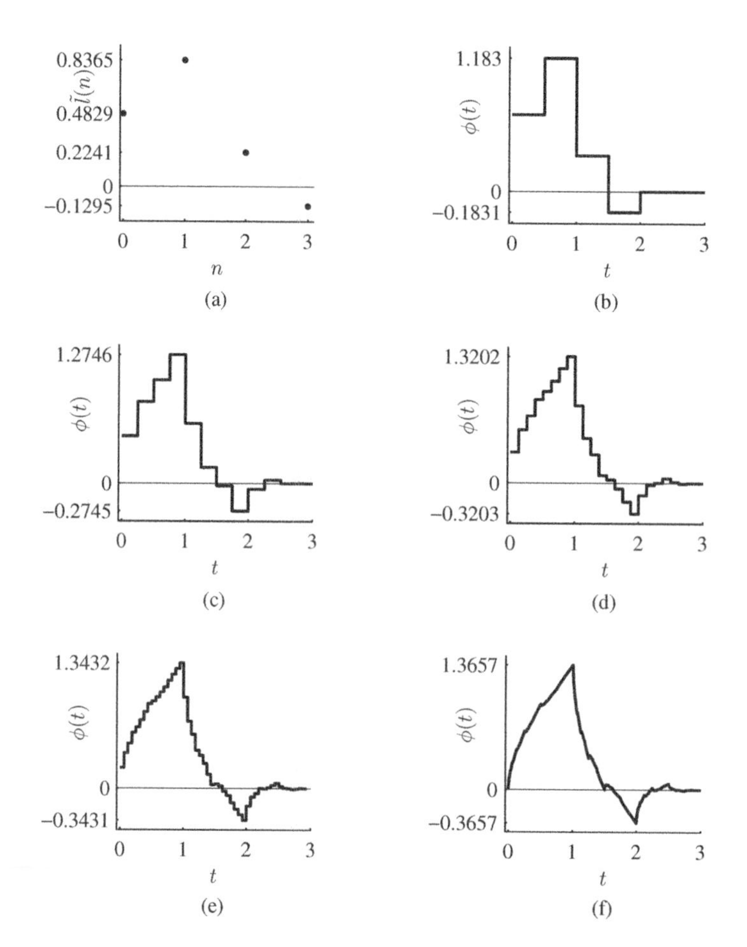

Figure 9.5 Approximations of the D4 scaling function derived from the synthesis lowpass filter. (a) Filter coefficients; (b) after one stage; (c) after two cascade stages; (d) after three cascade stages; (e) after four cascade stages; (f) after 10 cascade stages

The 8×8 inverse transform matrix is

$$\widetilde{W}_8 = \begin{bmatrix} 0.4830 & 0 & 0 & 0.2241 & -0.1294 & 0 & 0 & 0.8365 \\ 0.8365 & 0 & 0 & -0.1294 & -0.2241 & 0 & 0 & -0.4830 \\ 0.2241 & 0.4830 & 0 & 0 & 0.8365 & -0.1294 & 0 & 0 \\ -0.1294 & 0.8365 & 0 & 0 & -0.4830 & -0.2241 & 0 & 0 \\ 0 & 0.2241 & 0.4830 & 0 & 0 & 0.8365 & -0.1294 & 0 \\ 0 & -0.1294 & 0.8365 & 0 & 0 & -0.4830 & -0.2241 & 0 \\ 0 & 0 & 0.2241 & 0.4830 & 0 & 0 & 0.8365 & -0.1294 \\ 0 & 0 & -0.1294 & 0.8365 & 0 & 0 & -0.4830 & -0.2241 \end{bmatrix}$$

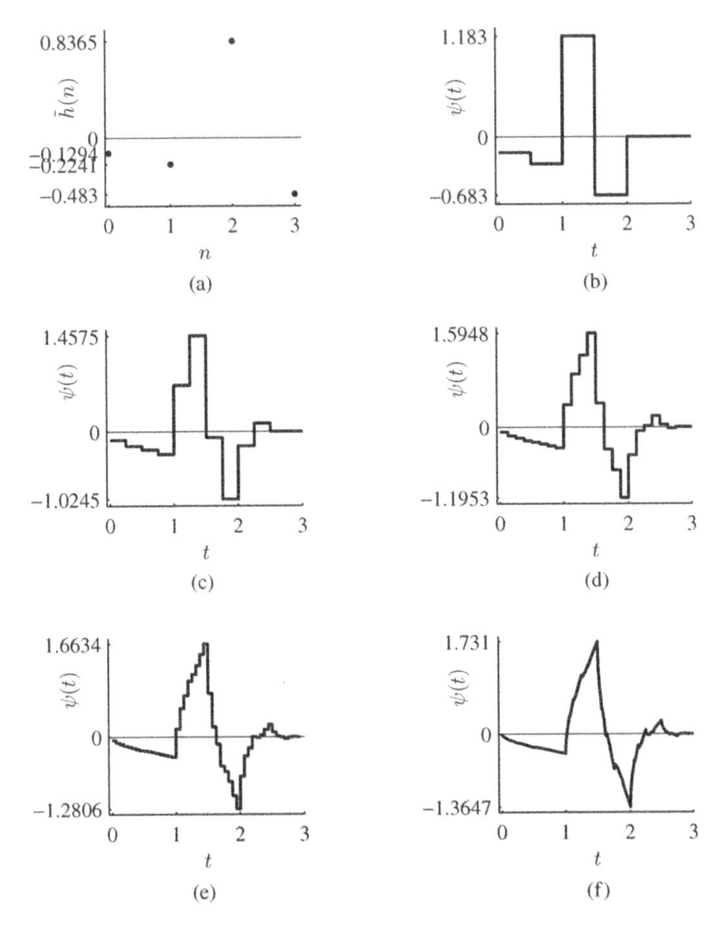

Figure 9.6 Approximations of the D4 wavelet function derived from the synthesis highpass filter. (a) Filter coefficients; (b) after one stage; (c) after two cascade stages; (d) after three cascade stages; (e) after four cascade stages; (f) after 10 cascade stages

Example 9.1 Compute the 1-level DWT of $x(n)$ using D4 filters. Assume periodic extension at the borders. Compute the IDWT and verify that the input $x(n)$ is reconstructed. Verify Parseval's theorem.

$$x(n) = \{1, 3, -2, 4, 2, 2, 3, -1\}$$

Solution

The DWT coefficients

$$\{2.0266, 2.5696, 3.4408, 0.4483, -4.4067, 0.0694, 2.2854, -0.7765\}$$

are obtained using the 8×8 transform matrix, assuming periodic extension of the input data.

$$\frac{\sqrt{2}}{8}\begin{bmatrix} 8+2\sqrt{3} \\ 18-2\sqrt{3} \\ 16+2\sqrt{3} \\ 6-2\sqrt{3} \\ -18-4\sqrt{3} \\ -10+6\sqrt{3} \\ 6+4\sqrt{3} \\ 6-6\sqrt{3} \end{bmatrix} =$$

$$\frac{\sqrt{2}}{8}\begin{bmatrix} 1+\sqrt{3} & 3+\sqrt{3} & 3-\sqrt{3} & 1-\sqrt{3} & 0 & 0 & 0 & 0 \\ 0 & 0 & 1+\sqrt{3} & 3+\sqrt{3} & 3-\sqrt{3} & 1-\sqrt{3} & 0 & 0 \\ 0 & 0 & 0 & 0 & 1+\sqrt{3} & 3+\sqrt{3} & 3-\sqrt{3} & 1-\sqrt{3} \\ 3-\sqrt{3} & 1-\sqrt{3} & 0 & 0 & 0 & 0 & 1+\sqrt{3} & 3+\sqrt{3} \\ 1-\sqrt{3} & -3+\sqrt{3} & 3+\sqrt{3} & -(1+\sqrt{3}) & 0 & 0 & 0 & 0 \\ 0 & 0 & 1-\sqrt{3} & -3+\sqrt{3} & 3+\sqrt{3} & -(1+\sqrt{3}) & 0 & 0 \\ 0 & 0 & 0 & 0 & 1-\sqrt{3} & -3+\sqrt{3} & 3+\sqrt{3} & -(1+\sqrt{3}) \\ 3+\sqrt{3} & -(1+\sqrt{3}) & 0 & 0 & 0 & 0 & 1-\sqrt{3} & -3+\sqrt{3} \end{bmatrix}\begin{bmatrix} 1 \\ 3 \\ -2 \\ 4 \\ 2 \\ 2 \\ 3 \\ -1 \end{bmatrix}$$

As an orthogonal transform, Parseval's theorem holds.

$$(1^2+3^2+(-2)^2+4^2+2^2+2^2+3^2+(-1)^2) = 48$$

$$((8+2\sqrt{3})^2+(18-2\sqrt{3})^2+(16+2\sqrt{3})^2+(6-2\sqrt{3})^2$$
$$+(-18-4\sqrt{3})^2+(-10+6\sqrt{3})^2+(6+4\sqrt{3})^2+(6-6\sqrt{3})^2)/32 = 48$$

The input can be obtained from the coefficients using the 8×8 inverse transform matrix.

$$\begin{bmatrix} 1 \\ 3 \\ -2 \\ 4 \\ 2 \\ 2 \\ 3 \\ -1 \end{bmatrix} = \frac{1}{32}$$

$$\begin{bmatrix} 1+\sqrt{3} & 0 & 0 & 3-\sqrt{3} & 1-\sqrt{3} & 0 & 0 & 3+\sqrt{3} \\ 3+\sqrt{3} & 0 & 0 & 1-\sqrt{3} & -3+\sqrt{3} & 0 & 0 & -1-\sqrt{3} \\ 3-\sqrt{3} & 1+\sqrt{3} & 0 & 0 & 3+\sqrt{3} & 1-\sqrt{3} & 0 & 0 \\ 1-\sqrt{3} & 3+\sqrt{3} & 0 & 0 & -1-\sqrt{3} & -3+\sqrt{3} & 0 & 0 \\ 0 & 3-\sqrt{3} & 1+\sqrt{3} & 0 & 0 & 3+\sqrt{3} & 1-\sqrt{3} & 0 \\ 0 & 1-\sqrt{3} & 3+\sqrt{3} & 0 & 0 & -1-\sqrt{3} & -3+\sqrt{3} & 0 \\ 0 & 0 & 3-\sqrt{3} & 1+\sqrt{3} & 0 & 0 & 3+\sqrt{3} & 1-\sqrt{3} \\ 0 & 0 & 1-\sqrt{3} & 3+\sqrt{3} & 0 & 0 & -1-\sqrt{3} & -3+\sqrt{3} \end{bmatrix}\begin{bmatrix} 8+2\sqrt{3} \\ 18-2\sqrt{3} \\ 16+2\sqrt{3} \\ 6-2\sqrt{3} \\ -18-4\sqrt{3} \\ -10+6\sqrt{3} \\ 6+4\sqrt{3} \\ 6-6\sqrt{3} \end{bmatrix}$$ ■

While the DWT is normally computed using convolution in the time domain, employing DFT/IDFT for computing the convolution is very instructive and helps understanding.

Even if the border extension is other than periodic, the DWT computation can be verified using DFT/IDFT except in the vicinity of the borders.

Example 9.2 Compute the 1-level DWT of $x(n)$ using D4 filters. Assume periodic extension at the borders and use DFT/IDFT. Compute the IDWT and verify that the input $x(n)$ is reconstructed.

$$x(n) = \{1, 3, -2, 4, 2, 2, 3, -1\}$$

Solution

The DFT of $x(n)$ is

$$X(k) = \{12, -3.8284 + j0.7574, 2 - j2, 1.8284 - j9.2426, -4, 1.8284 + j9.2426, 2 + j2, - 3.8284 - j0.7574\}$$

The DFT of the analysis lowpass filter $l(n)$ appended by four zeros is

$$L(k) = \{1.4142, -0.3124 - j1.3365, -0.9659 + j0.2588, 0.0536 + j0.3365, 0, 0.0536 - j0.3365, -0.9659 - j0.2588, -0.3124 + j1.3365\}$$

The product $X(k)L(k)$ is

$$X(k)L(k) = \{16.9706, 2.2083 + j4.8801, -1.4142 + j2.4495, 3.2083 + j0.1199, 0, 3.2083 - j0.1199, -1.4142 - j2.4495, 2.2083 - j4.8801\}$$

The IDFT of $X(k)L(k)$ is

$$\{3.1219, 0.4483, 1.2848, 2.0266, 0.4136, 2.5696, 3.6649, 3.4408\}$$

Note that the coefficients are shifted. The first coefficient is the fourth one. The approximation coefficients are obtained by downsampling by a factor of 2.

$$X_\phi(2, k) = \{2.0266, 2.5696, 3.4408, 0.4483\}$$

The DFT of the analysis highpass filter $h(n)$ appended by four zeros is

$$H(k) =\{0, 0.2000 - j0.2759 - 0.2588 - j0.9659, -1.1660 - j0.7241, - 1.4142, -1.1660 + j0.7241, -0.2588 + j0.9659, 0.2000 + j0.2759\}$$

The product of $X(k)H(k)$ is

$$X(k)H(k) =\{0, -0.5570 + j1.2076, -2.4495 - j1.4142, -8.8249 + j9.4526, 5.6569, -8.8249 - j9.4526, -2.4495 + j1.4142, -0.5570 - j1.2076\}$$

The IDFT of $X(k)H(k)$ is

$$\{-2.2507, -0.7765, 3.3807, -4.4067, 2.4402, 0.0694, -0.7418, 2.2854\}$$

The detail coefficients are,

$$X_{\psi}(2,k) = \{-4.4067, 0.0694, 2.2854, -0.7765\}$$

The IDWT: The DFT of the synthesis lowpass filter $\widetilde{l}(n)$ appended by four zeros is

$$\widetilde{L}(k) = \{1.4142, 1.1660 - j0.7241, 0.2588 - j0.9659, -0.2000 - j0.2759, \\ 0, -0.2000 + j0.2759, 0.2588 + j0.9659, 1.1660 + j0.7241\}$$

The DFT of the upsampled approximation coefficients

$$\{2.0266, 0, 2.5696, 0, 3.4408, 0, 0.4483, 0\}$$

is

$$\{8.4853, -1.4142 - j2.1213, 2.4495, -1.4142 + j2.1213, \\ 8.4853, -1.4142 - j2.1213, 2.4495, -1.4142 + j2.1213\}$$

Note the duplication of the spectrum. The product of the two DFTs is

$$SL = \{12, -3.1850 - j1.4493, 0.6340 - j2.3660, 0.8681 - j0.0342, \\ 0, 0.8681 + j0.0342, 0.6340 + j2.3660, -3.1850 + j1.4493\}$$

The DFT of the synthesis highpass filter $\widetilde{h}(n)$ appended by four zeros is

$$\widetilde{H}(k) = \{0, 0.0536 - j0.3365, -0.9659 - j0.2588, -0.3124 + j1.3365, \\ 1.4142, -0.3124 - j1.3365, -0.9659 + j0.2588, 0.0536 + j0.3365\}$$

The DFT of the upsampled detail coefficients

$$\{-4.4067, 0, 0.0694, 0, 2.2854, 0, -0.7765, 0\}$$

is

$$\{-2.8284, -6.6921 - j0.8459, -1.4142, -6.6921 + j0.8459, \\ -2.8284, -6.6921 - j0.8459, -1.4142, -6.6921 + j0.8459\}$$

The product of the two DFTs is

$$SH = \{0, -0.6434 + j2.2067, 1.3660 + j0.3660, 0.9602 - j9.2084, \\ -4, 0.9602 + j9.2084, 1.3660 - j0.3660, -0.6434 - j2.2067\}$$

The sum of SL and SH is

$$(SL + SH) = \{12, -3.8284 + j0.7574, 2 - j2, 1.8283 - j9.2426, \\ -4, 1.8283 + j9.2426, 2 + j2, -3.8284 - j0.7574\}$$

The IDFT of this sequence is the reconstructed input

$$\hat{x} = \{1, 3, -2, 4, 2, 2, 3, -1\}$$

■

Example 9.3 Compute the 1-level 2-D DWT of $x(n_1, n_2)$ using D4 filters. Assume periodic extension at the borders. Compute the IDWT and verify that the input $x(n_1, n_2)$ is reconstructed. Verify Parseval's theorem.

Solution

$$\boldsymbol{x} = \begin{bmatrix} 1 & 2 & -1 & 3 & 3 & 1 & 2 & 2 \\ 3 & 4 & 3 & -2 & 2 & 1 & 4 & 3 \\ 2 & 1 & 2 & 3 & -1 & 4 & 2 & 2 \\ 4 & 3 & -1 & 2 & 4 & 2 & 1 & 3 \\ 3 & 2 & -1 & 2 & 3 & 1 & 3 & 2 \\ 2 & -1 & 3 & -2 & 2 & 3 & 4 & 3 \\ 1 & 4 & 2 & 1 & -1 & 4 & 2 & 2 \\ 4 & 3 & -1 & 2 & 3 & 2 & 1 & 3 \end{bmatrix}$$

The result of computing the 1-D DWT of the rows of x yields

$$\boldsymbol{xr} = \boldsymbol{x}\boldsymbol{W}_8^T$$

$$= \begin{bmatrix} 1.5436 & 2.5696 & 2.4749 & 2.6043 & -2.8631 & 1.4836 & 0.0947 & -0.8365 \\ 5.7262 & 0.0947 & 2.3108 & 4.5962 & 2.1907 & 1.2501 & 1.4142 & -0.6124 \\ 1.8625 & 2.7337 & 3.0526 & 2.9578 & -0.2588 & -3.6996 & -0.0601 & 0.4830 \\ 3.9584 & 1.8278 & 3.4408 & 3.5009 & -2.9925 & 2.0613 & -1.5783 & 1.0953 \\ 2.6390 & 1.7331 & 2.6990 & 3.5355 & -2.6390 & 1.7077 & 0.9313 & 0.7071 \\ 1.0607 & -0.1641 & 3.9838 & 5.0191 & 3.4408 & 0.2842 & 0.9659 & 0.9659 \\ 4.1479 & 1.0607 & 3.0526 & 2.3455 & 0.1641 & -3.2513 & -0.0601 & -1.8024 \\ 3.9584 & 1.6037 & 2.9578 & 3.5009 & -2.9925 & 1.2247 & -1.4489 & 1.0953 \end{bmatrix}$$

The result of computing the 1-D DWT of the columns of $\boldsymbol{xr}$ yields the 2-D DWT.

$$\boldsymbol{X} = \boldsymbol{W}_8\boldsymbol{xr}$$

$$= \begin{bmatrix} 5.4408 & 1.6965 & 3.3672 & 5.3125 & 0.7790 & 0.6663 & 1.4196 & -0.9498 \\ 4.6651 & 3.2590 & 4.4420 & 4.5000 & -3.6651 & 0.2835 & -1.2655 & 1.1830 \\ 2.5792 & 0.7300 & 4.9375 & 5.9788 & 2.0278 & 0.1752 & 1.4318 & 0.6038 \\ 4.9196 & 2.4175 & 4.2042 & 4.0502 & -3.3493 & -0.3750 & -1.4028 & -0.0625 \\ -1.8370 & 1.0502 & 0.0535 & -0.5837 & 1.1083 & -4.5625 & 0.3828 & 0.1205 \\ 0.5670 & 0.7655 & -0.8325 & -0.6340 & -3.1651 & 1.3080 & 0.6740 & -0.1830 \\ 0.9788 & -0.0748 & -0.1172 & -1.3113 & 1.1528 & -3.5960 & 0.3125 & -2.3448 \\ -2.8983 & 1.6071 & -0.1038 & -1.1295 & -2.8035 & 0.7835 & -0.2712 & -0.4163 \end{bmatrix}$$

By premultiplying $\boldsymbol{X}$ by $\boldsymbol{W}_8^T$, we get $\boldsymbol{xr}$. By postmultiplying $\boldsymbol{xr}$ by $\boldsymbol{W}_8$, we get back the original input. ■

The 2-D D4 filters are shown in Figure 9.7.

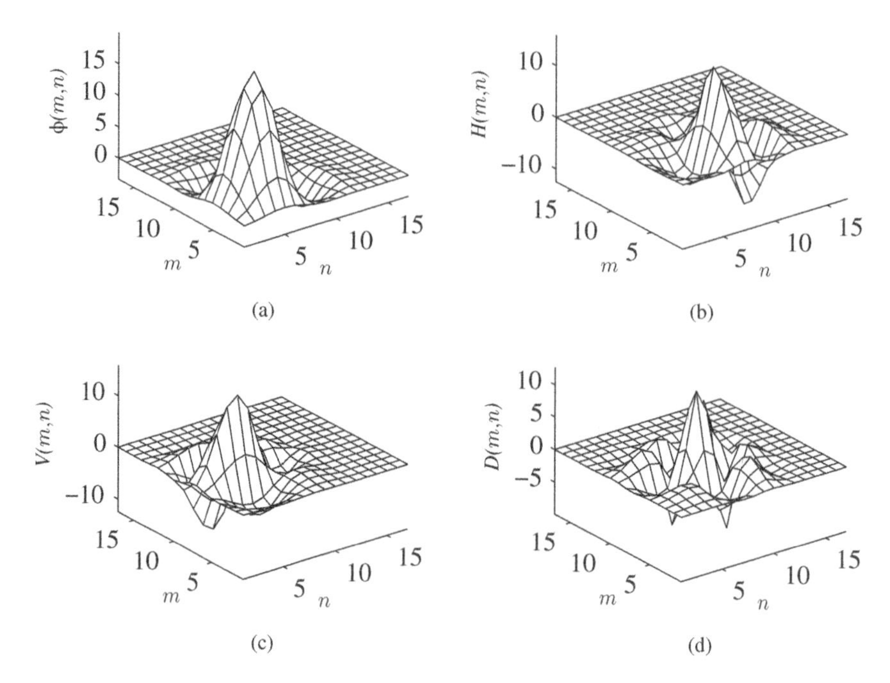

Figure 9.7 (a) The D4 2-D scaling filter; (b) the 2-D wavelet filter $H(m,n)$; (c) the 2-D wavelet filter $V(m,n)$; (d) the 2-D wavelet filter $D(m,n)$

9.3 Orthogonality Conditions

Orthogonality of two real discrete sequences is the property of having the sum of products being one or zero under specified conditions. In this section, we consider the orthogonality condition in the frequency domain. In the time domain, we use matrices, while we use functions of the frequency variable ω in the frequency domain. From the Haar lowpass filter frequency response, we get

$$|L(e^{j\omega})| = \left|\frac{1}{\sqrt{2}}\left(1+e^{-j\omega}\right)\right| = \left|\frac{1}{\sqrt{2}}e^{-j\frac{\omega}{2}}\left(e^{j\frac{\omega}{2}}+e^{-j\frac{\omega}{2}}\right)\right| = \sqrt{2}\cos\left(\frac{\omega}{2}\right)$$

$$|L(e^{j(\omega+\pi)})|^2 = \left(\sqrt{2}\cos\left(\frac{\omega+\pi}{2}\right)\right)^2 = 2\left(\cos\left(\frac{\omega}{2}+\frac{\pi}{2}\right)\right)^2 = 2\sin^2\left(\frac{\omega}{2}\right)$$

$$|L(e^{j\omega})|^2 + |L(e^{j(\omega+\pi)})|^2 = 2\cos^2\left(\frac{\omega}{2}\right) + 2\sin^2\left(\frac{\omega}{2}\right) = 2$$

The last equation is a sufficient and necessary condition, in the frequency domain, for the orthogonality condition to hold. Consider the orthogonality condition

$$|L(e^{j\omega})|^2 + |L(e^{j(\omega+\pi)})|^2 = L(e^{j\omega})L^*(e^{j\omega}) + L(e^{j(\omega+\pi)})L^*(e^{j(\omega+\pi)}) = 2$$

Let the coefficients of $L(e^{j\omega})$ be $\{a,b,c,d\}$. Then, the coefficients of $L(e^{j(\omega+\pi)})$ are $\{a,-b,c,-d\}$. As the terms making up the orthogonality condition are the product of a transform and its conjugate, they correspond to autocorrelation in the time domain.

It is so because the problem is concerned with the similarity of a signal with itself and its double-shifted versions. If two different filters are involved, autocorrelation becomes cross-correlation. The autocorrelation of $\{1,1\}/\sqrt{2}$ is $\{1,2,1\}/2$. The autocorrelation function is even-symmetric. The autocorrelation of $\{1,-1\}/\sqrt{2}$ is $\{-1,2,-1\}/2$. The sum of the two autocorrelations is 2, as required by the orthogonality condition. Note that the orthogonality condition we derived and required in the DWT is for double shifts of the filter coefficients.

The physical interpretation of the frequency-domain version of the orthogonality conditions is as follows. The autocorrelation operation, which is a similarity measure, of a sequence $l(n)$ with itself is defined as

$$r_{ll}(k) = \sum_{n=-\infty}^{\infty} l(n)l(n-k),$$

where the time separation k, called the lag, is the independent variable of the autocorrelation function. The autocorrelation of the impulse response $l(n) = \{l(0), l(1)\}$ is

$$\{l(0)l(1), l^2(0) + l^2(1), l(0)l(1)\}$$

The DTFT of this autocorrelation sequence is given by

$$L(e^{j\omega})L^*(e^{j\omega}) = |L(e^{j\omega})|^2 = (l^2(0) + l^2(1)) + 2l(0)l(1)\cos(\omega)$$

Similarly, the DTFT of the autocorrelation of $(-1)^n l(n) = \{l(0), -l(1)\}$ is given by

$$|L(e^{j(\omega+\pi)})|^2 = (l^2(0) + l^2(1)) - 2l(0)l(1)\cos(\omega)$$

Now,

$$|L(e^{j\omega})|^2 + |L(e^{j(\omega+\pi)})|^2 = 2(l^2(0) + l^2(1)) = 2$$

as $(l^2(0) + l^2(1)) = 1$ for orthogonal filters. This relation holds for any even-length orthogonal filter. This condition ensures the double-shift orthogonality

$$\sum_{n=-\infty}^{\infty} l(n)l(n-2m) = \delta(m), \quad m = 0, \pm 1, \pm 2, \ldots \tag{9.22}$$

Consider the autocorrelation of four-term impulse response. The DTFT of the autocorrelation of $l(n) = \{l(0), l(1), l(2), l(3)\}$ is given by

$$\begin{aligned}(l^2(0) + l^2(1) + l^2(2) + l^2(3)) + 2(l(0)l(1) + l(1)l(2) + l(2)l(3))\cos(\omega)\\ + 2(l(0)l(2) + l(1)l(3))\cos(2\omega) + 2(l(0)l(3))\cos(3\omega)\end{aligned}$$

The DTFT of the autocorrelation of $(-1)l(n) = \{l(0), -l(1), l(2), -l(3)\}$ is given by

$$\begin{aligned}(l^2(0) + l^2(1) + l^2(2) + l^2(3)) - 2(l(0)l(1) + l(1)l(2) + l(2)l(3))\cos(\omega)\\ + 2(l(0)l(2) + l(1)l(3))\cos(2\omega) - 2(l(0)l(3))\cos(3\omega)\end{aligned}$$

Now, the sum of the two DTFT expressions is

$$2(l^2(0) + l^2(1) + l^2(2) + l^2(3)) + 4(l(0)l(2) + l(1)l(3))\cos(2\omega) \tag{9.23}$$

To satisfy the orthogonality conditions, the last expression must be equal to 2.

$$2(l^2(0) + l^2(1) + l^2(2) + l^2(3)) + 4(l(0)l(2) + l(1)l(3))\cos(2\omega) = 2 \tag{9.24}$$

$$(l^2(0) + l^2(1) + l^2(2) + l^2(3)) + 2(l(0)l(2) + l(1)l(3))\cos(2\omega) = 1 \tag{9.25}$$

This reduces to

$$(l^2(0) + l^2(1) + l^2(2) + l^2(3)) = 1 \quad \text{and} \quad (l(0)l(2) + l(1)l(3)) = 0 \tag{9.26}$$

The odd-indexed terms in the sum of the two DTFT expressions get canceled, and the even-indexed terms add up. The condition that the sum of the even-indexed terms (except that with zero index) is zero implies that each of the other autocorrelation coefficients constituting the sum is zero, as the cosines are linearly independent. Similarly, for highpass filters,

$$|H(e^{j\omega})|^2 + |H(e^{j(\omega+\pi)})|^2 = 2$$

Establishing the orthogonality of lowpass and highpass filters is essentially the same as that of lowpass filters except that the operation involved is cross-correlation. Writing the lowpass and highpass filter coefficients and the overlapping double-shifted versions, we get

$$\begin{array}{cccccc} l(0) & l(1) & l(2) & l(3) & & \\ h(0) & h(1) & h(2) & h(3) & & \\ & & h(0) & h(1) & h(2) & h(3) \end{array}$$

The orthogonality conditions are

$$h(0)l(0) + h(1)l(1) + h(2)l(2) + h(3)l(3) = 0 \tag{9.27}$$

$$h(0)l(2) + h(1)l(3) = 0 \text{ and } h(2)l(0) + h(3)l(1) = 0 \tag{9.28}$$

The DTFT of the impulse responses $l(n)$ and $h(n)$ is given by

$$L(e^{j\omega}) = \sum_{n=0}^{N-1} l(n)e^{-j\omega n} \quad \text{and} \quad H(e^{j\omega}) = \sum_{n=0}^{N-1} h(n)e^{-j\omega n}$$

Then, the orthogonality condition is given by

$$L(e^{j\omega})H^*(e^{j\omega}) + L(e^{j(\omega+\pi)})H^*(e^{j(\omega+\pi)}) = 0$$

if and only if

$$\sum_{n=-\infty}^{\infty} l(n)h(n-2m) = 0, \quad m = 0, \pm 1, \pm 2, \ldots$$

Remember that the highpass analysis filter is obtained using the rule

$$h(n) = -(-1)^n l(N-1-n)$$

The DTFT of this expression is

$$H(e^{j\omega}) = e^{-j(N-1)\omega} L^*(e^{j(\omega+\pi)})$$

Shifting the transform by π radians corresponds to multiplying the sequence in the time domain by $(-1)^n$. Complex conjugation operation corresponds to time reversal of the sequence. Multiplying the transform of a sequence by $e^{-(N-1)j\omega}$ in the frequency domain corresponds to shifting the sequence in the time domain by $(N-1)$ sample intervals to the right. Then, the orthogonality condition is verified as

$$L(e^{j\omega})e^{j(N-1)\omega}L(e^{j(\omega+\pi)}) + L(e^{j(\omega+\pi)})e^{j(N-1)(\omega+\pi)}L(e^{j\omega}) = 0$$

Let $l(n) = \{a, b, c, d\}$. Then,

$$h(n) = \{-(-1)^n l(3-n), n = 0, 1, 2, 3\} = \{-d, c, -b, a\}$$

In the frequency domain,

$$\begin{aligned} H(e^{j\omega}) &= e^{-j3\omega}L^*(e^{j(\omega+\pi)}) \\ L(e^{j(\omega+\pi)}) &= a - be^{-j\omega} + ce^{-j2\omega} - de^{-j3\omega} \\ L^*(e^{j(\omega+\pi)}) &= a - be^{j\omega} + ce^{j2\omega} - de^{j3\omega} \\ e^{-j3\omega}L^*(e^{j(\omega+\pi)}) &= H(e^{j\omega}) = ae^{-j3\omega} - be^{-j2\omega} + ce^{-j\omega} - d \end{aligned}$$

The sum of the cross-correlation of $\{a, b, c, d\}$ and $\{-d, c, -b, a\}$, and that of $\{a, -b, c, -d\}$ and $\{-d, -c, -b, -a\}$, is zero.

Let $l(n) = \{2, -3, 1, 4\}$. Then,

$$h(n) = \{-(-1)^n l(3-n), \quad n = 0, 1, 2, 3\} = \{-4, 1, 3, 2\}$$

The cross-correlation of $\{2, -3, 1, 4\}$ and $\{-4, 1, 3, 2\}$ is

$$\{4, 0, -5, 0, 25, 0, -16\}$$

The cross-correlation of $(-1)^n l(n) = \{2, 3, 1, -4\}$ and $(-1)^n h(n) = \{-4, -1, 3, -2\}$ is

$$\{-4, 0, 5, 0, -25, 0, 16\}$$

The sum of the two cross-correlations is zero.

9.3.1 Characteristics of Daubechies Lowpass Filters

A set of linear and quadratic equations characterizing Daubechies lowpass filters for any even length is given as follows.
Orthogonality conditions

Time domain	Frequency domain
$\sum_{n=-\infty}^{\infty} l(n)l(n-2m) = \delta(m), m = 0, \pm 1, \pm 2, \ldots,$	$\lvert L(e^{j(\omega)})\rvert^2 + \lvert L(e^{j(\omega+\pi)})\rvert^2 = 2$

Lowpass and additional lowpass filter conditions

Time domain	Frequency domain
$\sum_{n=0}^{N-1} l(n) = \sqrt{2}$	$L(e^{j0}) = \sqrt{2}$
$\sum_{n=0}^{N-1} (-1)^n l(n) = 0$	$L(e^{j\pi}) = 0$
$\sum_{n=1}^{N-1} (-1)^n n^m l(n) = 0, m = 1, 2, \ldots, \frac{N}{2} - 1$	$L^m(e^{j\pi}) = 0$

9.4 Coiflet Filter

In this type of filters, we use the derivative conditions both at $\omega = 0$ and at $\omega = \pi$ in designing the filter. This flattens out the frequency response of the filters at both the ends. The frequency response of a filter with six coefficients is given by

$$L(e^{j\omega}) = l(-2)e^{j2\omega} + l(-1)e^{j\omega} + l(0) + l(1)e^{-j\omega} + l(2)e^{-j2\omega} + l(3)e^{-j3\omega} \tag{9.29}$$

Setting the frequency response at $\omega = 0$ equal to $\sqrt{2}$, we get

$$L(e^{j0}) = l(-2) + l(-1) + l(0) + l(1) + l(2) + l(3) = \sqrt{2} \tag{9.30}$$

Setting the frequency response at $\omega = \pi$ equal to zero, we get

$$L(e^{j\pi}) = l(-2) - l(-1) + l(0) - l(1) + l(2) - l(3) = 0 \tag{9.31}$$

Differentiating the frequency response, we get

$$j2l(-2)e^{j2\omega} + jl(-1)e^{j\omega} - jl(1)e^{-j\omega} - j2l(2)e^{-j2\omega} - j3l(3)e^{-j3\omega} \tag{9.32}$$

Setting Equation (9.32) to zero, with $\omega = 0$ and $\omega = \pi$, we get

$$L'(e^{j0}) = 2l(-2) + l(-1) - l(1) - 2l(2) - 3l(3) = 0 \tag{9.33}$$

$$L'(e^{j\pi}) = 2l(-2) - l(-1) + l(1) - 2l(2) + 3l(3) = 0 \tag{9.34}$$

Solving Equations (9.30), (9.31), (9.33), and (9.34) in terms of $l(2)$ and $l(3)$, we get

$$\begin{bmatrix} 1 & 1 & 1 & 1 \\ 1 & -1 & 1 & -1 \\ 2 & 1 & 0 & -1 \\ 2 & -1 & 0 & 1 \end{bmatrix} \begin{bmatrix} l(-2) \\ l(-1) \\ l(0) \\ l(1) \end{bmatrix} = \begin{bmatrix} \sqrt{2} - l(2) - l(3) \\ -l(2) + l(3) \\ 2l(2) + 3l(3) \\ 2l(2) - 3l(3) \end{bmatrix} \tag{9.35}$$

$$\left[\begin{array}{cccc|c} 1 & 1 & 1 & 1 & \sqrt{2} - l(2) - l(3) \\ 0 & -2 & 0 & -2 & -\sqrt{2} + 2l(3) \\ 0 & -1 & -2 & -3 & -2\sqrt{2} + 4l(2) + 5l(3) \\ 0 & -3 & -2 & -1 & -2\sqrt{2} + 4l(2) - l(3) \end{array}\right] \tag{9.36}$$

$$\left[\begin{array}{cccc|c} 1 & 1 & 1 & 1 & \sqrt{2} - l(2) - l(3) \\ 0 & -2 & 0 & -2 & -\sqrt{2} + 2l(3) \\ 0 & 0 & -2 & -2 & \frac{1}{\sqrt{2}} - 2\sqrt{2} + 4l(2) + 4l(3) \\ 0 & 0 & -2 & 2 & \frac{3}{\sqrt{2}} - 2\sqrt{2} + 4l(2) - 4l(3) \end{array}\right] \tag{9.37}$$

$$l(-2) = l(2), \quad l(-1) = \frac{1}{2\sqrt{2}} + l(3), \quad l(0) = \frac{1}{\sqrt{2}} - 2l(2), \quad l(1) = \frac{1}{2\sqrt{2}} - 2l(3) \tag{9.38}$$

Now, the orthogonality conditions are

$$l^2(-2) + l^2(-1) + l^2(0) + l^2(1) + l^2(2) + l^2(3) = 1 \tag{9.39}$$

$$l(-2)l(0) + l(-1)l(1) + l(0)l(2) + l(1)l(3) = 0 \tag{9.40}$$

$$l(-2)l(2) + l(-1)l(3) = 0 \tag{9.41}$$

Substituting for $l(-2)$ and $l(-1)$ in Equation (9.41), we get

$$l(2) + \frac{1}{2\sqrt{2}}l(3) + l^2(3) = 0 \tag{9.42}$$

Substituting for $l(-2), l(-1), l(0)$, and $l(1)$ in Equation (9.40), we get

$$\sqrt{2}l(2) - 4l^2(2) - 4l^2(3) = -\frac{1}{8} \tag{9.43}$$

From the last two equations,

$$l(2) = -\left(l(3) + \frac{1}{8\sqrt{2}}\right) \tag{9.44}$$

Substituting for $l(-2), l(-1), l(0)$, and $l(1)$ in Equation (9.39), we get

$$-2\sqrt{2}l(2) + 6l^2(2) - \frac{1}{\sqrt{2}}l(3) + 6l^2(3) = \frac{1}{4} \tag{9.45}$$

$$12l^2(3) + \frac{9}{2\sqrt{2}}l(3) + \frac{3}{64} = 0 \tag{9.46}$$

Solving the quadratic equation, we get

$$l(3) = \frac{1}{16\sqrt{2}}(-3 \pm \sqrt{7}) \tag{9.47}$$

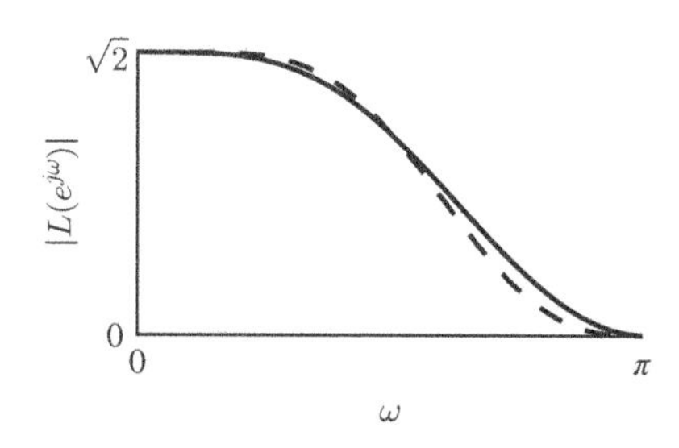

Figure 9.8 The magnitude of the frequency responses of the C6 analysis filter (solid line) and that of the Daubechies filter (dashed line) of length six

Using the root with $(-3+\sqrt{7})$, we get

$$l(-2) = \frac{1}{16\sqrt{2}}(1-\sqrt{7}), \quad l(-1) = \frac{1}{16\sqrt{2}}(5+\sqrt{7}), \quad l(0) = \frac{1}{8\sqrt{2}}(7+\sqrt{7}) \tag{9.48}$$

$$l(1) = \frac{1}{8\sqrt{2}}(7-\sqrt{7}), \quad l(2) = \frac{1}{16\sqrt{2}}(1-\sqrt{7}), \quad l(3) = \frac{1}{16\sqrt{2}}(-3+\sqrt{7}) \tag{9.49}$$

The other filter coefficients of the C6 filters are found as they were found for D4 filters. The magnitude of the frequency responses of the C6 analysis filter (solid line) and that of the Daubechies filter (dashed line) of length six are shown in Figure 9.8.

Example 9.4 Compute the 1-level DWT of $x(n)$ using the reflections of the C6 filters given as follows.

$$l(n) = \{-0.0157, -0.0727, 0.3849, 0.8526, 0.3379, -0.0727\}$$
$$h(n) = \{0.0727, 0.3379, -0.8526, 0.3849, 0.0727, -0.0157\}$$

Assume periodic extension at the borders. Verify Parseval's theorem.

$$x(n) = \{1, 3, -2, 4, 2, 2, 3, -1\}$$

Solution

The DWT coefficients

$$\{1.5339, 0.5985, 3.7694, 2.5835, -2.6774, -3.1562, 0.3278, 2.6774\}$$

can be obtained using the 8×8 transform matrix, assuming periodic extension. The first coefficient is

$$\begin{aligned} l(2)x(-2) + l(1)x(-1) + l(0)x(0) + l(-1)x(1) + l(-2)x(2) + l(-3)x(3) &= (-0.0727)3 \\ +(0.3379)(-1) + (0.8526)1 + (0.3849)3 + (-0.0727)(-2) \\ +(-0.0157)4 = 1.5339 \end{aligned}$$

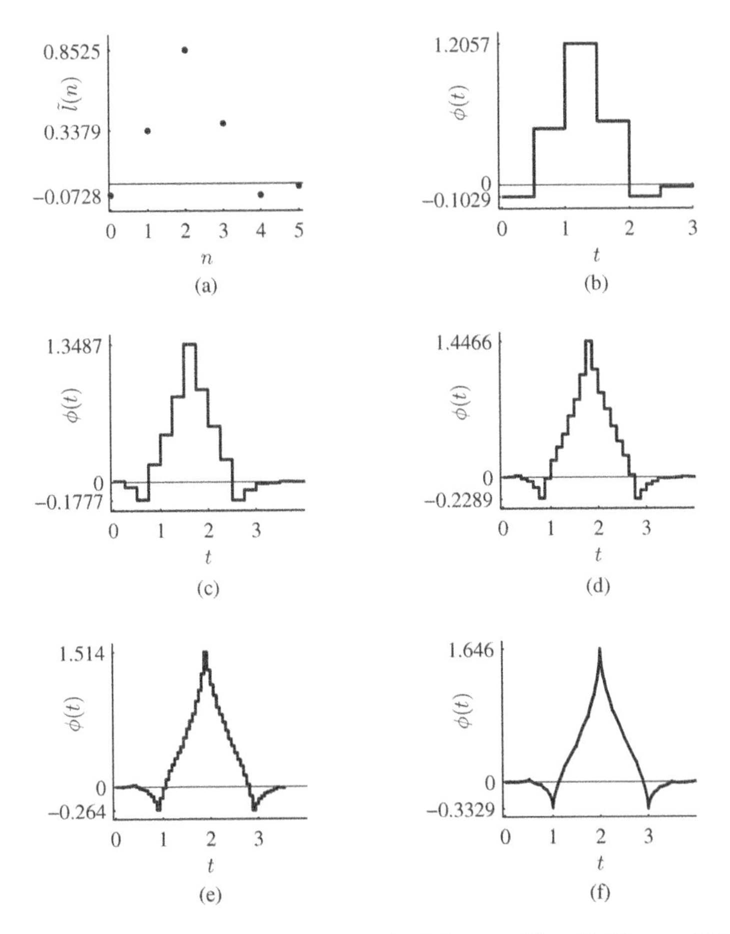

Figure 9.9 C6 scaling function derived from the synthesis lowpass filter. (a) Filter coefficients; (b) after one stage; (c) after two cascade stages; (d) after three cascade stages; (e) after four cascade stages; (f) after 10 cascade stages

The other coefficients are found by double-shifting the input and finding the sum of products. As an orthogonal transform, Parseval's theorem holds.

$$\{1^2 + 3^2 + (-2)^2 + 4^2 + 2^2 + 2^2 + 3^2 + (-1)^2\} = 48$$

$$\{1.5339^2 + 0.5985^2 + 3.7694^2 + 2.5835^2 + (-2.6774)^2 + (-3.1562)^2 + 0.3278^2 + 2.6774^2 = 48\}$$

■

Approximations of the scaling function are shown in Figure 9.9. This function is derived using Equation (8.13). Approximations of the wavelet function are shown in Figure 9.10. This function is derived using Equation (8.15).

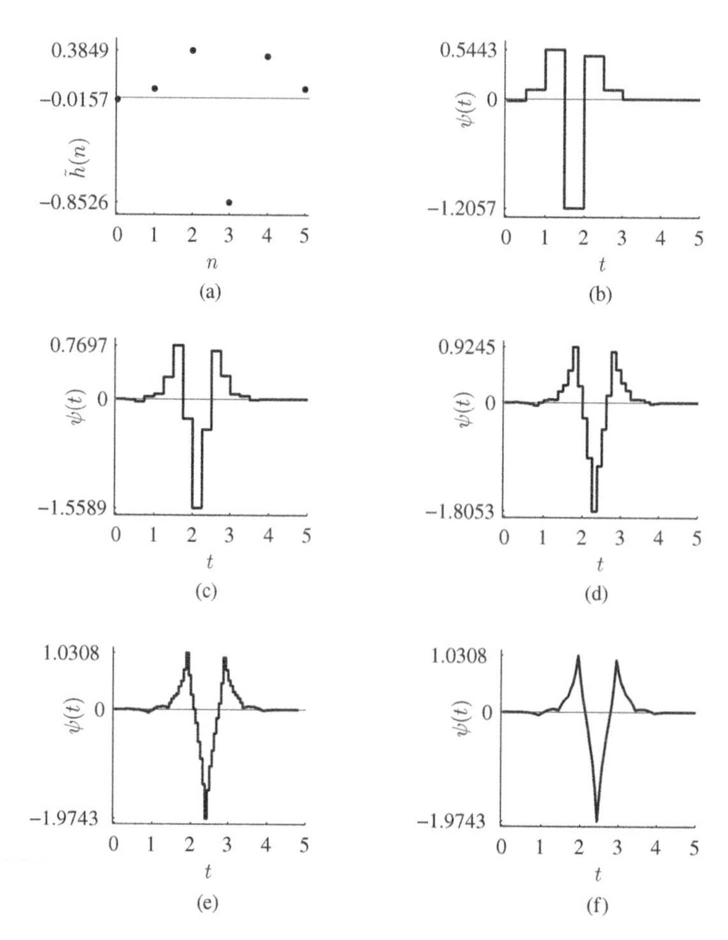

Figure 9.10 C6 wavelet function derived from the synthesis highpass filter. (a) Filter coefficients; (b) after one stage; (c) after two cascade stages; (d) after three cascade stages; (e) after four cascade stages; (f) after 10 cascade stages

9.5 Summary

- In this chapter, the design of commonly used DWT orthogonal filters is described.
- While the DWT filters are typically FIR type, their design methods are different from those of the conventional FIR filters.
- The DWT transform matrices are constituted using the double-shifted versions of the impulse responses of a set of lowpass and highpass filters.
- If the forward and the inverse transform matrices are inverses as well as transposes of each other, then the corresponding filter bank is called orthogonal.
- The basic orthogonal filter design method is to use the orthogonality and lowpass filter conditions to solve for the lowpass analysis filter. The other three filters are derived from the lowpass filter.
- The advantage of orthogonal filter banks is that they preserve the energy of the signal.
- Only the Haar DWT filters are orthogonal and symmetric.

- With longer filter lengths, the approximation of a signal becomes better. However, the computational cost also increases. In DWT applications, it is important to select the appropriate type of filter, its length, and the number of levels of decomposition that best suits the specific application.

Exercises

9.1 Verify the orthogonality conditions for highpass, and lowpass and highpass Haar analysis filters in both the frequency domain and time domain. The orthogonality conditions are

$$|H(e^{j\omega})|^2 + |H(e^{j(\omega+\pi)})|^2 = 2$$
$$L(e^{j\omega})H^*(e^{j\omega}) + L(e^{j(\omega+\pi)})H^*(e^{j(\omega+\pi)}) = 0$$

9.2 Verify the orthogonality conditions for lowpass, highpass, and lowpass and highpass D4 analysis filters. The orthogonality conditions are

$$|L(e^{j\omega})|^2 + |L(e^{j(\omega+\pi)})|^2 = 2$$
$$|H(e^{j\omega})|^2 + |H(e^{j(\omega+\pi)})|^2 = 2$$
$$L(e^{j\omega})H^*(e^{j\omega}) + L(e^{j(\omega+\pi)})H^*(e^{j(\omega+\pi)}) = 0$$

9.3 If $\{a, b, c, d\}$ is a solution to Equations (9.16–9.19), show that $\{-a, -b, -c, -d\}$ is also a solution to them with a factor of -1.

9.4 If $\{a, b, c, d\}$ is a solution to Equations (9.16–9.19), show that $\{d, c, b, a\}$ is also a solution to them.

9.5 Verify the orthogonality conditions for lowpass, highpass, and lowpass and highpass D6 analysis filters.

$$l(n) = \{0.0352, -0.0854, -0.1350, 0.4599, 0.8069, 0.3327\}$$
$$h(n) = \{-0.3327, 0.8069, -0.4599, -0.1350, 0.0854, 0.0352\}$$

9.6 Verify perfect reconstruction conditions for the D6 filters given as follows.

$$l(n) = \{0.0352, -0.0854, -0.1350, 0.4599, 0.8069, 0.3327\}$$
$$h(n) = \{-0.3327, 0.8069, -0.4599, -0.1350, 0.0854, 0.0352\}$$

9.7 Verify the orthogonality conditions for lowpass, highpass, and lowpass and highpass C6 analysis filters given as follows.

$$l(n) = \{-0.0157, -0.0727, 0.3849, 0.8526, 0.3379, -0.0727\}$$
$$h(n) = \{0.0727, 0.3379, -0.8526, 0.3849, 0.0727, -0.0157\}$$

9.8 Verify perfect reconstruction conditions for the C6 filters given as follows.

$$l(n) = \{-0.0157, -0.0727, 0.3849, 0.8526, 0.3379, -0.0727\}$$
$$h(n) = \{0.0727, 0.3379, -0.8526, 0.3849, 0.0727, -0.0157\}$$

9.9 Compute the 1-level DWT of $x(n)$ using D4 filters. Assume periodic extension at the borders. Compute the IDWT and verify that the input $x(n)$ is reconstructed. Verify Parseval's theorem.

9.9.1

$$x(n) = \{1, 4, 2, 3, 2, 3, 2, -3\}$$

***9.9.2**

$$x(n) = \{3, 4, 1, -2, 1, 2, -3, 2\}$$

9.9.3

$$x(n) = \{4, 8, 3, 1, 3, 5, -1, 1\}$$

9.10 Compute the 1-level 2-D DWT of $x(n_1, n_2)$ using D4 filters. Assume periodic extension at the borders. Compute the IDWT and verify that the input $x(n_1, n_2)$ is reconstructed. Verify Parseval's theorem.

***9.10.1**

$$\boldsymbol{x} = \begin{bmatrix} 2 & 2 & -1 & -3 & 3 & 1 & 1 & 2 \\ 4 & 3 & -1 & 2 & -3 & 2 & 1 & 3 \\ 2 & 1 & 2 & 3 & -1 & 4 & 2 & 2 \\ 3 & 2 & -1 & 2 & 4 & 1 & 3 & 2 \\ 4 & 3 & -1 & 2 & -1 & 2 & 1 & 3 \\ 2 & -1 & 3 & -2 & 2 & 3 & -2 & 3 \\ 1 & 4 & 2 & 1 & -1 & -3 & 2 & 2 \\ 3 & -2 & 3 & -2 & 2 & 1 & 3 & 3 \end{bmatrix}$$

9.10.2

$$\boldsymbol{x} = \begin{bmatrix} 1 & 4 & 2 & 1 & -1 & -3 & 2 & 2 \\ 4 & 3 & -1 & 2 & 3 & 2 & -1 & 3 \\ 4 & 3 & -1 & 2 & -1 & 2 & 1 & 3 \\ 3 & -2 & -1 & 2 & -4 & 1 & 1 & 2 \\ 2 & -1 & 2 & -3 & -1 & 4 & 2 & 2 \\ 2 & -1 & 4 & -2 & 2 & 3 & -2 & 3 \\ 2 & 2 & -1 & -3 & 3 & 1 & 1 & 2 \\ 3 & -2 & 3 & -3 & 2 & 1 & 4 & 3 \end{bmatrix}$$

9.10.3

$$\boldsymbol{x} = \begin{bmatrix} 3 & 2 & 2 & 1 & -1 & -3 & 2 & 2 \\ 4 & 3 & -1 & 4 & 3 & 2 & -1 & 3 \\ 4 & 3 & -4 & 2 & -1 & 2 & 1 & -3 \\ 3 & 4 & -3 & 2 & -2 & 1 & 1 & 2 \\ 1 & 2 & 3 & -3 & -1 & 4 & 2 & 2 \\ 3 & 1 & 4 & -2 & 2 & 3 & -2 & 3 \\ 4 & 2 & -1 & -3 & 2 & 1 & 1 & 2 \\ 1 & 2 & -3 & -3 & 2 & 1 & 4 & -2 \end{bmatrix}$$

9.11 Find the samples of the D4 scaling function after two stages of reconstruction. Use both the z -transform and IDFT methods.

The impulse responses of the synthesis filters are

$$\widetilde{l}(n) = \{0.4830, 0.8365, 0.2241, -0.1294\}$$

$$\widetilde{h}(n) = \{-0.1294, -0.2241, 0.8365, -0.4830\}$$

***9.12** Find the samples of the D4 wavelet function after two stages of reconstruction. Use both the z-transform and IDFT methods.

***9.13** Find the samples of the D6 scaling function after two stages of reconstruction. Use both the z-transform and IDFT methods.

The impulse responses of the synthesis filters are

$$\widetilde{l}(n) = \{0.3327, 0.8069, 0.4599, -0.1350, -0.0854, 0.0352\}$$

$$\widetilde{h}(n) = \{0.0352, 0.0854, -0.1350, -0.4599, 0.8069, -0.3327\}$$

9.14 Find the samples of the D6 wavelet function after two stages of reconstruction. Use both the z-transform and IDFT methods.

10

Biorthogonal Filter Banks

An even-symmetric filter is a filter whose coefficients about a central axis through the filter are mirror images of each other. For example, filters with coefficients $\{1, 1\}$ and $\{1, 2, 1\}$ are symmetric. One advantage of symmetric filters is that the phase response is linear, which is essential in applications such as image processing. Another advantage is that symmetric filters provide an effective solution to the boundary problem. Orthogonal DWT filters, except the Haar filters, are not symmetric. A DWT lowpass filter frequency response and that of its derivatives must evaluate to zero at $\omega = \pi$. For example, consider the frequency response

$$e^{j\omega} + 2 + e^{-j\omega}$$

and its derivative with respective ω

$$je^{j\omega} - je^{-j\omega}$$

Both of these expressions evaluate to zero at $\omega = \pi$. But the filter does not satisfy the double-shift orthogonality condition. Biorthogonal DWT filters of longer lengths can be symmetric. The orthogonality condition involves two different filters. Although the energy preservation property is lost, the biorthogonal DWT with some filters is quite close to the orthogonal DWT in that respect.

The DTFT of the coefficients of the filter is its frequency response. Therefore, the required coefficients can be obtained from the frequency response expression, if we place appropriate constraints on the frequency response. In Chapter 9, we derived the orthogonality conditions in the frequency domain. Biorthogonal DWT filters will be designed using these conditions.

In this chapter, the biorthogonal conditions are first stated. Then, the design of $5/3$ and $4/4$ spline filters and the CDF $9/7$ filter is presented. The first and last of these filters are recommended for image compression in JPEG2000 standard.

10.1 Biorthogonal Filters

Both the forward and inverse matrices of a biorthogonal DWT must be wavelet transform matrices, as an arbitrary inverse transform matrix may not have a fast algorithm for its computation. Furthermore, the transpose of the forward transform matrix is also used in transforming

Discrete Wavelet Transform: A Signal Processing Approach, First Edition. D. Sundararajan.

Companion Website: www.wiley.com/go/sundararajan/wavelet

2-D signals, and it has to be the inverse of another wavelet transform matrix. Therefore, the solution is to design two lowpass filters and derive the highpass filters from them so that the inverse of one transform matrix is the transpose of the other. In the orthogonal DWT, the inverse of the transform matrix is the transpose of itself. The two lowpass filters are called a biorthogonal filter pair. The design can proceed in the same way as that for the orthogonal filters, with the difference that the resulting set of equations must be solved for two filters.

Let $\boldsymbol{W}_N$ and $\tilde{\boldsymbol{W}}_N$ be two $N \times N$ wavelet transform matrices so that

$$\boldsymbol{W}_N^{-1} = \tilde{\boldsymbol{W}}_N^T$$

Let

$$\boldsymbol{W}_N = \begin{bmatrix} \boldsymbol{L} \\ \boldsymbol{H} \end{bmatrix}$$

where $\boldsymbol{L}$ and $\boldsymbol{H}$ are, respectively, the $\frac{N}{2} \times N$ matrices of lowpass and highpass filter portions. Similarly, let

$$\tilde{\boldsymbol{W}}_N = \begin{bmatrix} \tilde{\boldsymbol{L}} \\ \tilde{\boldsymbol{H}} \end{bmatrix}$$

Then,

$$\boldsymbol{W}_N \tilde{\boldsymbol{W}}_N^T = \begin{bmatrix} \boldsymbol{L} \\ \boldsymbol{H} \end{bmatrix} \begin{bmatrix} \tilde{\boldsymbol{L}}^T & \tilde{\boldsymbol{H}}^T \end{bmatrix} = \begin{bmatrix} \boldsymbol{L}\tilde{\boldsymbol{L}}^T & \boldsymbol{L}\tilde{\boldsymbol{H}}^T \\ \boldsymbol{H}\tilde{\boldsymbol{L}}^T & \boldsymbol{H}\tilde{\boldsymbol{H}}^T \end{bmatrix} = \begin{bmatrix} \boldsymbol{I}_{\frac{N}{2}} & \boldsymbol{0}_{\frac{N}{2}} \\ \boldsymbol{0}_{\frac{N}{2}} & \boldsymbol{I}_{\frac{N}{2}} \end{bmatrix} = \boldsymbol{I}_N \tag{10.1}$$

where $\boldsymbol{I}_{\frac{N}{2}}$ and $\boldsymbol{0}_{\frac{N}{2}}$ are, respectively, $\frac{N}{2} \times \frac{N}{2}$ identity and zero matrices. In the case of biorthogonal DWT filters, the two lowpass filters are designed, and the two highpass filters are derived from them. The orthogonality conditions that the analysis and reconstruction filters must satisfy are stated as follows. All of these equations involve cross-correlation as the similarity of different filters is determined. For lowpass filters,

$$\tilde{L}(e^{j\omega})L^*(e^{j\omega}) + \tilde{L}(e^{j(\omega+\pi)})L^*(e^{j(\omega+\pi)}) = 2 \tag{10.2}$$

if and only if, for all integers m,

$$\sum_n \tilde{l}(n)l(n-2m) = \delta(m)$$

Equation (10.2) is called the biorthogonality condition. The two filters with impulse responses $\boldsymbol{l}$ and $\tilde{\boldsymbol{l}}$ and frequency responses $L(e^{j\omega})$ and $\tilde{L}(e^{j\omega})$ are called a biorthogonal filter pair. For highpass filters,

$$\tilde{H}(e^{j\omega})H^*(e^{j\omega}) + \tilde{H}(e^{j(\omega+\pi)})H^*(e^{j(\omega+\pi)}) = 2 \tag{10.3}$$

if and only if, for all integers m,

$$\sum_n \tilde{h}(n)h(n-2m) = \delta(m)$$

For lowpass and highpass filters,

$$\tilde{H}(e^{j\omega})L^*(e^{j\omega}) + \tilde{H}(e^{j(\omega+\pi)})L^*(e^{j(\omega+\pi)}) = 0 \tag{10.4}$$

if and only if, for all integers m,

$$\sum_n \tilde{h}(n)l(n-2m) = 0$$

$$\tilde{L}(e^{j\omega})H^*(e^{j\omega}) + \tilde{L}(e^{j(\omega+\pi)})H^*(e^{j(\omega+\pi)}) = 0 \tag{10.5}$$

if and only if, for all integers m,

$$\sum_n \tilde{l}(n)h(n-2m) = 0$$

The frequency responses of the four filters must meet these conditions so that

$$\boldsymbol{W}_N \tilde{\boldsymbol{W}}_N^T = \boldsymbol{I}_N$$

The design of filters reduces to solving Equation (10.1) for the two filters $l(n)$ and $\tilde{l}(n)$. In practice, it is very difficult. One practical solution is to select a symmetric FIR lowpass filter $\tilde{l}(n)$. Then, the set of equations to be solved becomes linear. The frequency response of the filter $\tilde{l}(n)$, for a class of filters called spline filters, is given by

$$\tilde{L}(e^{j\omega}) = \sqrt{2}\cos^{(\tilde{N}-1)}\left(\frac{\omega}{2}\right) \tag{10.6}$$

where $\tilde{N}$ is odd.

$$\cos^{(\tilde{N}-1)}\left(\frac{\omega}{2}\right) = \sum_{n=-\frac{(\tilde{N}-1)}{2}}^{\frac{(\tilde{N}-1)}{2}} a(n)e^{-j\omega n}$$

where

$$a(n) = \frac{1}{2^{(\tilde{N}-1)}}\binom{(\tilde{N}-1)}{\frac{(\tilde{N}-1)}{2}-n}$$

$$= \frac{(\tilde{N}-1)!}{2^{(\tilde{N}-1)}\left(\frac{(\tilde{N}-1)}{2}-n\right)!\left(\frac{(\tilde{N}-1)}{2}+n\right)!}, \quad n = 0, \pm 1, \ldots, \pm\frac{\tilde{N}-1}{2}$$

Note that

$$\binom{n}{k} = \frac{n!}{k!(n-k)!}$$

Then,

$$\tilde{L}(e^{j\omega}) = \sqrt{2}\cos^{(\tilde{N}-1)}\left(\frac{\omega}{2}\right) = \sqrt{2}\sum_{n=-\frac{(\tilde{N}-1)}{2}}^{\frac{(\tilde{N}-1)}{2}} \frac{1}{2^{(\tilde{N}-1)}}\binom{(\tilde{N}-1)}{\frac{(\tilde{N}-1)}{2}-n} e^{-j\omega n}$$

Let $\tilde{N}=3$. Then, with $n=-1,0,1$ and $0!=1$, we get

$$a(n) = \frac{2!}{2^2(1-n)!(1+n)!} = \frac{1}{4}\{1,2,1\}$$

For even-length even-symmetric filters, the frequency response of the filter $\tilde{l}(n)$ is given by

$$\tilde{L}(e^{j\omega}) = \sqrt{2}\cos^{(\tilde{N}-1)}\left(\frac{\omega}{2}\right)$$

$$\cos^{(\tilde{N}-1)}\left(\frac{\omega}{2}\right) = e^{j\frac{\omega}{2}} \sum_{n=-\frac{\tilde{N}}{2}+1}^{\frac{\tilde{N}}{2}} a(n)e^{-j\omega n} \tag{10.7}$$

where

$$a(n) = \frac{1}{2^{(\tilde{N}-1)}}\begin{pmatrix}(\tilde{N}-1)\\ \frac{\tilde{N}}{2}-1+n\end{pmatrix} = \frac{(\tilde{N}-1)!}{2^{(\tilde{N}-1)}\left(\frac{\tilde{N}}{2}-1+n\right)!\left(\frac{\tilde{N}}{2}-n\right)!}, \quad n = -\frac{\tilde{N}}{2}+1, \ldots, \frac{\tilde{N}}{2}$$

Then,

$$\tilde{L}(e^{j\omega}) = \sqrt{2}\cos^{(\tilde{N}-1)}\left(\frac{\omega}{2}\right) = \sqrt{2}e^{j\frac{\omega}{2}} \sum_{n=-\frac{\tilde{N}}{2}+1}^{\frac{\tilde{N}}{2}} \frac{1}{2^{(\tilde{N}-1))}}\left(\begin{pmatrix}\tilde{N}-1\end{pmatrix}\\ \frac{\tilde{N}}{2}-1+n\right) e^{-j\omega n}$$

Let $\tilde{N} = 4$. Then, with $n = -1, 0, 1, 2$ and $0! = 1$, we get

$$a(n) = \frac{3!}{2^3(1+n)!(2-n)!} = \frac{1}{8}\{1, 3, 3, 1\}$$

Once the two lowpass filters are fixed, the two highpass filters are determined using the formulas

$$h(n) = (-1)^n\tilde{l}(1-n) \quad \text{and} \quad \tilde{h}(n) = (-1)^n l(1-n) \tag{10.8}$$

The DTFT of these equations is, respectively,

$$H(e^{j\omega}) = -e^{-j\omega}\tilde{L}^*(e^{j(\omega+\pi)}) \quad \text{and} \quad \tilde{H}(e^{j\omega}) = -e^{-j\omega}L^*(e^{j(\omega+\pi)}) \tag{10.9}$$

By substituting $n = n + 1$, the expression $h(n) = (-1)^n\tilde{l}(1-n)$ becomes

$$h(n+1) = (-1)^{(n+1)}\tilde{l}(-n) = -(-1)^n\tilde{l}(-n)$$

The DTFT of $h(n+1)$ is

$$-\tilde{L}^*(e^{j(\omega+\pi)})$$

The DTFT of $h(n)$ is

$$-e^{-j\omega}\tilde{L}^*(e^{j(\omega+\pi)})$$

For example, let $\tilde{l}(n), n = -1, 0, 1, 2 = \{1, 3, 3, 1\}$. Then,

$$\tilde{L}(e^{j\omega}) = e^{j\omega} + 3 + 3e^{-j\omega} + e^{-j2\omega}$$

$$\tilde{L}(e^{j(\omega+\pi)}) = -e^{j\omega} + 3 - 3e^{-j\omega} + e^{-j2\omega}$$

$$\tilde{L}^*(e^{j(\omega+\pi)}) = -e^{-j\omega} + 3 - 3e^{j\omega} + e^{j2\omega}$$

$$-e^{-j\omega}\tilde{L}^*(e^{j(\omega+\pi)}) = e^{-j2\omega} - 3e^{-j\omega} + 3 - e^{j\omega}$$

The inverse DTFT of the last expression is $\{-1, 3, -3, 1\}$.

Consider the equation

$$\tilde{H}(e^{j\omega})H^*(e^{j\omega}) + \tilde{H}(e^{j(\omega+\pi)})H^*(e^{j(\omega+\pi)}) = 2 \tag{10.10}$$

Writing the frequency responses of highpass filters in terms of those of lowpass filters, we get

$$(-1)e^{-j\omega}L^*(e^{j(\omega+\pi)})(-1)e^{j\omega}\tilde{L}(e^{j(\omega+\pi)}) + (-1)e^{-j\omega}L^*(e^{j\omega})(-1)e^{j\omega}\tilde{L}(e^{j\omega}) = 2$$

$$L^*(e^{j(\omega+\pi)})\tilde{L}(e^{j(\omega+\pi)}) + L^*(e^{j\omega})\tilde{L}(e^{j\omega}) = 2$$

Similarly,

$$\tilde{L}(e^{j\omega})H^*(e^{j\omega}) + \tilde{L}(e^{j(\omega+\pi)})H^*(e^{j(\omega+\pi)}) = 0$$

$$\tilde{L}(e^{j\omega})(-1)e^{j\omega}\tilde{L}(e^{j(\omega+\pi)}) + \tilde{L}(e^{j(\omega+\pi)})e^{j\omega}\tilde{L}(e^{j\omega}) = 0$$

$$\tilde{H}(e^{j\omega})L^*(e^{j\omega}) + \tilde{H}(e^{j(\omega+\pi)})L^*(e^{j(\omega+\pi)}) = 0 \tag{10.11}$$

$$(-1)e^{-j\omega}L^*(e^{j(\omega+\pi)})L^*(e^{j\omega}) + e^{-j\omega}L^*(e^{j\omega})L^*(e^{j(\omega+\pi)}) = 0 \tag{10.12}$$

10.2 5/3 Spline Filter

Let us consider the 5/3 lowpass filter design with $\tilde{N} = 3$. Then,

$$\tilde{L}(e^{j\omega}) = \sqrt{2}\cos^2\left(\frac{\omega}{2}\right) = \sqrt{2}\left(\frac{1+\cos(\omega)}{2}\right) = \frac{\sqrt{2}}{4}e^{j\omega} + \frac{\sqrt{2}}{2} + \frac{\sqrt{2}}{4}e^{-j\omega}$$

That is,

$$\tilde{\boldsymbol{l}} = \{\tilde{l}(-1), \tilde{l}(0), \tilde{l}(1)\} = \left\{\frac{\sqrt{2}}{4}, \frac{\sqrt{2}}{2}, \frac{\sqrt{2}}{4}\right\}$$

We have determined these coefficients earlier using a formula. However, as the filter order is small, we used a trigonometric identity to get the same result. The other filter with length 5 is

$$\boldsymbol{l} = \{l(-2), l(-1), l(0), l(1), l(2)\}$$

Writing the $\tilde{\boldsymbol{l}}$ filter in the first row and the overlapping double-shifted versions of the $\boldsymbol{l}$ filter in the next two rows, we get

$$\begin{array}{ccccccc} & \tilde{l}(-1) & \tilde{l}(0) & \tilde{l}(1) & & & \\ l(-2) & l(-1) & l(0) & l(1) & l(2) & & \\ & & l(-2) & l(-1) & l(0) & l(1) & l(2) \end{array}$$

Now, the orthogonality conditions can be written as

$$\sum_{n=-1}^{1}\tilde{l}(n)l(n) = 1 \quad \text{and} \quad \sum_{n=0}^{1}\tilde{l}(n)l(n-2) = 0$$

With the even symmetry of the coefficients, we get

$$\tilde{l}(0)l(0) + 2\tilde{l}(1)l(1) = 1 \quad \text{and} \quad \tilde{l}(0)l(2) + \tilde{l}(1)l(1) = 0$$

Substituting the values for $\tilde{l}_0$ and $\tilde{l}_1$ and simplifying, we get

$$\frac{\sqrt{2}}{2}l(0) + 2\frac{\sqrt{2}}{4}l(1) = 1 \quad \text{and} \quad \frac{\sqrt{2}}{2}l(2) + \frac{\sqrt{2}}{4}l(1) = 0$$

$$l(0) + l(1) = \sqrt{2} \quad \text{and} \quad l(1) + 2l(2) = 0$$

As there are three variables and two equations, we use the lowpass constraint to get

$$L(e^{j\pi}) = l(0) - 2l(1) + 2l(2) = 0$$

The filter coefficients are obtained by solving the following set of linear equations.

$$\begin{aligned} l(0) - 2l(1) + 2l(2) &= 0 \\ l(0) + l(1) + 0l(2) &= \sqrt{2} \\ 0l(0) + l(1) + 2l(2) &= 0 \end{aligned}$$

By elimination and back substitution, we get

$$\begin{bmatrix} 1 & -2 & 2 & 0 \\ 1 & 1 & 0 & \sqrt{2} \\ 0 & 1 & 2 & 0 \end{bmatrix} \rightarrow \begin{bmatrix} 1 & -2 & 2 & 0 \\ 0 & 3 & -2 & \sqrt{2} \\ 0 & 1 & 2 & 0 \end{bmatrix} \rightarrow \begin{bmatrix} 1 & -2 & 2 & 0 \\ 0 & 3 & -2 & \sqrt{2} \\ 0 & 0 & 1 & -\frac{\sqrt{2}}{8} \end{bmatrix}$$

$$\boldsymbol{l} = \{l(-2), l(-1), l(0), l(1), l(2)\} = \left\{-\frac{\sqrt{2}}{8}, \frac{\sqrt{2}}{4}, \frac{3\sqrt{2}}{4}, \frac{\sqrt{2}}{4}, -\frac{\sqrt{2}}{8}\right\}$$

The highpass filters are obtained using the formulas

$$h(n) = (-1)^n \tilde{l}(1-n) \quad \text{and} \quad \tilde{h}(n) = (-1)^n l(1-n) \tag{10.13}$$

Using the formula

$$h(n) = (-1)^n \tilde{l}(1-n), \quad n = 2, 1, 0 \tag{10.14}$$

we get

$$h(2) = \tilde{l}(-1) = \frac{\sqrt{2}}{4}, h(1) = -\tilde{l}(0) = -\frac{\sqrt{2}}{2}, h(0) = \tilde{l}(1) = \frac{\sqrt{2}}{4} \tag{10.15}$$

Using the formula

$$\tilde{h}(n) = (-1)^n l(1-n), \quad n = 3, 2, 1, 0, -1 \tag{10.16}$$

we get

$$\tilde{h}(3) = -l(-2) = \frac{\sqrt{2}}{8}, \tilde{h}(2) = l(-1) = \frac{\sqrt{2}}{4}, \tilde{h}(1) = -l(0) = -3\frac{\sqrt{2}}{4}, \tag{10.17}$$

$$\tilde{h}(0) = l(1) = \frac{\sqrt{2}}{4}, \tilde{h}(-1) = -l(2) = \frac{\sqrt{2}}{8} \tag{10.18}$$

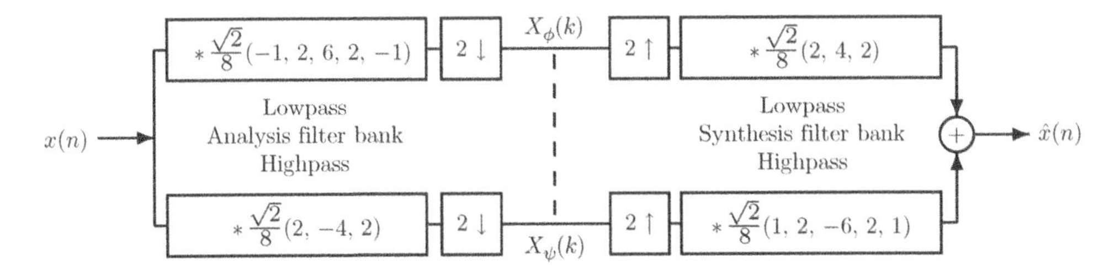

Figure 10.1 A two-channel analysis and synthesis filter bank with 5/3 spline filters

A two-channel filter bank with biorthogonal 5/3 spline filters is shown in Figure 10.1. The frequency responses of the noncausal filters are given by

$$L(e^{j\omega}) = \frac{\sqrt{2}}{8}(-e^{j2\omega} + 2e^{j\omega} + 6 + 2e^{-j\omega} - e^{-j2\omega})$$

$$H(e^{j\omega}) = \frac{\sqrt{2}}{8}(2 - 4e^{-j\omega} + 2e^{-j2\omega})$$

$$\tilde{L}(e^{j\omega}) = \frac{\sqrt{2}}{8}(2e^{j\omega} + 4 + 2e^{-j\omega})$$

$$\tilde{H}(e^{j\omega}) = \frac{\sqrt{2}}{8}(e^{j\omega} + 2 - 6e^{-j\omega} + 2e^{-j2\omega} + e^{-j3\omega})$$

The magnitude of the frequency response of the lowpass analysis filter is

$$|L(e^{j\omega})| = \frac{\sqrt{2}}{8}(6 + 4\cos(\omega) - 2\cos(2\omega))$$

The phase response for a causal filter is

$$\angle(L(e^{j\omega})) = -2\omega$$

To compute the output value corresponding to an input sample $x(n)$, the set of $\{x(n-2), x(n-1), x(n), x(n+1), x(n+2)\}$ values is required. There is a delay of two sample intervals to compute the output. Remember that if the transform of $x(n)$ is $X(e^{j\omega})$, then the transform of $x(n-2)$ is $e^{-j2\omega}X(e^{j\omega})$. The phase of the frequency component with frequency ω is changed by -2ω radians. Verify that the magnitude response is $\sqrt{2}$ at $\omega = 0$ and 0 at $\omega = \pi$ for lowpass filters and $\sqrt{2}$ at $\omega = \pi$ and 0 at $\omega = 0$ for highpass filters. The magnitude of the frequency responses of the 5/3 spline analysis filters is shown in Figure 10.2(a). Figure 10.2(b) shows the phase response of the lowpass filter, which is linear and is equal to -2ω. At $\omega = \pi$, the phase is -2π radians.

Approximations of the 5/3 spline scaling function are shown in Figure 10.3. The approximation is obtained using Equation (8.13). As can be clearly seen from Figure 10.3(f), the 5/3 scaling function is a triangular wave. This can be obtained from the convolution of the Haar scaling function with itself. The FT of the Haar scaling function, as derived in Chapter 8, is

$$\phi(j\omega) = \frac{1 - e^{-j\omega}}{j\omega}$$

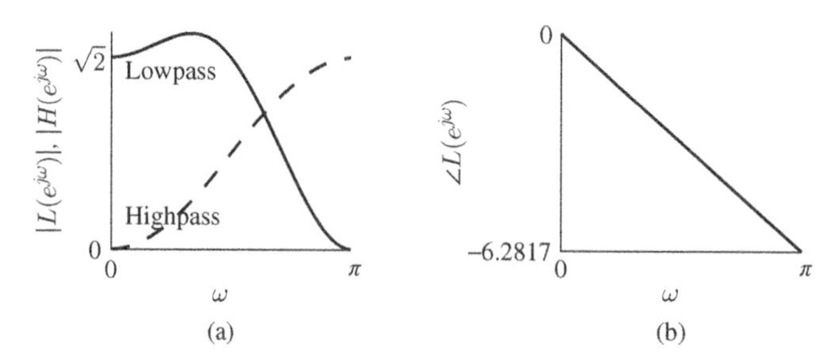

Figure 10.2 (a) The magnitude of the frequency responses of the 5/3 spline analysis filters; (b) the phase response of the lowpass filter

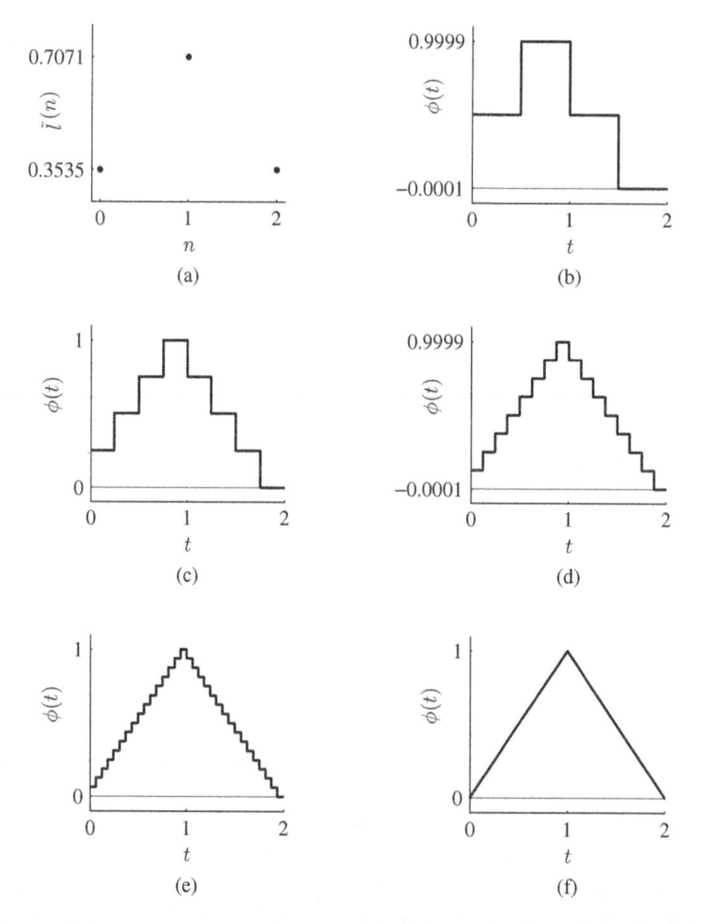

Figure 10.3 Approximation of the 5/3 spline scaling function derived from the synthesis lowpass filter. (a) Filter coefficients; (b) after one stage; (c) after two cascade stages; (d) after three cascade stages; (e) after four cascade stages; (f) after 10 cascade stages

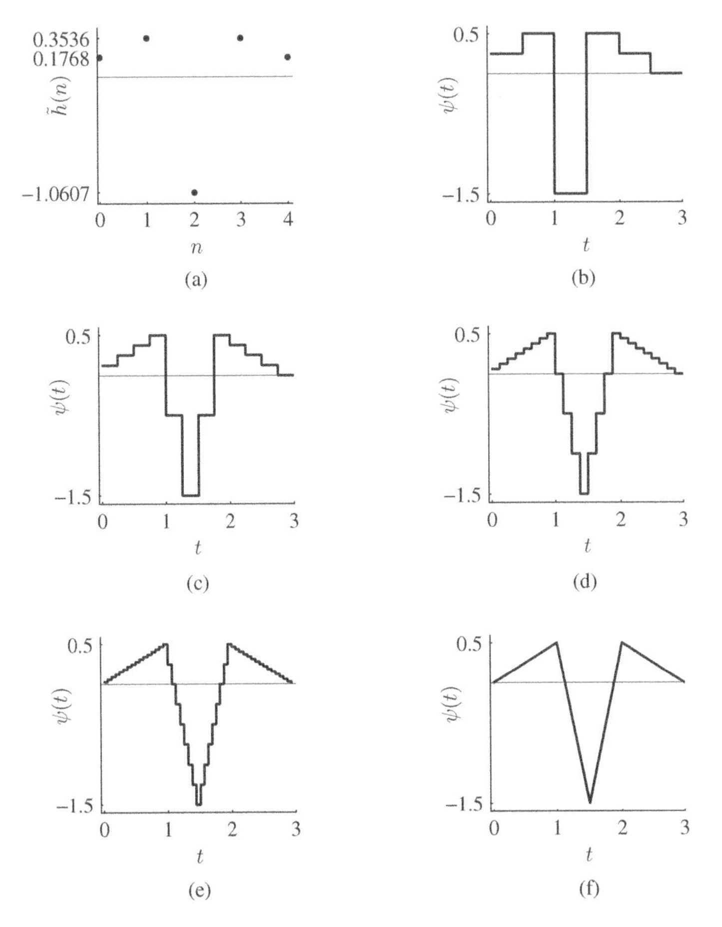

Figure 10.4 Approximation of the 5/3 spline wavelet function derived from the synthesis highpass filter. (a) Filter coefficients; (b) after one stage; (c) after two cascade stages; (d) after three cascade stages; (e) after four cascade stages; (f) after 10 cascade stages

Therefore, the FT of the 5/3 scaling function, from the FT convolution theorem, is the square of that of the Haar scaling function, and it is

$$\phi(j\omega) = \left(\frac{1 - e^{-j\omega}}{j\omega}\right)^2$$

Approximations of the wavelet function are shown in Figure 10.4. This function is derived using Equation (8.15).

In order to reduce the boundary problems, the filters are centered in the transform matrix. That is, $l(0)$ is placed in the first column of the first row of the transform matrix. Coefficient $h(0)$ is placed in the first column of the first row of the highpass portion. The other rows are formed by double-shift circular rotation. The inverse transform matrix is also formed by the same procedure using its filter coefficients. The 8×8 5/3 spline filter transform matrix is

$$\boldsymbol{W}_8 = \frac{\sqrt{2}}{8}\begin{bmatrix} 6 & 2 & -1 & 0 & 0 & 0 & -1 & 2 \\ -1 & 2 & 6 & 2 & -1 & 0 & 0 & 0 \\ 0 & 0 & -1 & 2 & 6 & 2 & -1 & 0 \\ -1 & 0 & 0 & 0 & -1 & 2 & 6 & 2 \\ 2 & -4 & 2 & 0 & 0 & 0 & 0 & 0 \\ 0 & 0 & 2 & -4 & 2 & 0 & 0 & 0 \\ 0 & 0 & 0 & 0 & 2 & -4 & 2 & 0 \\ 2 & 0 & 0 & 0 & 0 & 0 & 2 & -4 \end{bmatrix}$$

The 8×8 5/3 spline filter inverse transform matrix is

$$\tilde{\boldsymbol{W}}_8 = \frac{\sqrt{2}}{8}\begin{bmatrix} 4 & 2 & 0 & 0 & 0 & 0 & 0 & 2 \\ 0 & 2 & 4 & 2 & 0 & 0 & 0 & 0 \\ 0 & 0 & 0 & 2 & 4 & 2 & 0 & 0 \\ 0 & 0 & 0 & 0 & 0 & 2 & 4 & 2 \\ 2 & -6 & 2 & 1 & 0 & 0 & 0 & 1 \\ 0 & 1 & 2 & -6 & 2 & 1 & 0 & 0 \\ 0 & 0 & 0 & 1 & 2 & -6 & 2 & 1 \\ 2 & 1 & 0 & 0 & 0 & 1 & 2 & -6 \end{bmatrix}$$

Note that the **transpose** of either matrix, $\boldsymbol{W}_8$ or $\tilde{\boldsymbol{W}}_8$, is the inverse of the other. All the orthogonality conditions can be verified.

Example 10.1 Compute the 1-level DWT of $x(n)$ using 5/3 spline filters. Assume periodic extension at the borders. Compute the IDWT and verify that the input $x(n)$ is reconstructed. Find the energy of the signal and its DWT coefficients.

$$x(n) = \{1, 3, -2, 4, 2, 2, 3, -1\}$$

Solution

$$\frac{\sqrt{2}}{8}\begin{bmatrix} 9 \\ -1 \\ 23 \\ 17 \\ -14 \\ -16 \\ 2 \\ 12 \end{bmatrix} = \frac{\sqrt{2}}{8}\begin{bmatrix} 6 & 2 & -1 & 0 & 0 & 0 & -1 & 2 \\ -1 & 2 & 6 & 2 & -1 & 0 & 0 & 0 \\ 0 & 0 & -1 & 2 & 6 & 2 & -1 & 0 \\ -1 & 0 & 0 & 0 & -1 & 2 & 6 & 2 \\ 2 & -4 & 2 & 0 & 0 & 0 & 0 & 0 \\ 0 & 0 & 2 & -4 & 2 & 0 & 0 & 0 \\ 0 & 0 & 0 & 0 & 2 & -4 & 2 & 0 \\ 2 & 0 & 0 & 0 & 0 & 0 & 2 & -4 \end{bmatrix}\begin{bmatrix} 1 \\ 3 \\ -2 \\ 4 \\ 2 \\ 2 \\ 3 \\ -1 \end{bmatrix}$$

For biorthogonal DWT, Parseval's theorem does not hold.

$$(1^2 + 3^2 + (-2)^2 + 4^2 + 2^2 + 2^2 + 3^2 + (-1)^2) = 48$$

$$\frac{(9^2 + (-1)^2 + 23^2 + 17^2 + (-14)^2 + (-16)^2 + 2^2 + 12^2)}{32} = 46.8750$$

The input can be obtained from the coefficients using the **transpose** of the 8×8 inverse transform matrix.

$$\begin{bmatrix} 1 \\ 3 \\ -2 \\ 4 \\ 2 \\ 2 \\ 3 \\ -1 \end{bmatrix} = \frac{\sqrt{2}}{8} \begin{bmatrix} 4 & 0 & 0 & 0 & 2 & 0 & 0 & 2 \\ 2 & 2 & 0 & 0 & -6 & 1 & 0 & 1 \\ 0 & 4 & 0 & 0 & 2 & 2 & 0 & 0 \\ 0 & 2 & 2 & 0 & 1 & -6 & 1 & 0 \\ 0 & 0 & 4 & 0 & 0 & 2 & 2 & 0 \\ 0 & 0 & 2 & 2 & 0 & 1 & -6 & 1 \\ 0 & 0 & 0 & 4 & 0 & 0 & 2 & 2 \\ 2 & 0 & 0 & 2 & 1 & 0 & 1 & -6 \end{bmatrix} \frac{\sqrt{2}}{8} \begin{bmatrix} 9 \\ -1 \\ 23 \\ 17 \\ -14 \\ -16 \\ 2 \\ 12 \end{bmatrix}$$

■

The forward transform of an $N \times N$ image $\boldsymbol{x}$ is given by

$$\boldsymbol{X} = \boldsymbol{W}_N \boldsymbol{x} \boldsymbol{W}_N^T$$

and the inverse transform is given by

$$x = \tilde{\boldsymbol{W}}_N^T \boldsymbol{X} \tilde{\boldsymbol{W}}_N = (\tilde{\boldsymbol{W}}_N^T \boldsymbol{W}_N)\boldsymbol{x}(\boldsymbol{W}_N^T \tilde{\boldsymbol{W}}_N) = \boldsymbol{I}_N \boldsymbol{x} \boldsymbol{I}_N = \boldsymbol{x}$$

Example 10.2 Compute the 1-level 2-D DWT of $x(n_1, n_2)$ using 5/3 spline filters. Assume periodic extension at the borders. Compute the IDWT and verify that the input $x(n_1, n_2)$ is reconstructed. Find the energy of the signal and its DWT coefficients.

$$\boldsymbol{x} = \begin{bmatrix} 1 & 2 & -1 & 3 & 3 & 1 & 2 & 2 \\ 3 & 4 & 3 & -2 & 2 & 1 & 4 & 3 \\ 2 & 1 & 2 & 3 & -1 & 4 & 2 & 2 \\ 4 & 3 & -1 & 2 & 4 & 2 & 1 & 3 \\ 3 & 2 & -1 & 2 & 3 & 1 & 3 & 2 \\ 2 & -1 & 3 & -2 & 2 & 3 & 4 & 3 \\ 1 & 4 & 2 & 1 & -1 & 4 & 2 & 2 \\ 4 & 3 & -1 & 2 & 3 & 2 & 1 & 3 \end{bmatrix}$$

Solution

The result of computing the 1-D DWT of the rows of $\boldsymbol{x}$ yields

$$\boldsymbol{xr} = \boldsymbol{x}\boldsymbol{W}_8^T = \frac{\sqrt{2}}{8} \begin{bmatrix} 13 & 0 & 25 & 14 & -8 & -8 & 6 & -2 \\ 25 & 17 & 3 & 27 & -4 & 18 & 8 & 2 \\ 14 & 19 & 4 & 23 & 4 & -10 & -14 & 0 \\ 36 & -4 & 32 & 8 & -6 & -2 & 2 & -2 \\ 24 & -4 & 22 & 18 & -4 & -4 & 8 & 4 \\ 9 & 8 & 7 & 32 & 14 & 18 & 0 & 0 \\ 14 & 22 & 0 & 24 & -10 & -2 & -14 & -2 \\ 36 & -3 & 26 & 9 & -6 & -4 & 0 & -2 \end{bmatrix}$$

The result of computing the 1-D DWT of the columns of xr yields the 2-D DWT.

$$X = W_8 xr = \frac{1}{32}\begin{bmatrix} 172 & -13 & 204 & 109 & -62 & -8 & 80 & -10 \\ 169 & 144 & 47 & 176 & 16 & -16 & -78 & -2 \\ 206 & -57 & 206 & 141 & -2 & 20 & 80 & 22 \\ 137 & 146 & 19 & 194 & -32 & 28 & -98 & -18 \\ -46 & -30 & 46 & -34 & 8 & -108 & -48 & -12 \\ -68 & 46 & -76 & 50 & 24 & -20 & -20 & 16 \\ 40 & 4 & 16 & -44 & -84 & -84 & -12 & 4 \\ -90 & 56 & -54 & 40 & -12 & -4 & -16 & 0 \end{bmatrix}$$

You can notice that the magnitudes of the coefficients in other than the top left quarter are relatively small.

The inverse is computed by reversing the steps. By premultiplying X by $\tilde{W}_N^T$, we get xr. By postmultiplying xr by $\tilde{W}_N$, we get back the original input. ■

10.2.1 Daubechies Formulation

Daubechies has derived an explicit formula to find $L(e^{j\omega})$, given $\tilde{L}(e^{j\omega})$. Let $\tilde{L}(e^{j\omega})$ be defined by

$$\tilde{L}(e^{j\omega}) = \sqrt{2}\cos^{(\tilde{N}-1)}\left(\frac{\omega}{2}\right)$$

where $\tilde{N}$ is odd. Then, $L(e^{j\omega})$ of a filter of length $2N + \tilde{N} - 2$, for a given even N, is defined by

$$L(e^{j\omega}) = \sqrt{2}\cos^{N}\left(\frac{\omega}{2}\right)\sum_{m=0}^{k+\tilde{k}-1}\binom{k+\tilde{k}-1+m}{m}\sin^{2m}\left(\frac{\omega}{2}\right), k = \frac{N}{2}, \tilde{k} = \frac{\tilde{N}-1}{2}$$

Using this formulation, the 5/3 spline filter is derived again as follows with $N = 2$.

$$\tilde{L}(e^{j\omega}) = \sqrt{2}\cos^2\left(\frac{\omega}{2}\right) = \frac{\sqrt{2}}{4}e^{j\omega} + \frac{\sqrt{2}}{2} + \frac{\sqrt{2}}{4}e^{-j\omega}$$

$$L(e^{j\omega}) = \sqrt{2}\cos^2\left(\frac{\omega}{2}\right)\sum_{m=0}^{1}\binom{1+m}{m}\sin^{2m}\left(\frac{\omega}{2}\right) = \sqrt{2}\cos^2\left(\frac{\omega}{2}\right)\left(1 + 2\sin^2\left(\frac{\omega}{2}\right)\right)$$

$$= \sqrt{2}\left(\frac{1+\cos(\omega)}{2}\right)(1 + (1 - \cos(\omega))) = \frac{\sqrt{2}}{2}((1 + \cos(\omega))(2 - \cos(\omega)))$$

$$= -\frac{\sqrt{2}}{8}e^{j2\omega} + \frac{\sqrt{2}}{4}e^{j\omega} + \frac{3\sqrt{2}}{4} + \frac{\sqrt{2}}{4}e^{-j\omega} - \frac{\sqrt{2}}{8}e^{-j2\omega}$$

10.3 4/4 Spline Filter

Let us consider the 4/4 lowpass filter design with $\tilde{N} = 4$. Using the formula,

$$\cos^3\left(\frac{\omega}{2}\right) = \frac{1}{4}\left(3\cos\left(\frac{\omega}{2}\right) + \cos\left(\frac{3\omega}{2}\right)\right)$$

we get

$$\tilde{L}(e^{j\omega}) = \sqrt{2}\cos^3\left(\frac{\omega}{2}\right) = \frac{\sqrt{2}}{4}\left(3\cos\left(\frac{\omega}{2}\right) + \cos\left(\frac{3\omega}{2}\right)\right)$$
$$= \frac{\sqrt{2}}{8}\left(e^{j\left(\frac{3\omega}{2}\right)} + 3e^{j\left(\frac{\omega}{2}\right)} + 3e^{-j\left(\frac{\omega}{2}\right)} + e^{-j\left(\frac{3\omega}{2}\right)}\right)$$
$$= \frac{\sqrt{2}}{8}e^{j\left(\frac{\omega}{2}\right)}(e^{j\omega} + 3 + 3e^{-j\omega} + e^{-j2\omega})$$

Note that $\tilde{L}(e^{j0}) = \sqrt{2}$ and $\tilde{L}(e^{j\pi}) = 0$. The filter coefficients are

$$\{\tilde{l}(-1), \tilde{l}(0), \tilde{l}(1), \tilde{l}(2)\} = \frac{\sqrt{2}}{8}\{1, 3, 3, 1\}$$

We have determined these coefficients earlier using a formula. However, as the filter order is small, we used a trigonometric identity to get the same result. The other filter with length 4 is

$$\{l(-1), l(0), l(1), l(2)\}$$

Writing the $\tilde{l}(n)$ filter in the first row and the overlapping double-shifted versions of the $l(n)$ filter in the next two rows, we get

$$\begin{array}{cccccc} \tilde{l}(-1) & \tilde{l}(0) & \tilde{l}(1) & \tilde{l}(2) & & \\ l(-1) & l(0) & l(1) & l(2) & & \\ & & l(-1) & l(0) & l(1) & l(2) \end{array}$$

Now, the orthogonality conditions, with the even symmetry of the coefficients, can be written as

$$2\tilde{l}(0)l(0) + 2\tilde{l}(2)l(2) = 1 \quad \text{and} \quad \tilde{l}(0)l(2) + \tilde{l}(2)l(0) = 0$$

Substituting the values for $\tilde{l}(0)$ and $\tilde{l}(2)$ and simplifying, we get

$$6\frac{\sqrt{2}}{8}l(0) + 2\frac{\sqrt{2}}{8}l(2) = 1 \quad \text{and} \quad 3\frac{\sqrt{2}}{8}l(2) + \frac{\sqrt{2}}{8}l(0) = 0$$

From the second equation, $l(0) = -3l(2)$. Substituting for $l(0)$ in the first equation,

$$-18\frac{\sqrt{2}}{8}l(2) + 2\frac{\sqrt{2}}{8}l(2) = 1$$

Multiplying both sides by $-\frac{\sqrt{2}}{4}$,

$$18\frac{\sqrt{2}}{8}\frac{\sqrt{2}}{4}l(2) - 2\frac{\sqrt{2}}{8}\frac{\sqrt{2}}{4}l(2) = -\frac{\sqrt{2}}{4} = l(2) \quad \text{and} \quad l(0) = 3\frac{\sqrt{2}}{4}$$

$$\{l(-1), l(0), l(1), l(2)\} = \left\{-\frac{\sqrt{2}}{4}, 3\frac{\sqrt{2}}{4}, 3\frac{\sqrt{2}}{4}, -\frac{\sqrt{2}}{4}\right\}$$

The highpass filters are obtained using the formulas

$$h(n) = (-1)^n\tilde{l}(1-n) \quad \text{and} \quad \tilde{h}(n) = (-1)^n l(1-n) \tag{10.19}$$

Using the formula

$$h(n) = (-1)^n \tilde{l}(1-n), \quad n = 2, 1, 0, -1 \tag{10.20}$$

we get

$$h(2) = \tilde{l}(-1) = \frac{\sqrt{2}}{8}, h(1) = -\tilde{l}(0) = -\frac{3\sqrt{2}}{8}, h(0) = \tilde{l}(1) = \frac{3\sqrt{2}}{8}, h(-1) = -\tilde{l}(2) = -\frac{\sqrt{2}}{8} \tag{10.21}$$

Using the formula

$$\tilde{h}(n) = (-1)^n l(1-n), \quad n = 2, 1, 0, -1 \tag{10.22}$$

we get

$$\tilde{h}(2) = l(-1) = -\frac{\sqrt{2}}{4}, \tilde{h}(1) = -l(0) = -\frac{3\sqrt{2}}{4}, \tilde{h}(0) = l(1) = \frac{3\sqrt{2}}{4}, \tilde{h}(-1) = -l(2) = \frac{\sqrt{2}}{4} \tag{10.23}$$

The convolution of $l(n)$ and $\tilde{l}(n)$ with complete overlap is 1 and that of $h(n)$ and $\tilde{h}(n)$ is -1. The sum of the two partial sums is zero, rather than 2 as required by the PR condition. As multiplying the filter coefficients by -1 does not change the magnitude of the frequency response, the filter coefficients are changed to

$$\tilde{h}(-1) = -\frac{\sqrt{2}}{4}, \tilde{h}(0) = -\frac{3\sqrt{2}}{4}, \tilde{h}(1) = \frac{3\sqrt{2}}{4}, \tilde{h}(2) = \frac{\sqrt{2}}{4} \tag{10.24}$$

for filter bank implementation of the DWT.

A two-channel filter bank with 4/4 spline filters is shown in Figure 10.5. The frequency responses of the filters are given by

$$\begin{aligned}
L(e^{j\omega}) &= e^{j\frac{\omega}{2}} \frac{\sqrt{2}}{4}(-e^{j\omega} + 3 + 3e^{-j\omega} - e^{-j2\omega}) \\
&= \frac{\sqrt{2}}{4}\left(-e^{j\left(\frac{3}{2}\omega\right)} + 3e^{j\left(\frac{1}{2}\omega\right)} + 3e^{-j\left(\frac{1}{2}\omega\right)} - e^{-j\left(\frac{3}{2}\omega\right)}\right) \\
H(e^{j\omega}) &= e^{j\frac{\omega}{2}} \frac{\sqrt{2}}{8}(-e^{j\omega} + 3 - 3e^{-j\omega} + e^{-j2\omega}) \\
\tilde{L}(e^{j\omega}) &= e^{j\frac{\omega}{2}} \frac{\sqrt{2}}{8}(e^{j\omega} + 3 + 3e^{-j\omega} + e^{-j2\omega}) \\
\tilde{H}(e^{j\omega}) &= e^{j\frac{\omega}{2}} \frac{\sqrt{2}}{4}(-e^{j\omega} - 3 + 3e^{-j\omega} + e^{-j2\omega})
\end{aligned}$$

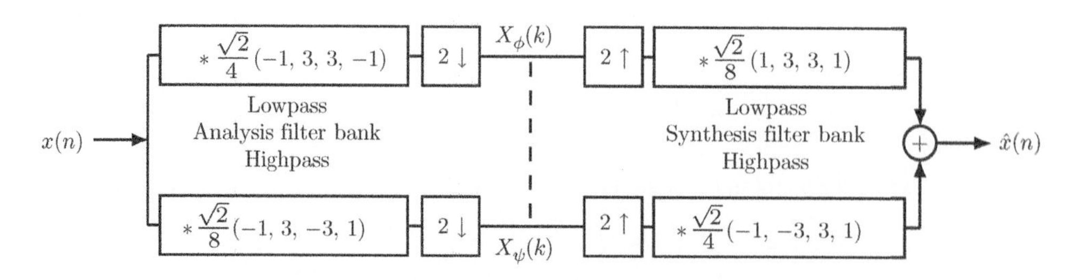

Figure 10.5 A two-channel analysis and synthesis filter bank with 4/4 spline filters

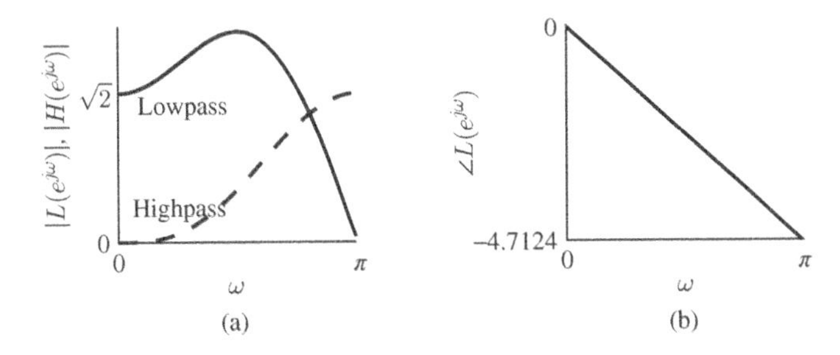

Figure 10.6 (a) The magnitude of the frequency responses of the 4/4 spline analysis filters; (b) the phase response of the lowpass filter

The magnitude of the frequency response of the lowpass filter is

$$|L(e^{j\omega})| = \frac{\sqrt{2}}{4}\left(-2\cos\left(\frac{3}{2}\omega\right) + 6\cos\left(\frac{1}{2}\omega\right)\right)$$

The phase response for a causal filter is

$$\angle(L(e^{j\omega})) = -3\omega/2$$

The magnitude of the frequency responses of the 4/4 spline analysis filters is shown in Figure 10.6(a). Verify that the magnitude response is $\sqrt{2}$ at $\omega = 0$ and 0 at $\omega = \pi$ for the lowpass filter and $\sqrt{2}$ at $\omega = \pi$ and 0 at $\omega = 0$ for the highpass filter. Figure 10.6(b) shows the phase response of the lowpass filter, which is linear and is equal to $-3\omega/2$ radians. At $\omega = \pi$, the phase is $-3\pi/2$ radians.

Approximations of the 4/4 spline scaling function are shown in Figure 10.7. This function is derived using the Equation (8.13). Approximations of the wavelet function are shown in Figure 10.8. This function is derived using Equation (8.15).

The 8 × 8 4/4 spline filter transform matrix is

$$\mathbf{W}_8 = \frac{\sqrt{2}}{8}\begin{bmatrix} 6 & 6 & -2 & 0 & 0 & 0 & 0 & -2 \\ 0 & -2 & 6 & 6 & -2 & 0 & 0 & 0 \\ 0 & 0 & 0 & -2 & 6 & 6 & -2 & 0 \\ -2 & 0 & 0 & 0 & 0 & -2 & 6 & 6 \\ -3 & 3 & -1 & 0 & 0 & 0 & 0 & 1 \\ 0 & 1 & -3 & 3 & -1 & 0 & 0 & 0 \\ 0 & 0 & 0 & 1 & -3 & 3 & -1 & 0 \\ -1 & 0 & 0 & 0 & 0 & 1 & -3 & 3 \end{bmatrix}$$

The 8 × 8 4/4 spline filter inverse transform matrix is

$$\tilde{\mathbf{W}}_8 = \frac{\sqrt{2}}{8}\begin{bmatrix} 3 & 3 & 1 & 0 & 0 & 0 & 0 & 1 \\ 0 & 1 & 3 & 3 & 1 & 0 & 0 & 0 \\ 0 & 0 & 0 & 1 & 3 & 3 & 1 & 0 \\ 1 & 0 & 0 & 0 & 0 & 1 & 3 & 3 \\ -6 & 6 & 2 & 0 & 0 & 0 & 0 & -2 \\ 0 & -2 & -6 & 6 & 2 & 0 & 0 & 0 \\ 0 & 0 & 0 & -2 & -6 & 6 & 2 & 0 \\ 2 & 0 & 0 & 0 & 0 & -2 & -6 & 6 \end{bmatrix}$$

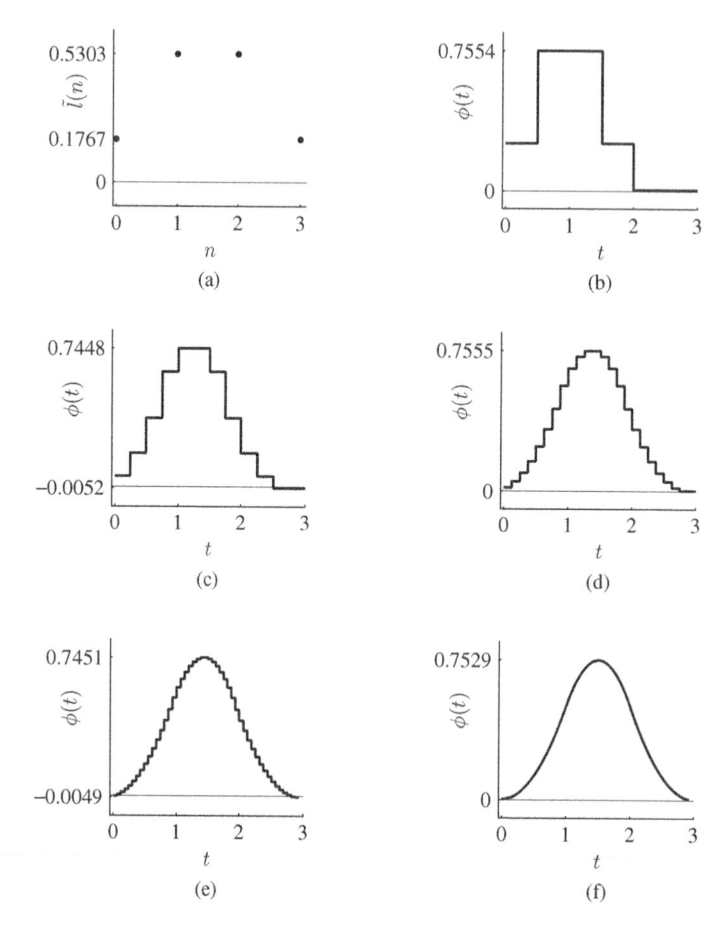

Figure 10.7 Approximation of the 4/4 spline scaling function derived from the synthesis lowpass filter. (a) Filter coefficients; (b) after one stage; (c) after two cascade stages; (d) after three cascade stages; (e) after four cascade stages; (f) after 10 cascade stages

Note that the **transpose** of either matrix, $\boldsymbol{W}_8$ or $\tilde{\boldsymbol{W}}_8$, is the inverse of the other. All the orthogonality conditions can be verified.

Example 10.3 Compute the 1-level DWT of $x(n)$ using 4/4 spline filters. Assume periodic extension at the borders. Compute the IDWT and verify that the input $x(n)$ is reconstructed. Find the energy of the signal and its DWT coefficients.

$$x(n) = \{1, 3, -2, 4, 2, 2, 3, -1\}$$

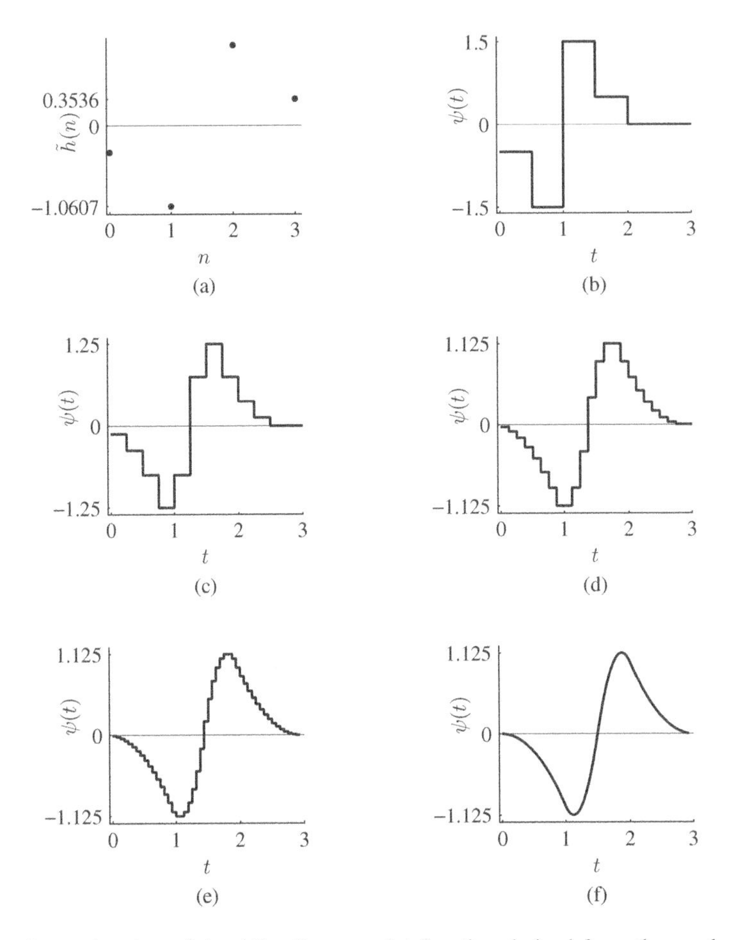

Figure 10.8 Approximation of the 4/4 spline wavelet function derived from the synthesis highpass filter. (a) Filter coefficients; (b) after one stage; (c) after two cascade stages; (d) after three cascade stages; (e) after four cascade stages; (f) after 10 cascade stages

Solution

$$\frac{\sqrt{2}}{8}\begin{bmatrix} 30 \\ 2 \\ 10 \\ 6 \\ 7 \\ 19 \\ 1 \\ -11 \end{bmatrix} = \frac{\sqrt{2}}{8}\begin{bmatrix} 6 & 6 & -2 & 0 & 0 & 0 & 0 & -2 \\ 0 & -2 & 6 & 6 & -2 & 0 & 0 & 0 \\ 0 & 0 & 0 & -2 & 6 & 6 & -2 & 0 \\ -2 & 0 & 0 & 0 & 0 & -2 & 6 & 6 \\ -3 & 3 & -1 & 0 & 0 & 0 & 0 & 1 \\ 0 & 1 & -3 & 3 & -1 & 0 & 0 & 0 \\ 0 & 0 & 0 & 1 & -3 & 3 & -1 & 0 \\ -1 & 0 & 0 & 0 & 0 & 1 & -3 & 3 \end{bmatrix}\begin{bmatrix} 1 \\ 3 \\ -2 \\ 4 \\ 2 \\ 2 \\ 3 \\ -1 \end{bmatrix}$$

For biorthogonal DWT, Parseval's theorem does not hold.

$$(1^2 + 3^2 + (-2)^2 + 4^2 + 2^2 + 2^2 + 3^2 + (-1)^2) = 48$$

$$\frac{(30^2 + 2^2 + 10^2 + 6^2 + 7^2 + 19^2 + 1^2 + (-11)^2)}{32} = 49.1250$$

The input can be obtained from the coefficients using the **transpose** of the 8×8 inverse transform matrix.

$$\begin{bmatrix} 1 \\ 3 \\ -2 \\ 4 \\ 2 \\ 2 \\ 3 \\ -1 \end{bmatrix} = \frac{\sqrt{2}}{8} \begin{bmatrix} 3 & 0 & 0 & 1 & -6 & 0 & 0 & 2 \\ 3 & 1 & 0 & 0 & 6 & -2 & 0 & 0 \\ 1 & 3 & 0 & 0 & 2 & -6 & 0 & 0 \\ 0 & 3 & 1 & 0 & 0 & 6 & -2 & 0 \\ 0 & 1 & 3 & 0 & 0 & 2 & -6 & 0 \\ 0 & 0 & 3 & 1 & 0 & 0 & 6 & -2 \\ 0 & 0 & 1 & 3 & 0 & 0 & 2 & -6 \\ 1 & 0 & 0 & 3 & -2 & 0 & 0 & 6 \end{bmatrix} \frac{\sqrt{2}}{8} \begin{bmatrix} 30 \\ 2 \\ 10 \\ 6 \\ 7 \\ 19 \\ 1 \\ -11 \end{bmatrix}$$

■

Example 10.4 Compute the 1-level 2-D DWT of $x(n_1, n_2)$ using 4/4 spline filters. Assume periodic extension at the borders. Compute the IDWT and verify that the input $x(n_1, n_2)$ is reconstructed. Find the energy of the signal and its DWT coefficients.

$$\boldsymbol{x} = \begin{bmatrix} 1 & 2 & -1 & 3 & 2 & 3 & 1 & 1 \\ 2 & 1 & 4 & 3 & 1 & 2 & -1 & 3 \\ 1 & 1 & 2 & 2 & 2 & 1 & 4 & 3 \\ 1 & 2 & 4 & 3 & 4 & 2 & 1 & 3 \\ 1 & 4 & -1 & 2 & 2 & 3 & -1 & 1 \\ 2 & 1 & 1 & -3 & 1 & 2 & -1 & 3 \\ 1 & 1 & 2 & 5 & 2 & -1 & 4 & 3 \\ 1 & -2 & 4 & 3 & 4 & -2 & 1 & 3 \end{bmatrix}$$

Solution

The result of computing the 1-D DWT of the rows of $\boldsymbol{x}$ yields

$$\boldsymbol{xr} = \boldsymbol{x}\boldsymbol{W}_8^T = \frac{\sqrt{2}}{8} \begin{bmatrix} 18 & 4 & 22 & 4 & 5 & 12 & 5 & 2 \\ 4 & 38 & 14 & 4 & -4 & -3 & 7 & 12 \\ 2 & 18 & 6 & 38 & 1 & -1 & -5 & -3 \\ 4 & 30 & 28 & 18 & 2 & -5 & -4 & 7 \\ 30 & -6 & 28 & -8 & 11 & 11 & 6 & 8 \\ 10 & -16 & 26 & 4 & -1 & -12 & 1 & 12 \\ 2 & 36 & -12 & 42 & 1 & 8 & -8 & -5 \\ -20 & 38 & 4 & 26 & -10 & -9 & -16 & 3 \end{bmatrix}$$

The result of computing the 1-D DWT of the columns of $\boldsymbol{xr}$ yields the 2-D DWT.

$$\boldsymbol{X} = \boldsymbol{W}_8\boldsymbol{xr} = \frac{1}{32}\begin{bmatrix} 168 & 140 & 196 & -80 & 24 & 74 & 114 & 84 \\ -32 & 224 & 120 & 344 & 4 & -52 & -80 & -16 \\ 228 & -264 & 292 & -144 & 54 & -12 & 66 & 116 \\ -164 & 468 & -144 & 392 & -62 & -6 & -156 & -40 \\ -64 & 122 & -26 & -12 & -38 & -53 & -5 & 36 \\ -20 & 80 & 52 & -48 & -12 & -26 & 4 & 34 \\ -58 & -36 & 34 & 12 & -35 & -82 & -11 & 24 \\ -74 & -14 & 52 & -48 & -39 & -75 & -28 & 34 \end{bmatrix}$$

You can notice that the magnitudes of the samples in other than the top left quarter are relatively small.

The inverse is obtained by reversing the steps. By premultiplying $\boldsymbol{X}$ by $\tilde{\boldsymbol{W}}_M^T$, we get $\boldsymbol{xr}$. By postmultiplying $\boldsymbol{xr}$ by $\tilde{\boldsymbol{W}}_M$, we get back the original input. ∎

10.3.1 Daubechies Formulation

Daubechies has derived an explicit formula to find $L(e^{j\omega})$, given $\tilde{L}(e^{j\omega})$. Let $\tilde{L}(e^{j\omega})$ be defined by

$$\tilde{L}(e^{j\omega}) = \sqrt{2}\cos^{(\tilde{N}-1)}\left(\frac{\omega}{2}\right)$$

where $\tilde{N}$ is even. Then, $L(e^{j\omega})$ of a filter of length $2N + \tilde{N} - 2$, for a given even N, is defined by

$$L(e^{j\omega}) = \sqrt{2}\cos^{N}\left(\frac{\omega}{2}\right)\sum_{m=0}^{k+\tilde{k}}\binom{k+\tilde{k}+m}{m}\sin^{2m}\left(\frac{\omega}{2}\right), k = \frac{N-1}{2}, \tilde{k} = \frac{\tilde{N}}{2} - 1$$

Using this formulation, the 4/4 spline filter is derived again as follows with $N = 1$.

$$\tilde{L}(e^{j\omega}) = \sqrt{2}\cos^{3}\left(\frac{\omega}{2}\right) = e^{j\frac{\omega}{2}}\frac{\sqrt{2}}{8}(e^{j\omega} + 3 + 3e^{-j\omega} + e^{-j2\omega})$$

$$L(e^{j\omega}) = \sqrt{2}\cos\left(\frac{\omega}{2}\right)\sum_{m=0}^{1}\binom{1+m}{m}\sin^{2m}\left(\frac{\omega}{2}\right) = \sqrt{2}\cos\left(\frac{\omega}{2}\right)\left(1 + 2\sin^2\left(\frac{\omega}{2}\right)\right)$$

$$= \sqrt{2}\cos\left(\frac{\omega}{2}\right)(1 + (1 - \cos(\omega))) = \sqrt{2}\cos\left(\frac{\omega}{2}\right)(2 - \cos(\omega))$$

$$= \frac{\sqrt{2}}{4}e^{j\frac{\omega}{2}}(1 + e^{-j\omega})(4 - e^{j\omega} - e^{-j\omega})$$

$$= \frac{\sqrt{2}}{4}e^{j\frac{\omega}{2}}(-e^{j\omega} + 3 + 3e^{-j\omega} - e^{-j2\omega})$$

10.4 CDF 9/7 Filter

The frequency response, $L(e^{j\omega})$, of an even-symmetric FIR filter with an odd length N (Equation (6.1)) is given by

$$L(e^{j\omega}) = l(0) + 2\sum_{n=1}^{N-1} l(n)\cos(\omega n)$$

The frequency response can be written as a linear combination of powers of $\cos(\omega)$, that is, a polynomial in $\cos(\omega)$. For the frequency response of the lowpass filter to be more flat in the vicinity of $\omega = \pi$, it should satisfy the condition that a number of its derivatives are equal to zero at $\omega = \pi$. Therefore, $(1 + \cos(\omega))^q$ must be a factor of $L(e^{j\omega})$. Then, the frequency response can be written as

$$L(e^{j\omega}) = \sqrt{2}(1 + \cos(\omega))^q p(\cos(\omega))$$

for some polynomial p, where $p(\cos(\pi)) = p(-1) \neq 0$. At $\omega = 0$,

$$L(e^{j0}) = \sqrt{2} = \sqrt{2}(1 + \cos(0))^q p(\cos(0)) = \sqrt{2}(2^q)p(1)$$

Therefore, $p(1) = 2^{-q}$. Using the identity $1 + \cos(\omega) = 2\cos^2\left(\frac{\omega}{2}\right)$, we get

$$L(e^{j\omega}) = \sqrt{2}\cos^{2q}\left(\frac{\omega}{2}\right)(2^q)p(\cos(\omega))$$

Let $s(\omega) = 2^q p(\omega)$. Then,

$$L(e^{j\omega}) = \sqrt{2}\cos^{2q}\left(\frac{\omega}{2}\right)s(\cos(\omega))$$

with $s(1) = 2^q p(1) = 2^q 2^{-q} = 1$, $s(-1) = 2^q p(-1) \neq 0$, and $L(e^{j0}) = \sqrt{2}$. Therefore, for an odd-length even-symmetric filter, the frequency response can be written as

$$L(e^{j\omega}) = \sqrt{2}\cos^{2q}\left(\frac{\omega}{2}\right)s(\cos(\omega)) \tag{10.25}$$

We get a similar expression for the other lowpass filter of the biorthogonal pair.

$$\tilde{L}(e^{j\omega}) = \sqrt{2}\cos^{2\tilde{q}}\left(\frac{\omega}{2}\right)\tilde{s}(\cos(\omega)) \tag{10.26}$$

with $\tilde{s}(1) = 1$ and $\tilde{s}(-1) \neq 0$. Given that the filters defined in Equations (10.25) and (10.26) solve Equation (10.2), the polynomials $\tilde{s}(\omega)$ and $s(\omega)$ must satisfy

$$s(\cos(\omega))\tilde{s}(\cos(\omega)) = \sum_{m=0}^{K-1}\binom{K-1+m}{m}\sin^{2m}\left(\frac{\omega}{2}\right)$$

where $K = q + \tilde{q}$ and the degree of $s(\cos(\omega))\tilde{s}(\cos(\omega))$ is less than K. We factorize the right-hand side and assign the factors to s and $\tilde{s}$ as evenly as possible. Choose K and factorize

$$S(t) = \sum_{m=0}^{K-1}\binom{K-1+m}{m}t^m$$

where $t = \sin^2\left(\frac{\omega}{2}\right)$. Construct one polynomial from the real roots of $S(t)$ and another from the complex roots. The first value of K, for which both real and complex roots exist, is 4. Taking $K = 4 = q + \tilde{q} = 2 + 2$, the 9/7 filter coefficients are obtained from these roots.

$$S(t) = \sum_{m=0}^{3} \binom{3+m}{m} t^m = 1 + 4t + 10t^2 + 20t^3$$

We used the fact that

$$\binom{n}{k} = \frac{n!}{k!(n-k)!}$$

The roots of $1 + 4t + 10t^2 + 20t^3$ are

$$s_1 = -0.342384, s_2 = -0.078808 - j0.373931, s_3 = -0.078808 + j0.373931$$

Now,

$$S(t) = 20(t - s_1)(t - s_2)(t - s_3)$$

Assigning the real root to $\tilde{s}$ and the complex roots to s, we get

$$\tilde{s}(t) = k(t - s_1) \quad \text{and} \quad s(t) = \frac{20}{k}(t - s_2)(t - s_3)$$

With $\tilde{q} = q = 2$ and $t = \sin^2\left(\frac{\omega}{2}\right)$,

$$\tilde{L}(e^{j\omega}) = \sqrt{2}\cos^4\left(\frac{\omega}{2}\right) k \left(\sin^2\left(\frac{\omega}{2}\right) + 0.342384\right)$$

and

$$L(e^{j\omega}) = \sqrt{2}\cos^4\left(\frac{\omega}{2}\right)$$
$$\left(\frac{20}{k}\right)\left(\sin^2\left(\frac{\omega}{2}\right) + 0.078808 - j0.373931\right)\left(\sin^2\left(\frac{\omega}{2}\right) + 0.078808 + j0.373931\right)$$

Applying the lowpass condition

$$\tilde{L}(e^{j0}) = L(e^{j0}) = \sqrt{2},$$

we get

$$\tilde{L}(e^{j0}) = \sqrt{2}\cos^4(0)k(\sin^2(0) + 0.342384) = \sqrt{2}(0.342384)k = \sqrt{2}$$

Solving for k, we get $k = 2.920696$. As

$$\tilde{L}^{(m)}(e^{j\pi}) = L^{(m)}(e^{j\pi}) = 0, \quad m = 0, 1, 2, 3$$

both the filters are equally flat at $\omega = \pi$. Using the formulas

$$\cos^2\left(\frac{\omega}{2}\right) = \frac{1 + \cos(\omega)}{2}, \quad \sin^2\left(\frac{\omega}{2}\right) = \frac{1 - \cos(\omega)}{2}, \quad \textit{and} \quad \cos(\omega) = \frac{e^{j\omega} + e^{-j\omega}}{2}$$

and expanding the frequency response expressions, we get

$$L(e^{j\omega}) = 0.037829e^{j4\omega} - 0.02385e^{j3\omega} - 0.110624e^{j2\omega} + 0.377403e^{j\omega} + 0.852699$$
$$+ 0.377403e^{-j\omega} - 0.110624e^{-j2\omega} - 0.02385e^{-j3\omega} + 0.037829e^{-j4\omega}$$

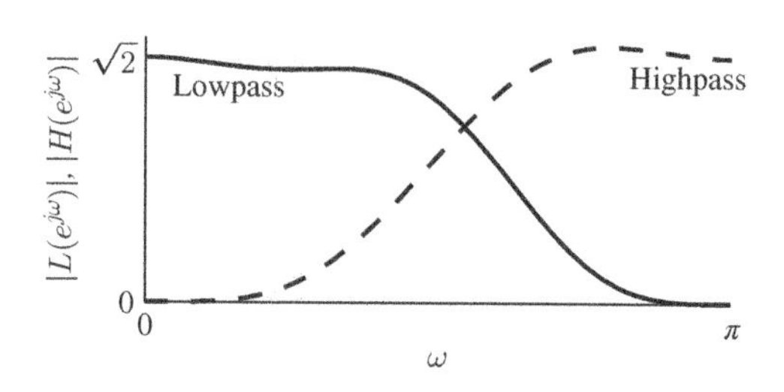

Figure 10.9 The magnitude of the frequency responses of the CDF 9/7 analysis filters

$$H(e^{j\omega}) = -0.064539e^{j2\omega} + 0.040689e^{j\omega} + 0.418092 - 0.788486e^{-j\omega} + 0.418092e^{-j2\omega} + 0.040689e^{-j3\omega} - 0.064539e^{-j4\omega}$$

$$\tilde{L}(e^{j\omega}) = -0.064539e^{j3\omega} - 0.040689e^{j2\omega} + 0.418092e^{j\omega} + 0.788486 + 0.418092e^{-j\omega} - 0.040689e^{-j2\omega} - 0.064539e^{-j3\omega}$$

$$\tilde{H}(e^{j\omega}) = -0.037829e^{j3\omega} - 0.02385e^{j2\omega} + 0.110624e^{j\omega} + 0.377403 - 0.852699e^{-j\omega} + 0.377403e^{-j2\omega} + 0.110624e^{-j3\omega} - 0.02385e^{-j4\omega} - 0.037829e^{-j5\omega}$$

The magnitude of the frequency responses of the 9/7 analysis filters is shown in Figure 10.9.

Approximations of the CDF 9/7 scaling function are shown in Figure 10.10. This function is derived using the Equation (8.13). Approximations of the CDF 9/7 wavelet function are shown in Figure 10.11. This function is derived using Equation (8.15). We list the impulse responses of all the four CDF 9/7 filters with a precision of six digits. Lowpass analysis filter

$$\begin{aligned} l(0) &= 0.852699 \\ l(1) = l(-1) &= 0.377403 \\ l(2) = l(-2) &= -0.110624 \\ l(3) = l(-3) &= -0.023850 \\ l(4) = l(-4) &= 0.037829 \end{aligned}$$

Highpass analysis filter

$$\begin{aligned} h(-2) = h(4) = \tilde{l}(3) &= -0.064539 \\ h(-1) = h(3) = -\tilde{l}(2) &= 0.040689 \\ h(0) = h(2) = \tilde{l}(1) &= 0.418092 \\ h(1) = -\tilde{l}(0) &= -0.788486 \end{aligned}$$

Lowpass synthesis filter

$$\begin{aligned} \tilde{l}_0 &= 0.788486 \\ \tilde{l}_{-1} = \tilde{l}_1 &= 0.418092 \end{aligned}$$

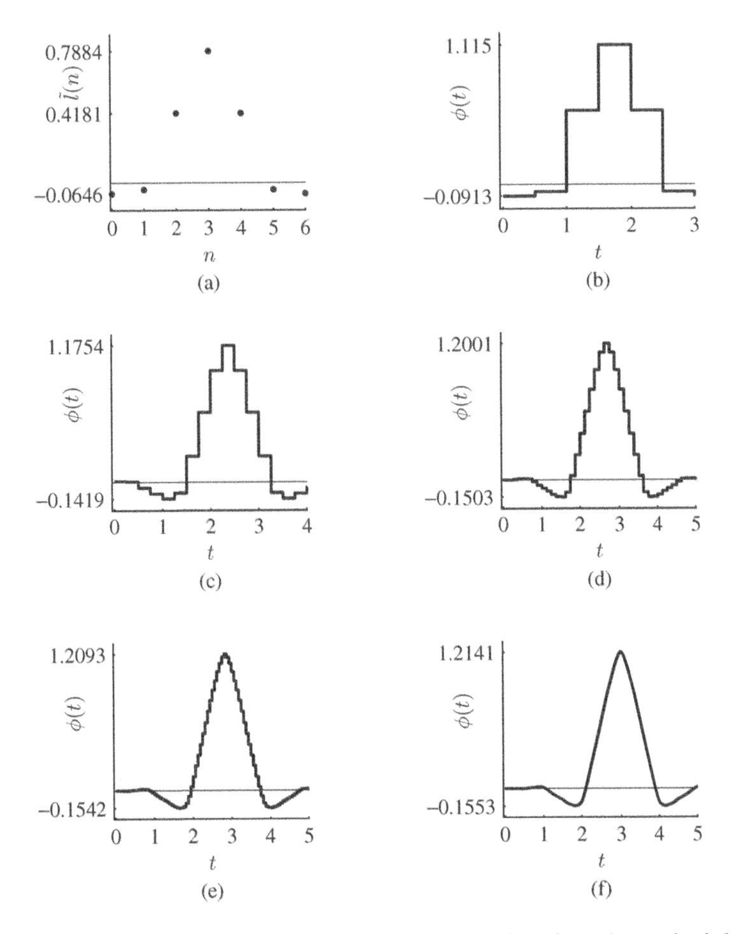

Figure 10.10 Approximation of the CDF 9/7 scaling function derived from the synthesis lowpass filter. (a) Filter coefficients; (b) after one stage; (c) after two cascade stages; (d) after three cascade stages; (e) after four cascade stages; (f) after 10 cascade stages

$$\tilde{l}_{-2} = \tilde{l}_2 = -0.040689$$
$$\tilde{l}_{-3} = \tilde{l}_3 = -0.064539$$

Highpass synthesis filter

$$\tilde{h}_{-3} = \tilde{h}_5 = -l(4) = -0.037829$$
$$\tilde{h}_{-2} = \tilde{h}_4 = l(3) = -0.023850$$
$$\tilde{h}_{-1} = \tilde{h}_3 = -l(2) = 0.110624$$
$$\tilde{h}_0 = \tilde{h}_2 = l(1) = 0.377403$$
$$\tilde{h}_1 = -l(0) = -0.852699$$

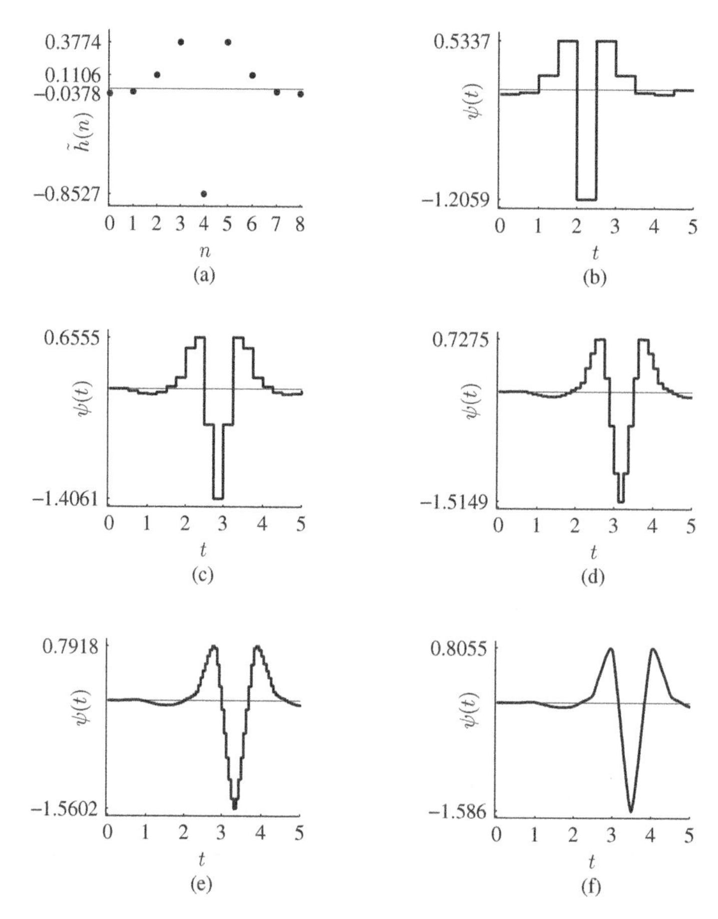

Figure 10.11 Approximation of the CDF 9/7 wavelet function derived from the synthesis highpass filter. (a) Filter coefficients; (b) after one stage; (c) after two cascade stages; (d) after three cascade stages; (e) after four cascade stages; (f) after 10 cascade stages

The 14×14 forward transform matrix $\boldsymbol{W}_{14}$ is

$$
\begin{bmatrix}
l(0) & l(1) & l(2) & l(3) & l(4) & 0 & 0 & 0 & 0 & 0 & l(-4) & l(-3) & l(-2) & l(-1) \\
l(-2) & l(-1) & l(0) & l(1) & l(2) & l(3) & l(4) & 0 & 0 & 0 & 0 & 0 & l(-4) & l(-3) \\
l(-4) & l(-3) & l(-2) & l(-1) & l(0) & l(1) & l(2) & l(3) & l(4) & 0 & 0 & 0 & 0 & 0 \\
0 & 0 & l(-4) & l(-3) & l(-2) & l(-1) & l(0) & l(1) & l(2) & l(3) & l(4) & 0 & 0 & 0 \\
0 & 0 & 0 & 0 & l(-4) & l(-3) & l(-2) & l(-1) & l(0) & l(1) & l(2) & l(3) & l(4) & 0 \\
l(4) & 0 & 0 & 0 & 0 & 0 & l(-4) & l(-3) & l(-2) & l(-1) & l(0) & l(1) & l(2) & l(3) \\
l(2) & l(3) & l(4) & 0 & 0 & 0 & 0 & 0 & l(-4) & l(-3) & l(-2) & l(-1) & l(0) & l(1) \\
h(0) & h(1) & h(2) & h(3) & h(4) & 0 & 0 & 0 & 0 & 0 & 0 & 0 & h(-2) & h(-1) \\
h(-2) & h(-1) & h(0) & h(1) & h(2) & h(3) & h(4) & 0 & 0 & 0 & 0 & 0 & 0 & 0 \\
0 & 0 & h(-2) & h(-1) & h(0) & h(1) & h(2) & h(3) & h(4) & 0 & 0 & 0 & 0 & 0 \\
0 & 0 & 0 & 0 & h(-2) & h(-1) & h(0) & h(1) & h(2) & h(3) & h(4) & 0 & 0 & 0 \\
0 & 0 & 0 & 0 & 0 & 0 & h(-2) & h(-1) & h(0) & h(1) & h(2) & h(3) & h(4) & 0 \\
h(4) & 0 & 0 & 0 & 0 & 0 & 0 & 0 & h(-2) & h(-1) & h(0) & h(1) & h(2) & h(3) \\
h(2) & h(3) & h(4) & 0 & 0 & 0 & 0 & 0 & 0 & 0 & h(-2) & h(-1) & h(0) & h(1)
\end{bmatrix}
$$

Example 10.5 Compute the 1-level DWT of $x(n)$ using CDF 9/7 spline filters. Assume periodic extension at the borders. Compute the IDWT and verify that the input $x(n)$ is reconstructed. Find the energy of the signal and its DWT coefficients.

$$x(n) = \{2, -4, 5, 3, 0, 1, 7, -5, 2, -6, 3, 1, -8, 2, 0, 3\}$$

Solution

The DWT coefficients

$$\{0.3530, 3.8342, 0.5481, 4.6122, -3.9026, 1.6712, -5.7984, 2.9249,$$
$$6.3247, -0.9779, 1.6050, 7.3082, 6.7232, -3.1708, -5.0816, -1.4170\}$$

can be obtained using the 16×16 transform matrix, assuming periodic extension. For biorthogonal transforms, Parseval's theorem does not hold.

$$\{2^2 + (-4)^2 + 5^2 + 3^2 + 0^2 + 1^2 + 7^2 + (-5)^2 2^2 + (-6)^2 + 3^2 + 1^2 + (-8)^2 + 2^2$$
$$+ 0^2 + 3^2\} = 256$$

$$\{0.3530^2 + 3.8342^2 + 0.5481^2 + 4.6122^2 + (-3.9026)^2 + 1.6712^2 + (-5.7984)^2$$
$$+ 2.9249^2 + 6.3247^2 + (-0.9779)^2 + 1.6050^2 + 7.3082^2 + 6.7232^2 + (-3.1708)^2$$
$$+ (-5.0816)^2 + (-1.4170)^2\} = 276.6287$$

The input can be obtained from the coefficients using the **transpose** of the 16×16 inverse transform matrix. ■

The 14×14 inverse transform matrix $\tilde{W}_{14}$ is

$$\begin{bmatrix}
\tilde{l}(0) & \tilde{l}(1) & \tilde{l}(2) & \tilde{l}(3) & 0 & 0 & 0 & 0 & 0 & 0 & 0 & \tilde{l}(-3) & \tilde{l}(-2) & \tilde{l}(-1) \\
\tilde{l}(-2) & \tilde{l}(-1) & \tilde{l}(0) & \tilde{l}(1) & \tilde{l}(2) & \tilde{l}(3) & 0 & 0 & 0 & 0 & 0 & 0 & 0 & \tilde{l}(-3) \\
0 & \tilde{l}(-3) & \tilde{l}(-2) & \tilde{l}(-1) & \tilde{l}(0) & \tilde{l}(1) & \tilde{l}(2) & \tilde{l}(3) & 0 & 0 & 0 & 0 & 0 & 0 \\
0 & 0 & 0 & \tilde{l}(-3) & \tilde{l}(-2) & \tilde{l}(-1) & \tilde{l}(0) & \tilde{l}(1) & \tilde{l}(2) & \tilde{l}(3) & 0 & 0 & 0 & 0 \\
0 & 0 & 0 & 0 & 0 & \tilde{l}(-3) & \tilde{l}(-2) & \tilde{l}(-1) & \tilde{l}(0) & \tilde{l}(1) & \tilde{l}(2) & \tilde{l}(3) & 0 & 0 \\
0 & 0 & 0 & 0 & 0 & 0 & 0 & \tilde{l}(-3) & \tilde{l}(-2) & \tilde{l}(-1) & \tilde{l}(0) & \tilde{l}(1) & \tilde{l}(2) & \tilde{l}(3) \\
\tilde{l}(2) & \tilde{l}(3) & 0 & 0 & 0 & 0 & 0 & 0 & 0 & \tilde{l}(-3) & \tilde{l}(-2) & \tilde{l}(-1) & \tilde{l}(0) & \tilde{l}(1) \\
\tilde{h}(0) & \tilde{h}(1) & \tilde{h}(2) & \tilde{h}(3) & \tilde{h}(4) & \tilde{h}(5) & 0 & 0 & 0 & 0 & 0 & \tilde{h}(-3) & \tilde{h}(-2) & \tilde{h}(-1) \\
\tilde{h}(-2) & \tilde{h}(-1) & \tilde{h}(0) & \tilde{h}(1) & \tilde{h}(2) & \tilde{h}(3) & \tilde{h}(4) & \tilde{h}(5) & 0 & 0 & 0 & 0 & 0 & \tilde{h}(-3) \\
0 & \tilde{h}(-3) & \tilde{h}(-2) & \tilde{h}(-1) & \tilde{h}(0) & \tilde{h}(1) & \tilde{h}(2) & \tilde{h}(3) & \tilde{h}(4) & \tilde{h}(5) & 0 & 0 & 0 & 0 \\
0 & 0 & 0 & \tilde{h}(-3) & \tilde{h}(-2) & \tilde{h}(-1) & \tilde{h}(0) & \tilde{h}(1) & \tilde{h}(2) & \tilde{h}(3) & \tilde{h}(4) & \tilde{h}(5) & 0 & 0 \\
0 & 0 & 0 & 0 & 0 & \tilde{h}(-3) & \tilde{h}(-2) & \tilde{h}(-1) & \tilde{h}(0) & \tilde{h}(1) & \tilde{h}(2) & \tilde{h}(3) & \tilde{h}(4) & \tilde{h}(5) \\
\tilde{h}(4) & \tilde{h}(5) & 0 & 0 & 0 & 0 & 0 & \tilde{h}(-3) & \tilde{h}(-2) & \tilde{h}(-1) & \tilde{h}(0) & \tilde{h}(1) & \tilde{h}(2) & \tilde{h}(3) \\
\tilde{h}(2) & \tilde{h}(3) & \tilde{h}(4) & \tilde{h}(5) & 0 & 0 & 0 & 0 & 0 & \tilde{h}(-3) & \tilde{h}(-2) & \tilde{h}(-1) & \tilde{h}(0) & \tilde{h}(1)
\end{bmatrix}$$

10.5 Summary

- In this chapter, the design of commonly used DWT biorthogonal filters is presented.
- While the DWT filters are typically FIR type, their design methods are different from those of the conventional FIR filter.

- The DWT transform matrices are constituted using the double-shifted versions of the impulse responses of a set of lowpass and highpass filters.
- If the transpose of the forward transform matrix is the inverse of the inverse transform matrix, they are called biorthogonal.
- In the biorthogonal spline filter design, one lowpass filter is arbitrarily selected, and the other lowpass filter is obtained by solving the linear equations derived from the orthogonality and lowpass filter conditions. The highpass filters are derived from the lowpass filters.
- Biorthogonal filter banks do not preserve the energy, but some of them are quite close to the orthogonal filter banks in that respect. The advantages of biorthogonal filter banks are that they provide linear phase response and handle the border problem effectively.
- With longer filter lengths, the approximation of a signal becomes better. However, the computation time also increases. In DWT applications, it is important to select the appropriate filter type, its length, and the number of iterations that best suits the specific application.

Exercises

10.1 Verify that the 5/3 spline filters satisfy the biorthogonal conditions in the frequency-domain and time-domain.

10.2 Verify that the CDF 9/7 spline filters satisfy the biorthogonal conditions in the frequency domain and time domain.

10.3 Check whether the filters satisfy the biorthogonal conditions using the IDFT. Write the DTFT biorthogonal conditions for the filter. For each of the four conditions, take 8 samples from 0 to 2π radians and find the IDFT of the samples.

$$\{l(-1), l(0), l(1), l(2)\} = \frac{\sqrt{2}}{8}\{-2, 6, 6, -2\}$$

$$\{h(-1), h(0), h(1), h(2)\} = \frac{\sqrt{2}}{8}\{-1, 3, -3, 1\}$$

$$\{\tilde{l}(-1), \tilde{l}0), \tilde{l}(1), \tilde{l}(2)\} = \frac{\sqrt{2}}{8}\{1, 3, 3, 1\}$$

$$\{\tilde{h}(-1), \tilde{h}0), \tilde{h}(1), \tilde{h}(2)\} = \frac{\sqrt{2}}{8}\{2, 6, -6, -2\}$$

10.4 Check whether the filters satisfy the biorthogonal conditions using the IDFT.

$$\{l(-1), l(0), l(1), l(2)\} = \frac{\sqrt{2}}{8}\{-2, 6, 6, -2\}$$

$$\{h(-1), h(0), h(1), h(2)\} = \frac{\sqrt{2}}{8}\{-1, 3, -3, 1\}$$

$$\{\tilde{l}(-1), \tilde{l}0), \tilde{l}(1), \tilde{l}(2)\} = \frac{\sqrt{2}}{8}\{1, 3, 3, 1\}$$

$$\{\tilde{h}(-1), \tilde{h}0), \tilde{h}(1), \tilde{h}(2)\} = \frac{\sqrt{2}}{8}\{-2, -6, 6, 2\}$$

10.5 Verify that the 5/3 spline filters satisfy the PR conditions in the frequency domain and time domain.

10.6 Verify that the CDF 9/7 spline filters satisfy the PR conditions in the frequency domain and time domain.

10.7 Check whether the filters satisfy the PR conditions using the IDFT.

$$\{l(-1), l(0), l(1), l(2)\} = \frac{\sqrt{2}}{8}\{-2, 6, 6, -2\}$$

$$\{h(-1), h(0), h(1), h(2)\} = \frac{\sqrt{2}}{8}\{-1, 3, -3, 1\}$$

$$\{\tilde{l}(-1), \tilde{l}0), \tilde{l}(1), \tilde{l}(2)\} = \frac{\sqrt{2}}{8}\{1, 3, 3, 1\}$$

$$\{\tilde{h}(-1), \tilde{h}0), \tilde{h}(1), \tilde{h}(2)\} = \frac{\sqrt{2}}{8}\{2, 6, -6, -2\}$$

10.8 Check whether the filters satisfy the PR conditions using the IDFT.

$$\{l(-1), l(0), l(1), l(2)\} = \frac{\sqrt{2}}{8}\{-2, 6, 6, -2\}$$

$$\{h(-1), h(0), h(1), h(2)\} = \frac{\sqrt{2}}{8}\{-1, 3, -3, 1\}$$

$$\{\tilde{l}(-1), \tilde{l}0), \tilde{l}(1), \tilde{l}(2)\} = \frac{\sqrt{2}}{8}\{1, 3, 3, 1\}$$

$$\{\tilde{h}(-1), \tilde{h}0), \tilde{h}(1), \tilde{h}(2)\} = \frac{\sqrt{2}}{8}\{-2, -6, 6, 2\}$$

10.9 Compute the 1-level DWT of $x(n)$ using 5/3 spline filters. Assume periodic extension at the borders. Compute the IDWT and verify that the input $x(n)$ is reconstructed. Find the energy of the signal and its DWT coefficients.

10.9.1

$$x(n) = \{1, 4, 2, 3, 2, 3, 2, -3\}$$

***10.9.2**

$$x(n) = \{3, 4, 1, -2, 1, 2, -3, 2\}$$

10.9.3

$$x(n) = \{4, 8, 3, 1, 3, 5, -1, 1\}$$

10.10 Compute the 1-level 2-D DWT of $x(n_1, n_2)$ using 5/3 spline filters. Assume periodic extension at the borders. Compute the IDWT and verify that the input $x(n_1, n_2)$ is reconstructed. Find the energy of the signal and its DWT coefficients.

***10.10.1**

$$x = \begin{bmatrix} 2 & 2 & -1 & -3 & 3 & 1 & 1 & 2 \\ 4 & 3 & -1 & 2 & -3 & 2 & 1 & 3 \\ 2 & 1 & 2 & 3 & -1 & 4 & 2 & 2 \\ 3 & 2 & -1 & 2 & 4 & 1 & 3 & 2 \\ 4 & 3 & -1 & 2 & -1 & 2 & 1 & 3 \\ 2 & -1 & 3 & -2 & 2 & 3 & -2 & 3 \\ 1 & 4 & 2 & 1 & -1 & -3 & 2 & 2 \\ 3 & -2 & 3 & -2 & 2 & 1 & 3 & 3 \end{bmatrix}$$

10.10.2

$$x = \begin{bmatrix} 1 & 4 & 2 & 1 & -1 & -3 & 2 & 2 \\ 4 & 3 & -1 & 2 & 3 & 2 & -1 & 3 \\ 4 & 3 & -1 & 2 & -1 & 2 & 1 & 3 \\ 3 & -2 & -1 & 2 & -4 & 1 & 1 & 2 \\ 2 & -1 & 2 & -3 & -1 & 4 & 2 & 2 \\ 2 & -1 & 4 & -2 & 2 & 3 & -2 & 3 \\ 2 & 2 & -1 & -3 & 3 & 1 & 1 & 2 \\ 3 & -2 & 3 & -3 & 2 & 1 & 4 & 3 \end{bmatrix}$$

10.10.3

$$x = \begin{bmatrix} 3 & 2 & 2 & 1 & -1 & -3 & 2 & 2 \\ 4 & 3 & -1 & 4 & 3 & 2 & -1 & 3 \\ 4 & 3 & -4 & 2 & -1 & 2 & 1 & -3 \\ 3 & 4 & -3 & 2 & -2 & 1 & 1 & 2 \\ 1 & 2 & 3 & -3 & -1 & 4 & 2 & 2 \\ 3 & 1 & 4 & -2 & 2 & 3 & -2 & 3 \\ 4 & 2 & -1 & -3 & 2 & 1 & 1 & 2 \\ 1 & 2 & -3 & -3 & 2 & 1 & 4 & -2 \end{bmatrix}$$

10.11 Compute the 1-level DWT of $x(n)$ using 4/4 spline filters. Assume periodic extension at the borders. Compute the IDWT and verify that the input $x(n)$ is reconstructed. Find the energy of the signal and its DWT coefficients.

10.11.1

$$x(n) = \{1, 4, 2, 3, 2, 3, 2, -3\}$$

10.11.2

$$x(n) = \{3, 4, 1, -2, 1, 2, -3, 2\}$$

***10.11.3**

$$x(n) = \{4, 8, 3, 1, 3, 5, -1, 1\}$$

10.12 Compute the 1-level 2-D DWT of $x(n_1, n_2)$ using 4/4 spline filters. Assume periodic extension at the borders. Compute the IDWT and verify that the input $x(n_1, n_2)$ is reconstructed. Find the energy of the signal and its DWT coefficients.

10.12.1

$$x = \begin{bmatrix} 2 & 2 & -1 & -3 & 3 & 1 & 1 & 2 \\ 4 & 3 & -1 & 2 & -3 & 2 & 1 & 3 \\ 2 & 1 & 2 & 3 & -1 & 4 & 2 & 2 \\ 3 & 2 & -1 & 2 & 4 & 1 & 3 & 2 \\ 4 & 3 & -1 & 2 & -1 & 2 & 1 & 3 \\ 2 & -1 & 3 & -2 & 2 & 3 & -2 & 3 \\ 1 & 4 & 2 & 1 & -1 & -3 & 2 & 2 \\ 3 & -2 & 3 & -2 & 2 & 1 & 3 & 3 \end{bmatrix}$$

***10.12.2**

$$x = \begin{bmatrix} 1 & 4 & 2 & 1 & -1 & -3 & 2 & 2 \\ 4 & 3 & -1 & 2 & 3 & 2 & -1 & 3 \\ 4 & 3 & -1 & 2 & -1 & 2 & 1 & 3 \\ 3 & -2 & -1 & 2 & -4 & 1 & 1 & 2 \\ 2 & -1 & 2 & -3 & -1 & 4 & 2 & 2 \\ 2 & -1 & 4 & -2 & 2 & 3 & -2 & 3 \\ 2 & 2 & -1 & -3 & 3 & 1 & 1 & 2 \\ 3 & -2 & 3 & -3 & 2 & 1 & 4 & 3 \end{bmatrix}$$

10.12.3

$$x = \begin{bmatrix} 3 & 2 & 2 & 1 & -1 & -3 & 2 & 2 \\ 4 & 3 & -1 & 4 & 3 & 2 & -1 & 3 \\ 4 & 3 & -4 & 2 & -1 & 2 & 1 & -3 \\ 3 & 4 & -3 & 2 & -2 & 1 & 1 & 2 \\ 1 & 2 & 3 & -3 & -1 & 4 & 2 & 2 \\ 3 & 1 & 4 & -2 & 2 & 3 & -2 & 3 \\ 4 & 2 & -1 & -3 & 2 & 1 & 1 & 2 \\ 1 & 2 & -3 & -3 & 2 & 1 & 4 & -2 \end{bmatrix}$$

10.13 Compute the 1-level DWT of $x(n)$ using CDF 9/7 spline filters. Assume periodic extension at the borders. Compute the IDWT and verify that the input $x(n)$ is reconstructed. Find the energy of the signal and its DWT coefficients.

***10.13.1**

$$x(n) = \{1, 4, 2, 3, 2, 3, 2, -3, 3, 4, 1, -2, 1, 2, -3, 2\}$$

10.13.2

$$x(n) = \{3, 4, 3, -2, 1, -2, -3, 2, 4, 8, 3, 1, 3, 5, -1, 1\}$$

10.13.3

$$x(n) = \{4, 1, 3, 4, 3, -5, -1, 1, 2, 1, 3, 4, 3, 1, 2, 2\}$$

10.14 Find the samples of the 5/3 spline scaling function after two stages of reconstruction. Use both the z-transform and IDFT methods.

***10.15** Find the samples of the 5/3 spline wavelet function after two stages of reconstruction. Use both the z-transform and IDFT methods.

***10.16** Find the samples of the 4/4 spline scaling function after two stages of reconstruction. Use both the z-transform and IDFT methods.

10.17 Find the samples of the 4/4 spline wavelet function after two stages of reconstruction. Use both the z-transform and IDFT methods.

11

Implementation of the Discrete Wavelet Transform

The DWT is essentially a set of bandpass filters and their efficient implementation. The filtering is achieved by repeatedly using a set of lowpass and highpass filters. Several filters are used in implementing the DWT with advantages and disadvantages in terms of the number of coefficients, the desired frequency response, errors in the representation of a signal, and so on. For efficient implementation of the DWT, several issues such as the execution time and memory trade-off, numerical errors, stability, and so on have to be considered. In this chapter, the implementation of the DWT using typical filters is described. Basically, there are two approaches to implement the DWT. The first approach is to evaluate the required convolutions directly. The other approach is to factorize the polyphase matrix into a product of a set of sparse matrices.

The DWT filters are linear time-invariant systems, and their outputs can be computed using the convolution relation, which relates the input, output, and impulse response of a system. In the implementation of the DWT, the output of the filter is downsampled. As the length of typical DWT filters is short, transform methods are not used in the implementation of the convolution operation. Using the polyphase realization of the filters, fast processing can be achieved if the operations of each phase of the filters are carried out in parallel. The boundary problem has to be tackled. The problem is how to extend the data at the boundaries to carry out the convolution at the boundary values of a finite signal. Matrix factorization method is well known in designing fast algorithms, for example, in fast Fourier and Walsh transform algorithms. This method is applicable in the DWT also. The convolution operation is evaluated in stages by factorizing the polyphase matrix of a filter bank. As usual, due to simplicity, the implementation aspects of the DWT are first presented using the Haar filter. Then, the implementation aspects of the DWT with other filters are described.

Unlike the DFT, for a given filter and input, more than one set of DWT coefficients is possible. In downsampling the filter output, we usually take the even-indexed values. Odd-indexed values are also valid DWT coefficients. Furthermore, the filter coefficients can be used in the given order or in reversed order. Border extensions of the input data can be carried out in different ways.

Discrete Wavelet Transform: A Signal Processing Approach, First Edition. D. Sundararajan.

Companion Website: www.wiley.com/go/sundararajan/wavelet

11.1 Implementation of the DWT with Haar Filters

The basic operation in finding the transform of a signal is the determination of its correlation with each of the basis functions. For example, the determination of the Fourier series is finding the amplitudes of each of an infinite number of sinusoids by correlating the given waveform with each of its constituent sinusoids. While the functions of correlation and convolution are different, their computation is very similar. Carrying out the convolution operation without time-reversing either of the two functions is the correlation operation. The DFT operation is similar to that of a set of bandpass filters with a narrow passband. While individual sinusoids are filtered out in the DFT operation, the components of a signal corresponding to a set of subbands of the frequency spectrum are filtered out in the DWT. The required operation is the convolution of the signal with the impulse responses of the corresponding filters. The impulse responses can be derived from the time reversals of the 2-point Haar ($\boldsymbol{W}_{2,0}$) basis functions. As the spectra of the decomposed signal components are subbands, each signal component is represented by a reduced set of samples. Reducing the number of samples is called the downsampling operation (discarding every other sample). After processing the individual components of a signal, the complete processed signal has to be reconstructed. The reconstruction process involves upsampling (inserting a zero after every sample) and filtering operations. Furthermore, all these operations have to be carried out using efficient algorithms. Therefore, signal decomposition, downsampling, upsampling, and reconstruction constitute the essential multirate digital signal processing operations required for the implementation of the DWT.

11.1.1 1-Level Haar DWT

The linear convolution of sequences $x(n)$ and $h(n)$ is defined as

$$y(n) = \sum_{m=-\infty}^{\infty} x(m)h(n-m)$$

In the DWT decomposition of a signal, convolution (filtering) is followed by downsampling. For an input sequence $x(n)$ with N values and Haar filter lowpass and highpass impulse responses given, respectively, by $l(n)$ and $h(n)$, the 1-level DWT coefficients are given by

$$X_{\phi}(k) = \sum_{m=2k}^{2k+1} x(m)l((2k+1)-m), \quad k = 0, 1, \dots, \frac{N}{2} - 1 \tag{11.1}$$

$$X_{\psi}(k) = \sum_{m=2k}^{2k+1} x(m)h((2k+1)-m)), \quad k = 0, 1, \dots, \frac{N}{2} - 1 \tag{11.2}$$

The convolution of the input $\{x(0) = 3, x(1) = 4, x(2) = 3, x(3) = -2\}$ with the scaled lowpass filter impulse response $\{1, 1\}$ is shown in Figure 11.1(a). The computation of 1-level Haar DWT approximation coefficients alone is shown in Figure 11.1(b). The convolution of the input with the scaled highpass filter impulse response $\{-1, 1\}$ is shown in Figure 11.2(a). The computation of Haar DWT detail coefficients alone is shown in Figure 11.2(b). The computation of Haar IDWT using the convolution operation without upsampling is shown in Figure 11.3. Note that the division by the constant $\sqrt{2}$ of the impulse response values is not

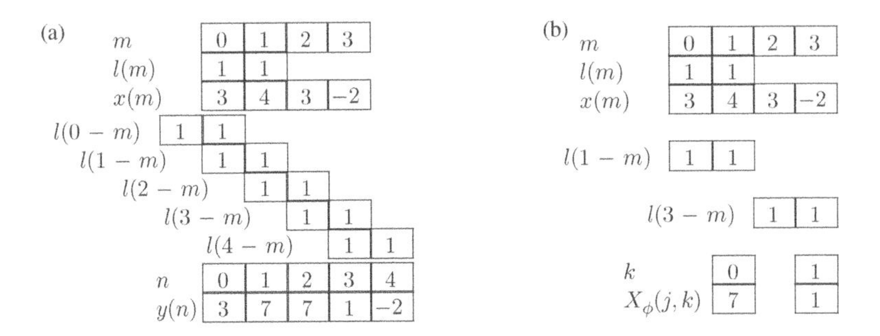

Figure 11.1 The 1-level Haar DWT approximation coefficients using the convolution operation

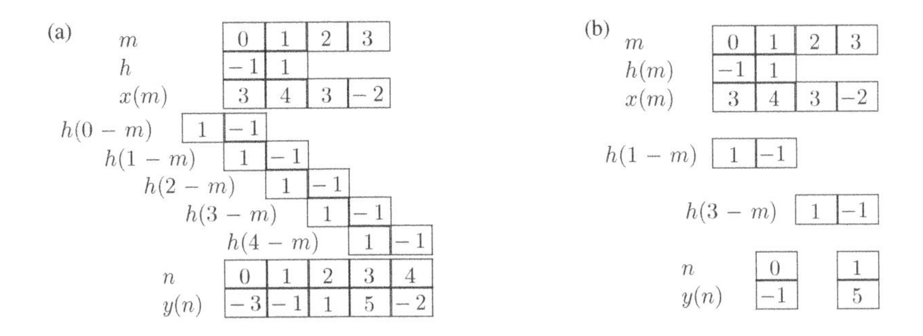

Figure 11.2 The 1-level Haar DWT detail coefficients using the convolution operation

shown in Figures 11.1–11.3, for simplicity. The DWT approximation coefficients $\{7, 1\}$ are alternately convolved with the scaled even- and odd-indexed impulse response $\{1\}$ and $\{1\}$ to get $x_{lp} = \{7, 7, 1, 1\}$. The DWT detail coefficients $\{-1, 5\}$ are alternately convolved with the scaled even- and odd-indexed impulse response $\{1\}$ and $\{-1\}$ to get $x_{hp} = \{-1, 1, 5, -5\}$. The summation of the two interpolated sequences

$$x_{lp}(n) + x_{hp}(n) = \{7, 7, 1, 1\} + \{-1, 1, 5, -5\} = \{6, 8, 6, -4\}$$

is just two times that of the given input sequence. The IDWT operation is given by

$$y(n) = 2x(n) = \sum_{m=0}^{(N/2)-1} (X_\phi(m)\tilde{l}(n-2m) + X_\psi(m)\tilde{h}(n-2m)), \quad n = 0, 1, \dots, N-1 \tag{11.3}$$

where $\tilde{l}(n) = \{1, 1\}$ and $\tilde{h}(n) = \{1, -1\}$. In the implementation, the two convolutions need to be executed only for valid values of the impulse response.

11.1.2 2-Level Haar DWT

Consider the computation of a 2-level 4-point DWT using a two-stage two-channel Haar analysis filter bank, shown in Figure 11.4. The 4-point input is $\{2, 1, 3, 4\}$, which is also considered

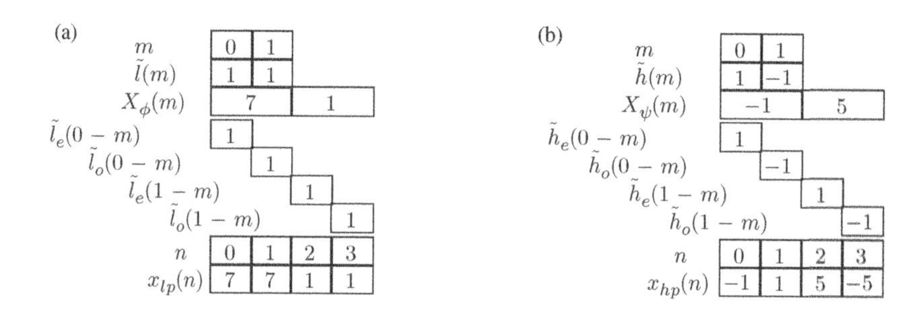

Figure 11.3 The 1-level Haar IDWT using the convolution operation without upsampling

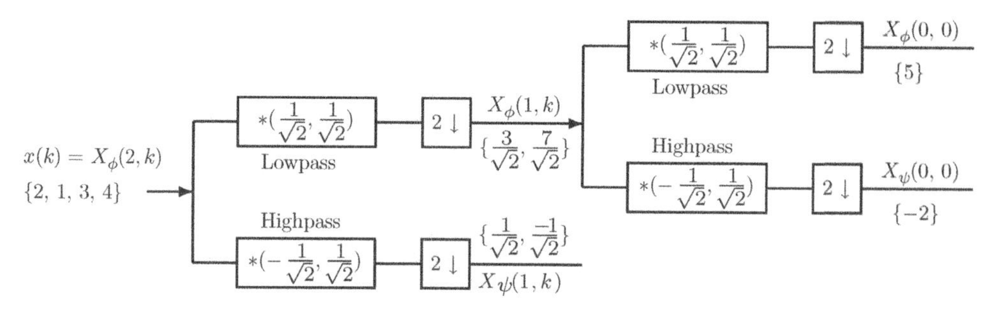

Figure 11.4 Computation of a 2-level 4-point DWT using a two-stage two-channel Haar analysis filter bank

as approximation of the input at scale 2, $X_\phi(2,k)$. The approximation or scaling coefficients at scale 1, $X_\phi(1,k)$, are computed by convolving the input $x(k)$ with the lowpass filter impulse response $\left\{\frac{1}{\sqrt{2}}, \frac{1}{\sqrt{2}}\right\}$ and then downsampling by a factor of 2. Note that the convolution output has five values. Of these five values, only the middle three values correspond to cases where both the impulse response values overlap with the given input. The first and the third values of the three middle values constitute the approximation output $X_\phi(1,k)$. These values correspond to those obtained from the definition (Equation 11.1). Similarly, the detail coefficients at scale 1, $X_\psi(1,k)$, are computed by convolving the input $x(k)$ with the highpass filter impulse response $\left\{-\frac{1}{\sqrt{2}}, \frac{1}{\sqrt{2}}\right\}$ and then downsampling by a factor of 2. The four coefficients obtained are the ones required to reconstruct the input. The approximation output $X_\phi(1,k)$ of the first stage again goes through the same process with half the values to produce $X_\phi(0,0)$ and $X_\psi(0,0)$ at the end of the second-stage analysis filter bank.

The computation of a 2-level 4-point IDWT using a two-stage two-channel Haar synthesis filter bank is shown in Figure 11.5. In the synthesis filter bank, convolution is preceded by upsampling. The outputs of the convolution in both the channels are added to reconstruct the signal at a higher scale. An upsampler inserts zeros after each sample so that its output contains double the number of samples in the input. Coefficients $X_\phi(0,0) = 5$ and $X_\psi(0,0) = -2$ are upsampled to yield $\{5, 0\}$ and $\{-2, 0\}$, respectively. These samples are convolved, respectively, with impulse responses $\left\{\frac{1}{\sqrt{2}}, \frac{1}{\sqrt{2}}\right\}$ and $\left\{\frac{1}{\sqrt{2}}, -\frac{1}{\sqrt{2}}\right\}$ to produce $\left\{\frac{5}{\sqrt{2}}, \frac{5}{\sqrt{2}}, 0\right\}$ and $\left\{\frac{-2}{\sqrt{2}}, \frac{2}{\sqrt{2}}, 0\right\}$. Adding the first two values of the last two sequences yields $\left\{\frac{3}{\sqrt{2}}, \frac{7}{\sqrt{2}}\right\}$. A similar process in the second filter bank reconstructs the input to the analysis filter bank. As shown in Figures 11.1–11.3, the operations shown in Figures 11.4 and 11.5 can be efficiently carried out.

11.1.3 1-Level Haar 2-D DWT

The computation of a 1-level 4 × 4 2-D DWT using a two-stage analysis filter bank is shown in Figure 11.6. Coefficients $\boldsymbol{X}_{\phi}$ are obtained by applying lowpass filtering and downsampling to each row of the 2-D data $\boldsymbol{x}$ followed by applying lowpass filtering and downsampling to each column of the resulting data. Coefficients $\boldsymbol{X}_{\psi}^{H}$ are obtained by applying highpass filtering and downsampling to each row of the 2-D data $\boldsymbol{x}$ followed by applying lowpass filtering and downsampling to each column of the resulting data. Coefficients $\boldsymbol{X}_{\psi}^{V}$ are obtained by applying

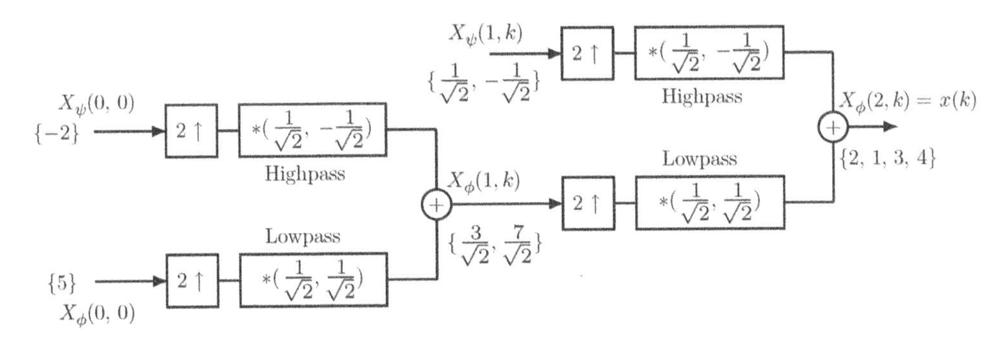

Figure 11.5 Computation of a 2-level 4-point IDWT using a two-stage two-channel Haar synthesis filter bank

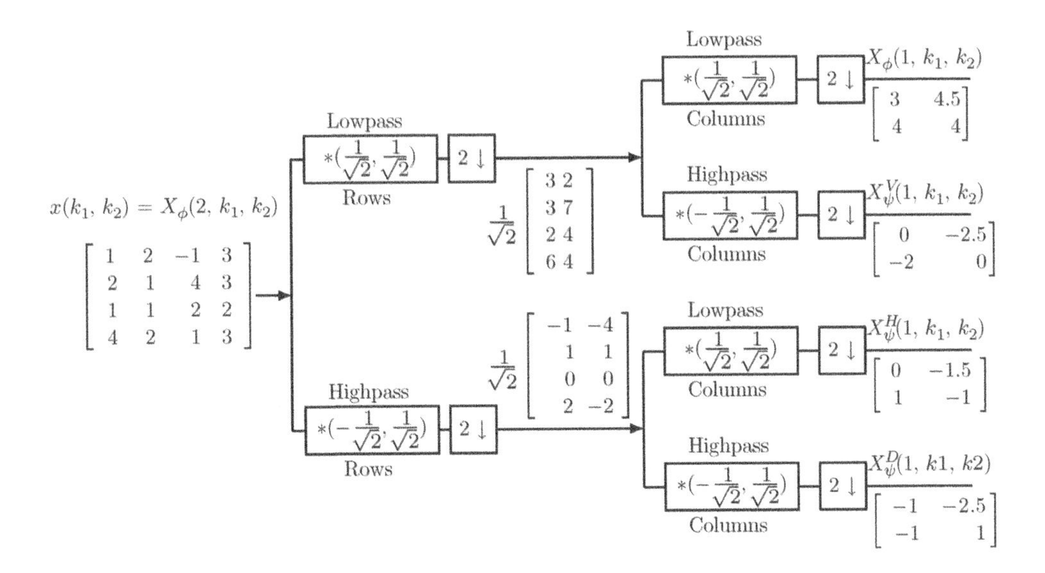

Figure 11.6 Computation of a 1-level 4 × 4 2-D Haar DWT using a two-stage filter bank

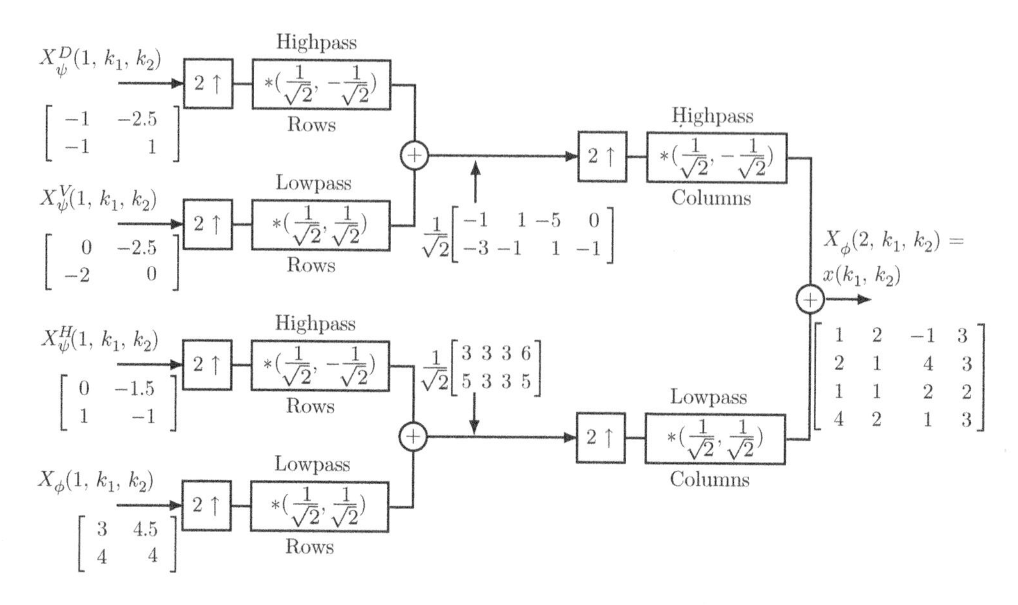

Figure 11.7 Computation of a 1-level 4×4 2-D IDWT using a two-stage filter bank

lowpass filtering and downsampling to each row of the 2-D data $\boldsymbol{x}$ followed by applying highpass filtering and downsampling to each column of the resulting data. Coefficients $\boldsymbol{X}_{\psi}^{D}$ are obtained by applying highpass filtering and downsampling to each row of the 2-D data $\boldsymbol{x}$ followed by applying highpass filtering and downsampling to each column of the resulting data. The computation of a 1-level 4×4 2-D IDWT using a two-stage synthesis filter bank is shown in Figure 11.7. The order of the computation can be changed.

11.1.4 The Signal-Flow Graph of the Fast Haar DWT Algorithms

The Haar filters constitute a two-band uniform DFT filter bank, and the implementation is similar to that of the DFT using fast algorithms. The signal-flow graph of the 8-point 3-level Haar DWT algorithm is shown in Figure 11.8. Multiplications by the factor $1/\sqrt{2}$ are merged as much as possible. The partial values at various stages of the 8-point Haar DWT algorithm are shown in Figure 11.9. For the same input, 1-level DWT coefficients

$$\frac{1}{\sqrt{2}}\{1, 5, 9, 13, -1, -1, -1, -1\}$$

are obtained by multiplying the first-half values of the stage 1 output by $1/\sqrt{2}$. 2-level DWT coefficients

$$\left\{3, 11, -2, -2, -\frac{1}{\sqrt{2}}, -\frac{1}{\sqrt{2}}, -\frac{1}{\sqrt{2}}, -\frac{1}{\sqrt{2}}\right\}$$

are obtained by multiplying the first-quarter values of the stage 2 output by $1/2$. For an N-point sequence, the number of real multiplications required for fast Haar DWT algorithm is

$$\frac{N}{2} + \frac{N}{4} + \cdots + 4 + 2 + 2 = N$$

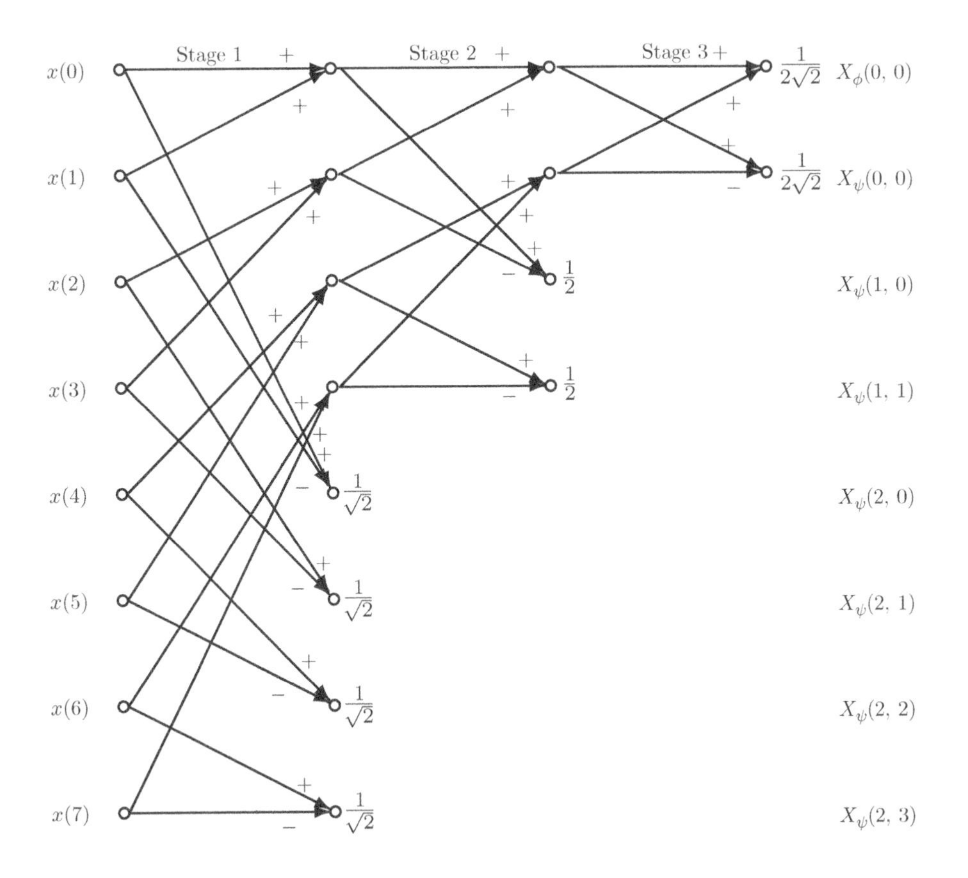

Figure 11.8 The signal-flow graph of the 1-D Haar 3-level DWT algorithm, with $N = 8$

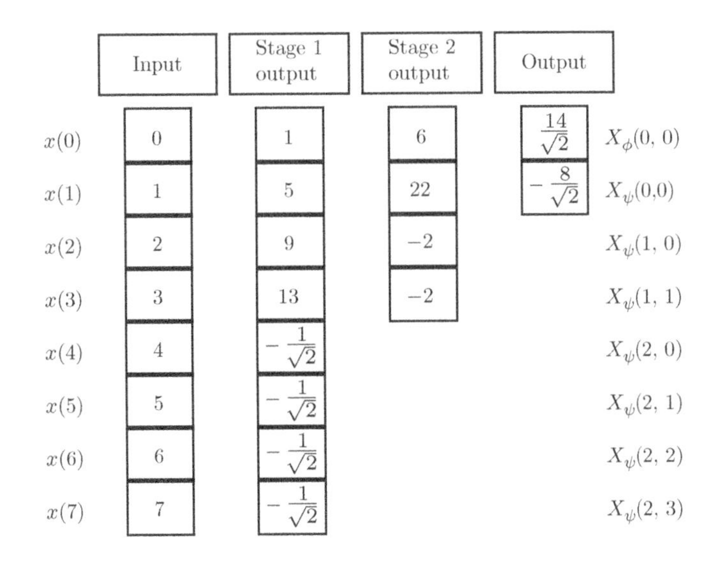

Figure 11.9 The trace of the 1-D Haar 3-level DWT algorithm, with $N = 8$

The number of real additions required is

$$N + \frac{N}{2} + \cdots + 4 + 2 = 2N - 2$$

Therefore, the computational complexity of evaluating the Haar DWT is $O(N)$. The IDWT algorithm is obtained by factorizing the inverse transform matrix. For example, with $N = 4$, the product of the two 4×4 matrices on the right-hand side equals the leftmost 4×4 matrix, which is the 2-level 4-point inverse transform matrix. The signal-flow graph of the 8-point 3-level Haar IDWT algorithm is shown in Figure 11.10. The partial values at various stages of the 8-point IDWT algorithm are shown in Figure 11.11.

$$\begin{bmatrix} \frac{1}{2} & \frac{1}{2} & \frac{1}{\sqrt{2}} & 0 \\ \frac{1}{2} & \frac{1}{2} & -\frac{1}{\sqrt{2}} & 0 \\ \frac{1}{2} & -\frac{1}{2} & 0 & \frac{1}{\sqrt{2}} \\ \frac{1}{2} & -\frac{1}{2} & 0 & -\frac{1}{\sqrt{2}} \end{bmatrix} = \begin{bmatrix} \frac{1}{\sqrt{2}} & 0 & \frac{1}{\sqrt{2}} & 0 \\ \frac{1}{\sqrt{2}} & 0 & -\frac{1}{\sqrt{2}} & 0 \\ 0 & \frac{1}{\sqrt{2}} & 0 & \frac{1}{\sqrt{2}} \\ 0 & \frac{1}{\sqrt{2}} & 0 & -\frac{1}{\sqrt{2}} \end{bmatrix} \begin{bmatrix} \frac{1}{\sqrt{2}} & \frac{1}{\sqrt{2}} & 0 & 0 \\ \frac{1}{\sqrt{2}} & -\frac{1}{\sqrt{2}} & 0 & 0 \\ 0 & 0 & 1 & 0 \\ 0 & 0 & 0 & 1 \end{bmatrix}$$

11.1.5 Haar DWT in Place

In implementing the DWT algorithm, due to the order the inputs are accessed and the outputs are stored, auxiliary memory of about half the data size is required. It is well known that from fast discrete Fourier and Walsh transform algorithms, data reordering is required to achieve an in-place computation. An in-place computation requires memory to store the given data and a few temporary variables. In this section, we present a data reordering algorithm that enables the implementation of the Haar DWT in place.

In computing the Haar DWT, sums and differences of adjacent pairs of numbers are formed (in addition to multiplying the values by the constant $1/\sqrt{2}$), and the sums are written in the top half of the array and the differences in the bottom half. A straightforward implementation of this procedure for an input with N values requires auxiliary memory to hold $N/2$ values in order to prevent overwriting of necessary input values. The bottom half output values have to be stored temporarily in the auxiliary memory in computing the DWT, as shown in the dashed box in Figure 11.12(a).

In implementing the DWT in place, the input values have to be reordered so that the even-indexed values are placed in the top half of the array and the odd-indexed values in the bottom half, as shown in Figure 11.12(b). The indices of the input values are reordered as

$$\{0, 1, 2, 3, 4, 5, 6, 7\} \rightarrow \{0, 2, 4, 6, 1, 3, 5, 7\}$$

Then, the transform can be computed in place. For example, the pair of input values $\{1, 7\}$ produces the output $\{8/\sqrt{2}, -6/\sqrt{2}\}$ in place. The reordering of N values can be implemented in a MATLAB assignment statement of the form

```
x(0:N-1) = [x(0:2:N-1),x(1:2:N-1)];
```

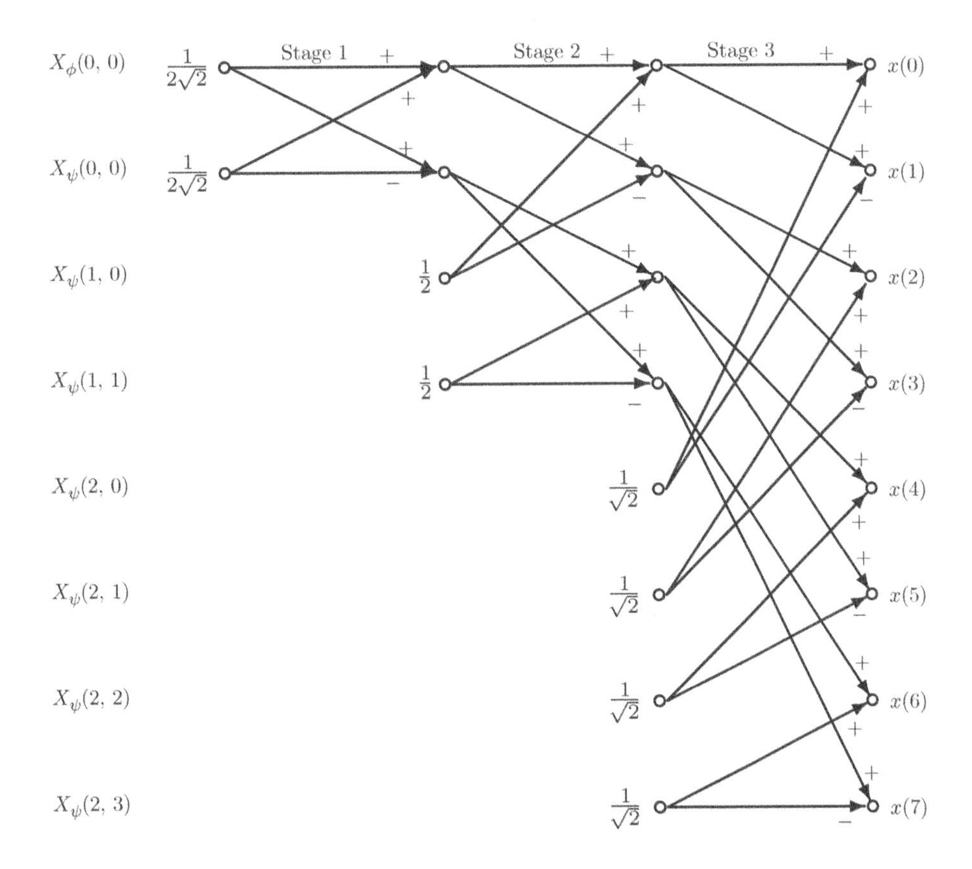

Figure 11.10 The signal-flow graph of the 1-D 3-level Haar IDWT algorithm, with $N = 8$

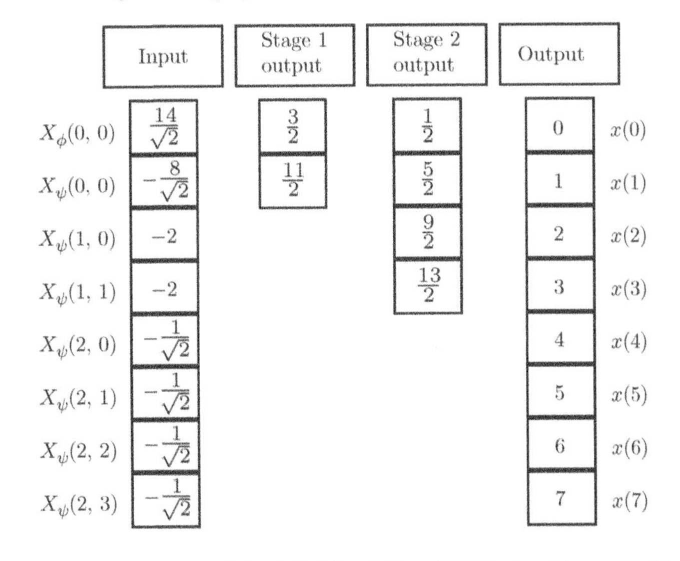

Figure 11.11 The trace of the 1-D 3-level Haar IDWT algorithm, with $N = 8$

(a)

	Input		Output	
$x(0)$	1		$\frac{8}{\sqrt{2}}$	$X_\phi(2,0)$
$x(1)$	7		$-\frac{3}{\sqrt{2}}$	$X_\phi(2,1)$
$x(2)$	−5		$-\frac{3}{\sqrt{2}}$	$X_\phi(2,2)$
$x(3)$	2		$-\frac{7}{\sqrt{2}}$	$X_\phi(2,3)$
$x(4)$	−6	$-\frac{6}{\sqrt{2}}$	$-\frac{6}{\sqrt{2}}$	$X_\psi(2,0)$
$x(5)$	3	$-\frac{7}{\sqrt{2}}$	$-\frac{7}{\sqrt{2}}$	$X_\psi(2,1)$
$x(6)$	1	$-\frac{9}{\sqrt{2}}$	$-\frac{9}{\sqrt{2}}$	$X_\psi(2,2)$
$x(7)$	−8	$\frac{9}{\sqrt{2}}$	$\frac{9}{\sqrt{2}}$	$X_\psi(2,3)$

(b)

	Input	Output
$x(0)$	1	$\frac{8}{\sqrt{2}}$
$x(2)$	−5	$-\frac{3}{\sqrt{2}}$
$x(4)$	−6	$-\frac{3}{\sqrt{2}}$
$x(6)$	1	$-\frac{7}{\sqrt{2}}$
$x(1)$	7	$-\frac{6}{\sqrt{2}}$
$x(3)$	2	$-\frac{7}{\sqrt{2}}$
$x(5)$	3	$-\frac{9}{\sqrt{2}}$
$x(7)$	−8	$\frac{9}{\sqrt{2}}$

Figure 11.12 The trace of the 1-D 1-level Haar DWT algorithm with (a) and without (b) auxiliary memory

11.2 Symmetrical Extension of the Data

The DWT coefficients are computed by convolving the filter coefficients with the input data. At either end of the data, the filter coefficients fall outside the defined data. Therefore, suitable extension of the data is required at the boundaries. Of course, the data can be assumed to be zero outside the defined values. This mode of data extension is called zero-padding. Another mode is periodic extension. The data can also be extended by boundary value replication. Other modes are also possible. The mode that suits the particular application should be selected.

Two symmetric type of extensions of data, which are simple and effective, are often used. In symmetric extension, the data are extended by mirror reflection. Assume that the input is

$$\{x(0) = 1, x(1) = 3, x(2) = 2, x(3) = 4\}$$

The two types of boundary extension of the signal, whole-point symmetry and half-point symmetry, are shown in Figure 11.13. The two methods differ in that in one case the first and last samples, $x(0)$ and $x(N-1)$, are repeated and in the other case they are not repeated. The first method of extension is to make $x(-1) = x(0)$ and $x(N) = x(N-1)$. In the case of half-point symmetry shown in Figure 11.13(a), the first and the last values of a sequence are replicated. The extension is carried out at both the ends to obtain a symmetrical signal of length $2N$ with symmetry about $n = -0.5$ and $n = N - 0.5$. For the specific example, the extended signal becomes

$$x_h(n) = \{x(-2) = 3, x(-1) = 1, \quad \boldsymbol{x(0) = 1, x(1) = 3, x(2) = 2, x(3) = 4}, \\ x(4) = 4, x(5) = 2\}$$

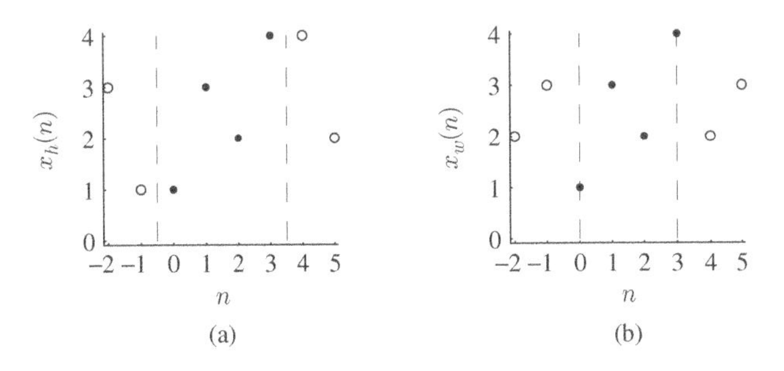

Figure 11.13 (a) A signal with half-point extension; (b) a signal with whole-point extension. The dots represent the given signal values and the unfilled circles are the extended signal values

In general, the half-point symmetry extension of the signal of length $N = 8$, where N is an even number,

$$\{x(0), x(1), x(2), x(3), x(4), x(5), x(6), x(7)\}$$

is

$$\boldsymbol{x}_h = \{x(0), x(1), x(2), x(3), x(4), x(5), x(6), x(7), x(7), x(6),$$
$$x(5), x(4), x(3), x(2), x(1), x(0)\}$$

of length $2N = (2)(8) = 16$. This input convolved with a symmetric lowpass filter impulse response, of an even length, yields a symmetric output of the form

$$\{y(0), y(1), y(2), y(3)|y(3), y(2), y(1), y(0)\} \tag{11.4}$$

The symmetry in the highpass filter output is of the form

$$\{w(0), w(1), w(2), w(3)| - w(3), -w(2), -w(1), -w(0)\} \tag{11.5}$$

The first four values of the output of the two symmetrical sequences constitute the output.

$$\{y(0), y(1), y(2), y(3)|w(0), w(1), w(2), w(3)\}$$

For even-length filters, half-point symmetry is preferred. The initial output values are computed using the first few elements of the input, which is a simpler and effective approximation at the borders. It is easy to verify the symmetry and periodicity of the output using $\boldsymbol{x}_h$ and the scaled Haar analysis filters

$$l(n) = \{1, 1\} \quad \text{and} \quad h(n) = \{-1, 1\}$$

The first and the last (8th) lowpass outputs are computed using the same values, $x(0), x(1)$, and the outputs are the same, $y(0)$. The fourth and the fifth lowpass outputs are computed using the same values, $x(6), x(7)$, and the outputs are the same, $y(3)$. Because of the sign change in the filter coefficients, although the input elements are the same, the highpass outputs are the negative of the other due to the change in the order of the input values.

The second method of extension is to make $x(-1) = x(1)$ and $x(N) = x(N-2)$. In the case of whole-point symmetry shown in Figure 11.13(b), the first and the last values of a sequence are not replicated. The extension is carried out at both the ends to obtain a symmetrical signal of length $2N - 2$ with symmetry about $n = 0$ and $n = N - 1$. For the specific example, the extended signal becomes

$$x_w(n) = \{x(-1) = 3, \quad \boldsymbol{x(0) = 1, x(1) = 3, x(2) = 2, x(3) = 4}, \quad x(4) = 2\}$$

In general, let the signal be

$$\{x(0), x(1), x(2), x(3), x(4), x(5), x(6), x(7)\}$$

of length $N = 8$, where N is an even number. Then, the whole-point symmetry extension of this signal is

$$\boldsymbol{x}_w = \{x(0), x(1), x(2), x(3), x(4), x(5), x(6), x(7), x(6), x(5), x(4), x(3), x(2), x(1)\}$$

of length $2N - 2 = (2)(8) - 2 = 14$. This input convolved with a symmetric lowpass filter impulse response, of an odd length, yields a symmetric output of the form

$$\{y(0), y(1), y(2), y(3) | y(3), y(2), y(1)\} \tag{11.6}$$

The symmetry in the highpass filter output is of the form

$$\{w(0), w(1), w(2), w(3) | w(2), w(1), w(0)\} \tag{11.7}$$

The first four values of the output of the two symmetrical sequences constitute the output.

$$\{y(0), y(1), y(2), y(3) | w(0), w(1), w(2), w(3)\}$$

For odd-length filters, whole-point symmetry is preferred. It is easy to verify the symmetry and periodicity of the output using the scaled 5/3 filters

$$l(n) = \{-1, 2, 6, 2, -1\} \quad \text{and} \quad h(n) = \{2, -4, 2\}$$

Input value $x(0)$ corresponds to the filter coefficient 6 for computing the first lowpass output, $y(0)$. Input value $x(0)$ corresponds to the first filter coefficient 2 for computing the first highpass output, $w(0)$.

11.3 Implementation of the DWT with the D4 Filter

Consider the implementation of the DWT using the Daubechies orthogonal filter of length 4. While Haar filters compute simple averages and differences, longer length filters compute weighted averages and differences. The impulse response of the lowpass analysis filter is

$$\{l_3, l_2, l_1, l_0\} = \{0.4830, 0.8365, 0.2241, -0.1294\}$$

For orthogonal DWT, only the lowpass filter is designed, and the other three are derived from it. Therefore, only one set of filters is stored in the implementation. The 8-point 1-level transform matrix, multiplied by the input, is

$$\begin{bmatrix} l_3 & l_2 & l_1 & l_0 & 0 & 0 & 0 & 0 \\ 0 & 0 & l_3 & l_2 & l_1 & l_0 & 0 & 0 \\ 0 & 0 & 0 & 0 & l_3 & l_2 & l_1 & l_0 \\ l_1 & l_0 & 0 & 0 & 0 & 0 & l_3 & l_2 \\ h_3 & h_2 & h_1 & h_0 & 0 & 0 & 0 & 0 \\ 0 & 0 & h_3 & h_2 & h_1 & h_0 & 0 & 0 \\ 0 & 0 & 0 & 0 & h_3 & h_2 & h_1 & h_0 \\ h_1 & h_0 & 0 & 0 & 0 & 0 & h_3 & h_2 \end{bmatrix} \begin{bmatrix} x(0) \\ x(1) \\ x(2) \\ x(3) \\ x(4) \\ x(5) \\ x(6) \\ x(7) \end{bmatrix} \tag{11.8}$$

Note the wrap-around in the last row of both the lowpass and highpass portions, which does not occur in the Haar transform matrix. The longer the filters, the more is the wrap-around. If the input data are periodic (not the usual case), there is no problem.

Assuming that the number of input samples is even, each of the two decompositions should have half the number of elements, and we should be able to reconstruct the input from its decomposed components. The problem is that we use the first four values of the input data to find the first value for each of the lowpass and highpass output portions,

$$x_0 l_3 + x_1 l_2 + x_2 l_1 + x_3 l_0 = X_\phi(2,0)$$
$$x_0 h_3 + x_1 h_2 + x_2 h_1 + x_3 h_0 = X_\psi(2,0)$$

Now, we need four equations to solve for four unknowns $\{x_0, x_1, x_2, x_3\}$ in the reconstruction process. Therefore, we have to insert two values at the beginning of the data by suitable extension. Then, we get the required four equations.

We need an extension at the end of the data as well to compute four output values with $N = 8$. With two values inserted at the beginning and end using half-point symmetry, we get

$$\begin{bmatrix} l_3 & l_2 & l_1 & l_0 & 0 & 0 & 0 & 0 & 0 & 0 & 0 & 0 \\ 0 & 0 & l_3 & l_2 & l_1 & l_0 & 0 & 0 & 0 & 0 & 0 & 0 \\ 0 & 0 & 0 & 0 & l_3 & l_2 & l_1 & l_0 & 0 & 0 & 0 & 0 \\ 0 & 0 & 0 & 0 & 0 & 0 & l_3 & l_2 & l_1 & l_0 & 0 & 0 \\ 0 & 0 & 0 & 0 & 0 & 0 & 0 & 0 & l_3 & l_2 & l_1 & l_0 \\ h_3 & h_2 & h_1 & h_0 & 0 & 0 & 0 & 0 & 0 & 0 & 0 & 0 \\ 0 & 0 & h_3 & h_2 & h_1 & h_0 & 0 & 0 & 0 & 0 & 0 & 0 \\ 0 & 0 & 0 & 0 & h_3 & h_2 & h_1 & h_0 & 0 & 0 & 0 & 0 \\ 0 & 0 & 0 & 0 & 0 & 0 & h_3 & h_2 & h_1 & h_0 & 0 & 0 \\ 0 & 0 & 0 & 0 & 0 & 0 & 0 & 0 & h_3 & h_2 & h_1 & h_0 \end{bmatrix} \begin{bmatrix} x(1) \\ x(0) \\ x(0) \\ x(1) \\ x(2) \\ x(3) \\ x(4) \\ x(5) \\ x(6) \\ x(7) \\ x(7) \\ x(6) \end{bmatrix} \tag{11.9}$$

Example 11.1 Compute the 1-level DWT of the input $x(n)$ using the D4 filter by the convolution approach. Assume half-point symmetry of the data. Compute the IDWT of the DWT coefficients and verify that the input $x(n)$ is reconstructed.

$$\boldsymbol{x} = \{1, 3, 2, 4\}$$

Solution

DWT: The input, with the half-point symmetry extension, is

$$\boldsymbol{x}_h = \{3, 1, \quad 1, 3, 2, 4, \quad 4, 2\}$$

The convolution of $\boldsymbol{x}_h$ with the analysis lowpass filter impulse response,

$$\{l_0, l_1, l_2, l_3\} = \{-0.1294, 0.2241, 0.8365, 0.4830\}$$

after downsampling, is

$$\{X_\phi(1, -1), X_\phi(1, 0), X_\phi(1, 1)\} = \{2.1213, 2.9232, 4.9497\}$$

The convolution of $\boldsymbol{x}_h$ with the analysis highpass filter impulse response,

$$\{h_0, h_1, h_2, h_3\} = \{-0.4830, 0.8365, -0.2241 - 0.1294\}$$

after downsampling, is

$$\{X_\psi(1, -1), X_\psi(1, 0), X_\psi(1, 1)\} = \{-1.2247, -1.0607, 1.2247\}$$

IDWT: The convolution of the lowpass output alternately with the even- and odd-indexed coefficients $\{0.4830, 0.2241\}$ and $\{0.8365, -0.1294\}$ of the synthesis lowpass filter impulse response

$$\{\tilde{l}_0, \tilde{l}_1, \tilde{l}_2, \tilde{l}_3\} = \{0.4830, 0.8365, 0.2241, -0.1294\}$$

is

$$\{1.8873, 2.1708, 3.0458, 3.7623\}$$

Note that the convolution with the even-indexed coefficients gives the even-indexed output values, and the other convolution gives the odd-indexed output values. The convolution of the highpass output alternately with the even- and odd-indexed coefficients $\{-0.1294, 0.8365\}$ and $\{-0.2241, -0.4830\}$ of the synthesis highpass filter impulse response

$$\{\tilde{h}_0, \tilde{h}_1, \tilde{h}_2, \tilde{h}_3\} = \{-0.1294, -0.2241, 0.8365, -0.4830\}$$

is

$$\{-0.8873, 0.8292, -1.0458, 0.2377\}$$

Adding the last two output sequences, we get the reconstructed input $\{1, 3, 2, 4\}$. ∎

11.4 Implementation of the DWT with Symmetrical Filters

11.4.1 5/3 Spline Filter

Symmetrical filters provide an effective solution to the boundary problem. With symmetrical filters and symmetrical extension of the data, the convolution output is also symmetrical and periodic. Therefore, the coefficient expansion problem is effectively tackled. Of course, symmetrical filters also provide linear phase response. This ensures that the image contents, such as ridges, do not get shifted between subbands. The 8×8 biorthogonal 5/3 spline filter transform matrix, multiplied by the input, is

$$\frac{\sqrt{2}}{8}\begin{bmatrix} 6 & 2 & -1 & 0 & 0 & 0 & -1 & 2 \\ -1 & 2 & 6 & 2 & -1 & 0 & 0 & 0 \\ 0 & 0 & -1 & 2 & 6 & 2 & -1 & 0 \\ -1 & 0 & 0 & 0 & -1 & 2 & 6 & 2 \\ 2 & -4 & 2 & 0 & 0 & 0 & 0 & 0 \\ 0 & 0 & 2 & -4 & 2 & 0 & 0 & 0 \\ 0 & 0 & 0 & 0 & 2 & -4 & 2 & 0 \\ 2 & 0 & 0 & 0 & 0 & 0 & 2 & -4 \end{bmatrix}\begin{bmatrix} x(0) \\ x(1) \\ x(2) \\ x(3) \\ x(4) \\ x(5) \\ x(6) \\ x(7) \end{bmatrix}$$

Using the whole-point symmetry extension, we get

$$\frac{\sqrt{2}}{8}\begin{bmatrix} -1 & 2 & 6 & 2 & -1 & 0 & 0 & 0 & 0 & 0 & 0 \\ 0 & 0 & -1 & 2 & 6 & 2 & -1 & 0 & 0 & 0 & 0 \\ 0 & 0 & 0 & 0 & -1 & 2 & 6 & 2 & -1 & 0 & 0 \\ 0 & 0 & 0 & 0 & 0 & 0 & -1 & 2 & 6 & 2 & -1 \\ 0 & 0 & 2 & -4 & 2 & 0 & 0 & 0 & 0 & 0 & 0 \\ 0 & 0 & 0 & 0 & 2 & -4 & 2 & 0 & 0 & 0 & 0 \\ 0 & 0 & 0 & 0 & 0 & 0 & 2 & -4 & 2 & 0 & 0 \\ 0 & 0 & 0 & 0 & 0 & 0 & 0 & 0 & 2 & -4 & 2 \end{bmatrix}\begin{bmatrix} x(2) \\ x(1) \\ x(0) \\ x(1) \\ x(2) \\ x(3) \\ x(4) \\ x(5) \\ x(6) \\ x(7) \\ x(6) \end{bmatrix} \tag{11.10}$$

Example 11.2 Compute the 1-level DWT of the input $x(n)$ using the $5/3$ spline filter by the convolution approach. Assume whole-point symmetry of the data. Compute the IDWT of the DWT coefficients and verify that the input $x(n)$ is reconstructed.

$$\boldsymbol{x} = \{1, 3, 2, 4, 3, -1, 4, 2\}$$

Solution

DWT: The input, with the whole-point symmetry extension, is

$$\boldsymbol{x}_w = \{2, 3, \quad 1, 3, 2, 4, 3, -1, 4, 2, \quad 4\}$$

While convolution operation is used in practical implementation, we have also used the transform matrix for clarity.

$$\begin{bmatrix} X_\phi(2,0) \\ X_\phi(2,1) \\ X_\phi(2,2) \\ X_\phi(2,3) \\ X_\psi(2,0) \\ X_\psi(2,0) \\ X_\psi(2,0) \\ X_\psi(2,0) \end{bmatrix} = \begin{bmatrix} -1 & 2 & 6 & 2 & -1 & 0 & 0 & 0 & 0 & 0 & 0 \\ 0 & 0 & -1 & 2 & 6 & 2 & -1 & 0 & 0 & 0 & 0 \\ 0 & 0 & 0 & 0 & -1 & 2 & 6 & 2 & -1 & 0 & 0 \\ 0 & 0 & 0 & 0 & 0 & 0 & -1 & 2 & 6 & 2 & -1 \\ 0 & 0 & 2 & -4 & 2 & 0 & 0 & 0 & 0 & 0 & 0 \\ 0 & 0 & 0 & 0 & 2 & -4 & 2 & 0 & 0 & 0 & 0 \\ 0 & 0 & 0 & 0 & 0 & 0 & 2 & -4 & 2 & 0 & 0 \\ 0 & 0 & 0 & 0 & 0 & 0 & 0 & 0 & 2 & -4 & 2 \end{bmatrix} \begin{bmatrix} 2 \\ 3 \\ 1 \\ 3 \\ 2 \\ 4 \\ 3 \\ -1 \\ 4 \\ 2 \\ 4 \end{bmatrix} = \begin{bmatrix} 14 \\ 22 \\ 18 \\ 19 \\ -6 \\ -6 \\ 18 \\ 8 \end{bmatrix} \tag{11.11}$$

Note that we have suppressed the factor $\frac{\sqrt{2}}{8}$ so that the example can be easily worked out by hand calculation. The convolution of $\boldsymbol{x}_w$ with the scaled analysis lowpass filter impulse response,

$$\{l_{-2}, l_{-1}, l_0, l_1, l_2\} = \{-1, 2, 6, 2, -1\}$$

after downsampling, is

$$\{X_\phi(2,0), X_\phi(2,1), X_\phi(2,2), X_\phi(2,3)\} = \{14, 22, 18, 19\}$$

With periodic extension (Equation 11.6), the lowpass output is

$$\{14, 22, 18, 19, \quad 19\}$$

The convolution of $\boldsymbol{x}_w$ with the scaled analysis highpass filter impulse response,

$$\{h_{-1}, h_0, h_1\} = \{2, -4, 2\}$$

after downsampling, is

$$\{X_\psi(2,0), X_\psi(2,1), X_\psi(2,2), X_\psi(2,3)\} = \{-6, -6, 18, 8\}$$

With periodic extension (Equation 11.7), the highpass output is

$$\{-6, \quad -6, -6, 18, 8, \quad 18\}$$

IDWT: The convolution of the lowpass output alternately with the even- and odd-indexed coefficients $\{4\}$ and $\{2, 2\}$ of the scaled synthesis lowpass filter impulse response $\tilde{\boldsymbol{l}} = \{2, 4, 2\}$ yields the even- and odd-indexed values of the output of the lowpass channel

$$\{56, 72, 88, 80, 72, 74, 76, 76\}$$

Using matrix formulation of convolution,

$$\begin{bmatrix} x_l(0) \\ x_l(1) \\ x_l(2) \\ x_l(3) \\ x_l(4) \\ x_l(5) \\ x_l(6) \\ x_l(7) \end{bmatrix} = \begin{bmatrix} 4 & 0 & 0 & 0 & 0 \\ 2 & 2 & 0 & 0 & 0 \\ 0 & 4 & 0 & 0 & 0 \\ 0 & 2 & 2 & 0 & 0 \\ 0 & 0 & 4 & 0 & 0 \\ 0 & 0 & 2 & 2 & 0 \\ 0 & 0 & 0 & 4 & 0 \\ 0 & 0 & 0 & 2 & 2 \end{bmatrix} \begin{bmatrix} 14 \\ 22 \\ 18 \\ 19 \\ 19 \end{bmatrix} = \begin{bmatrix} 56 \\ 72 \\ 88 \\ 80 \\ 72 \\ 74 \\ 76 \\ 76 \end{bmatrix} \tag{11.12}$$

This transform matrix is directly obtained from the columns 1,2,3, and 4 of the transpose of the inverse transform matrix of the 5/3 filter (Example 10.1).

The convolution of the highpass output alternately with the even- and odd-indexed coefficients $\{2, 2\}$ and $\{1, -6, 1\}$ of the scaled synthesis highpass filter impulse response $\tilde{\boldsymbol{h}} = \{1, 2, -6, 2, 1\}$ yields the even- and odd-indexed values of the output of the highpass channel

$$\{-24, 24, -24, 48, 24, -106, 52, -12\}$$

Using matrix formulation of convolution,

$$\begin{bmatrix} x_h(0) \\ x_h(1) \\ x_h(2) \\ x_h(3) \\ x_h(4) \\ x_h(5) \\ x_h(6) \\ x_h(7) \end{bmatrix} = \begin{bmatrix} 2 & 2 & 0 & 0 & 0 & 0 \\ 1 & -6 & 1 & 0 & 0 & 0 \\ 0 & 2 & 2 & 0 & 0 & 0 \\ 0 & 1 & -6 & 1 & 0 & 0 \\ 0 & 0 & 2 & 2 & 0 & 0 \\ 0 & 0 & 1 & -6 & 1 & 0 \\ 0 & 0 & 0 & 2 & 2 & 0 \\ 0 & 0 & 0 & 1 & -6 & 1 \end{bmatrix} \begin{bmatrix} -6 \\ -6 \\ -6 \\ 18 \\ 8 \\ 18 \end{bmatrix} = \begin{bmatrix} -24 \\ 24 \\ -24 \\ 48 \\ 24 \\ -106 \\ 52 \\ -12 \end{bmatrix} \tag{11.13}$$

This transform matrix is directly obtained from the columns 5,6,7, and 8 of the transpose of the inverse transform matrix of the 5/3 filter (Example 10.1). Adding the last two output sequences, we get the scaled original input $32\{1, 3, 2, 4, 3, -1, 4, 2\}$.

Figure 11.14(a) shows half cycle of the cosine wave $x(n) = \cos\left(\frac{2\pi}{32}n\right)$. Figures 11.14(b) and (c) shows the reconstructed signal, respectively, using its DWT approximation coefficients alone obtained using 5/3 spline filter with periodic border extension and with whole-point symmetry border extension. The least square errors with periodic border extension and with whole-point symmetry border extension are, respectively, 0.7473 and 0.2607. Evidently, the whole-point symmetry border extension results in a better approximation of the signal. ■

11.4.2 CDF 9/7 Filter

Example 11.3 Compute the 1-level DWT of the input $x(n)$ using the CDF 9/7 filter by the convolution approach. Assume whole-point symmetry of the data. Compute the IDWT of the

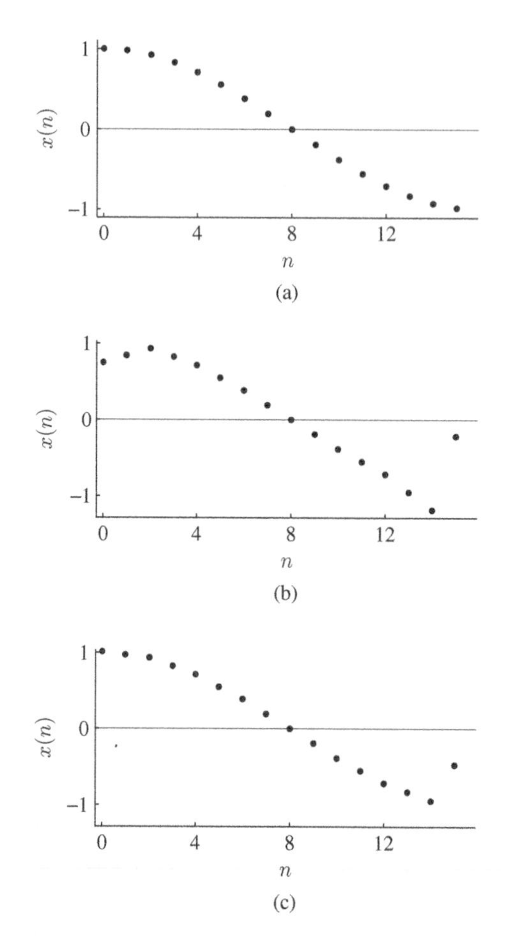

Figure 11.14 (a) Half cycle of the cosine wave $x(n) = \cos\left(\frac{2\pi}{32}n\right)$; (b) reconstructed signal from its DWT approximation coefficients alone obtained using 5/3 spline filter with periodic border extension; (c) reconstructed signal with whole-point symmetry border extension

DWT coefficients and verify that the input $x(n)$ is reconstructed.

$$\boldsymbol{x} = \{2, -4, 5, 3, 0, 1, 7, -5, 2, -6, 3, 1, -8, 2, 0, 3\}$$

Solution

DWT: The input, with the whole-point symmetry extension, is

$$\boldsymbol{x}_w = \{0, 3, 5, -4, \quad 2, -4, 5, 3, 0, 1, 7, -5, 2, -6, 3, 1, -8, 2, 0, 3, \quad 0, 2 - 8, 1\}$$

The convolution of $\boldsymbol{x}_w$ with the analysis lowpass filter impulse response,

$$\{l_{-4}, l_{-3}, l_{-2}, l_{-1}, l_0, l_1, l_2, l_3, l_4\}$$
$$= \{0.0378, -0.0238, -0.1106, 0.3774, 0.8527, 0.3774, -0.1106, -0.0238, 0.0378\}$$

after downsampling, is

$$\{X_\phi(3,0), X_\phi(3,1), X_\phi(3,2), X_\phi(3,3), X_\phi(3,4), X_\phi(3,5), X_\phi(3,6), X_\phi(3,7)\}$$
$$= \{-2.5632, 4.1903, 0.5481, 4.6122, -3.9026, 1.6712, -5.8740, 2.5113\}$$

With periodic extension, the lowpass output is

$$\{0.5481, 4.1903, -2.5632, 4.1903, 0.5481, 4.6122, -3.9026, 1.6712, -5.8740,$$
$$2.5113, 2.5113, -5.8740\}$$

The convolution of $\boldsymbol{x}_w$ with the analysis highpass filter impulse response,

$$\{h_{-2}, h_{-1}, h_0, h_1, h_2, h_3, h_4\}$$
$$= \{-0.0645, 0.0407, 0.4181, -0.7885, 0.4181, 0.0407, -0.0645\}$$

after downsampling, is

$$\{X_\psi(3,0), X_\psi(3,1), X_\psi(3,2), X_\psi(3,3)X_\psi(3,4), X_\psi(3,5), X_\psi(3,6), X_\psi(3,7)\}$$
$$= \{5.7172, -0.9779, 1.6050, 7.3082, 6.7232, -3.1708, -4.9526, -1.1701\}$$

With periodic extension, the highpass output is

$$\{-0.9779, 5.7172, 5.7172, -0.9779, 1.6050, 7.3082, 6.7232, -3.1708,$$
$$-4.9526, -1.1701, -4.9526, -3.1708\}$$

IDWT: The convolution of the lowpass output alternately with the even- and odd-indexed coefficients

$$\{-0.0407, 0.7885, -0.0407\}$$

and

$$\{-0.0645, 0.4181, 0.4181, -0.0645\}$$

of the synthesis lowpass filter impulse response

$$\tilde{l} = \{-0.0645, -0.0407, 0.4181, 0.7885, 0.4181, -0.0407, -0.0645\}$$

yields the even- and odd-indexed values of the output of the lowpass channel.

$$\{-2.3620, 0.3745, 3.3860, 1.8488, 0.0740, 2.1389, 3.7732, 0.1535,$$
$$-3.3328, -0.8515, 1.7155, -1.6674, -4.8018, -1.6759, 2.1170, 2.8581\}$$

The convolution of the highpass output alternately with the even- and odd-indexed coefficients

$$\{-0.0238, 0.3774, 0.3774, -0.0238,\}$$

and

$$\{-0.0378, 0.1106, -0.8527, 0.1106, -0.0378\}$$

of the synthesis highpass filter impulse response

$$\tilde{h} = \{-0.0378, -0.0238, 0.1106, 0.3774, -0.8527, 0.3774, 0.1106, -0.0238, -0.0378\}$$

yields the even- and indexed values of the output of the highpass channel.

$$\{4.3620, -4.3745, 1.6140, 1.1512, -0.0740, -1.1389, 3.2268, -5.1535,$$
$$5.3328, -5.1485, 1.2845, 2.6674, -3.1982, 3.6759, -2.1170, 0.1419\}$$

Adding the last two output sequences, we get back the reconstructed input

$$\{2, -4, 5, 3, 0, 1, 7, -5, 2, -6, 3, 1, -8, 2, 0, 3\}.$$

■

11.4.3 4/4 Spline Filter

The 8×8 biorthogonal 4/4 spline filter transform matrix, multiplied by the input, is

$$\frac{\sqrt{2}}{8}\begin{bmatrix} 6 & 6 & -2 & 0 & 0 & 0 & 0 & -2 \\ 0 & -2 & 6 & 6 & -2 & 0 & 0 & 0 \\ 0 & 0 & 0 & -2 & 6 & 6 & -2 & 0 \\ -2 & 0 & 0 & 0 & 0 & -2 & 6 & 6 \\ -3 & 3 & -1 & 0 & 0 & 0 & 0 & 1 \\ 0 & 1 & -3 & 3 & -1 & 0 & 0 & 0 \\ 0 & 0 & 0 & 1 & -3 & 3 & -1 & 0 \\ -1 & 0 & 0 & 0 & 0 & 1 & -3 & 3 \end{bmatrix} \begin{bmatrix} x(0) \\ x(1) \\ x(2) \\ x(3) \\ x(4) \\ x(5) \\ x(6) \\ x(7) \end{bmatrix}$$

Using the half-point symmetry extension, we get

$$\frac{\sqrt{2}}{8}\begin{bmatrix} -2 & 6 & 6 & -2 & 0 & 0 & 0 & 0 & 0 & 0 \\ 0 & 0 & -2 & 6 & 6 & -2 & 0 & 0 & 0 & 0 \\ 0 & 0 & 0 & 0 & -2 & 6 & 6 & -2 & 0 & 0 \\ 0 & 0 & 0 & 0 & 0 & 0 & -2 & 6 & 6 & -2 \\ 1 & -3 & 3 & -1 & 0 & 0 & 0 & 0 & 0 & 0 \\ 0 & 0 & 1 & -3 & 3 & -1 & 0 & 0 & 0 & 0 \\ 0 & 0 & 0 & 0 & 1 & -3 & 3 & -1 & 0 & 0 \\ 0 & 0 & 0 & 0 & 0 & 0 & 1 & -3 & 3 & -1 \end{bmatrix} \begin{bmatrix} x(0) \\ x(0) \\ x(1) \\ x(2) \\ x(3) \\ x(4) \\ x(5) \\ x(6) \\ x(7) \\ x(7) \end{bmatrix} \quad (11.14)$$

Example 11.4 Compute the 1-level DWT of the input $x(n)$ using the 4/4 spline filter by the convolution approach. Assume half-point symmetry of the data. Compute the IDWT of the DWT coefficients and verify that the input $x(n)$ is reconstructed.

$$\boldsymbol{x} = \{1, 3, 2, 4, 3, -1, 4, 2\}$$

Solution

DWT: The input, with the half-point symmetry extension, is

$$\boldsymbol{x}_h = \{1, \quad 1,3,2,4,3,-1,4,2, \quad 2\}$$

$$\begin{bmatrix} X_\phi(2,0) \\ X_\phi(2,1) \\ X_\phi(2,2) \\ X_\phi(2,3) \\ X_\psi(2,0) \\ X_\psi(2,0) \\ X_\psi(2,0) \\ X_\psi(2,0) \end{bmatrix} = \begin{bmatrix} -2 & 6 & 6 & -2 & 0 & 0 & 0 & 0 & 0 & 0 \\ 0 & 0 & -2 & 6 & 6 & -2 & 0 & 0 & 0 & 0 \\ 0 & 0 & 0 & 0 & -2 & 6 & 6 & -2 & 0 & 0 \\ 0 & 0 & 0 & 0 & 0 & 0 & -2 & 6 & 6 & -2 \\ 1 & -3 & 3 & -1 & 0 & 0 & 0 & 0 & 0 & 0 \\ 0 & 0 & 1 & -3 & 3 & -1 & 0 & 0 & 0 & 0 \\ 0 & 0 & 0 & 0 & 1 & -3 & 3 & -1 & 0 & 0 \\ 0 & 0 & 0 & 0 & 0 & 0 & 1 & -3 & 3 & -1 \end{bmatrix} \begin{bmatrix} 1 \\ 1 \\ 3 \\ 2 \\ 4 \\ 3 \\ -1 \\ 4 \\ 2 \\ 2 \end{bmatrix} = \begin{bmatrix} 18 \\ 24 \\ -4 \\ 34 \\ 5 \\ 6 \\ -12 \\ -9 \end{bmatrix}$$

Note that we have suppressed the factor $\frac{\sqrt{2}}{8}$ so that the example can be easily worked out by hand calculation. The convolution of $\boldsymbol{x}_h$ with the scaled analysis lowpass filter impulse response,

$$\{l(-1), l(0), l(1), l(2)\} = \{-2, 6, 6, -2\}$$

after downsampling, is

$$\{X_\phi(2,0), X_\phi(2,1), X_\phi(2,2), X_\phi(2,3)\} = \{18, 24, -4, 34\}$$

With periodic extension (Equation 11.4), the lowpass output is

$$\{18, \quad 18, 24, -4, 34, \quad 34\}$$

The convolution of $\boldsymbol{x}_h$ with the scaled analysis highpass filter impulse response,

$$\{h(-1), h(0), h(1), h(2)\} = \{-1, 3, -3, 1\}$$

after downsampling, is

$$\{X_\psi(2,0), X_\psi(2,1), X_\psi(2,2), X_\psi(2,3)\} = \{5, 6, -12, -9\}$$

With periodic extension (Equation 11.5), the highpass output is

$$\{-5, \quad 5, 6, -12, -9, \quad 9\}$$

IDWT: The convolution of the lowpass output alternately with the even- and odd-indexed coefficients $\{3, 1\}$ and $\{1, 3\}$ of the scaled synthesis lowpass filter impulse response

$$\{\tilde{l}(-1), \tilde{l}(0), \tilde{l}(1), \tilde{l}(2)\} = \{1, 3, 3, 1\}$$

yields the even- and indexed values of the output of the lowpass channel

$$\{72, 78, 90, 68, 12, 22, 98, 136\}$$

Using matrix formulation of convolution,

$$\begin{bmatrix} x_l(0) \\ x_l(1) \\ x_l(2) \\ x_l(3) \\ x_l(4) \\ x_l(5) \\ x_l(6) \\ x_l(7) \end{bmatrix} = \begin{bmatrix} 1 & 3 & 0 & 0 & 0 & 0 \\ 0 & 3 & 1 & 0 & 0 & 0 \\ 0 & 1 & 3 & 0 & 0 & 0 \\ 0 & 0 & 3 & 1 & 0 & 0 \\ 0 & 0 & 1 & 3 & 0 & 0 \\ 0 & 0 & 0 & 3 & 1 & 0 \\ 0 & 0 & 0 & 1 & 3 & 0 \\ 0 & 0 & 0 & 0 & 3 & 1 \end{bmatrix} \begin{bmatrix} 18 \\ 18 \\ 18 \\ 24 \\ -4 \\ 34 \\ 34 \end{bmatrix} = \begin{bmatrix} 72 \\ 78 \\ 90 \\ 68 \\ 12 \\ 22 \\ 98 \\ 136 \end{bmatrix} \tag{11.15}$$

The convolution of the highpass output alternately with the even- and odd-indexed coefficients $\{-6, 2\}$ and $\{-2, 6\}$ of the scaled synthesis highpass filter impulse response

$$\{\tilde{h}(-1), \tilde{h}(0), \tilde{h}(1), \tilde{h}(2)\} = \{-2, -6, 6, 2\}$$

yields the even- and indexed values of the output of the highpass channel

$$\{-40, 18, -26, 60, 84, -54, 30, -72\}$$

Using matrix formulation of convolution,

$$\begin{bmatrix} x_h(0) \\ x_h(1) \\ x_h(2) \\ x_h(3) \\ x_h(4) \\ x_h(5) \\ x_h(6) \\ x_h(7) \end{bmatrix} = \begin{bmatrix} 2 & -6 & 0 & 0 & 0 & 0 \\ 0 & 6 & -2 & 0 & 0 & 0 \\ 0 & 2 & -6 & 0 & 0 & 0 \\ 0 & 0 & 6 & -2 & 0 & 0 \\ 0 & 0 & 2 & -6 & 0 & 0 \\ 0 & 0 & 0 & 6 & -2 & 0 \\ 0 & 0 & 0 & 2 & -6 & 0 \\ 0 & 0 & 0 & 0 & 6 & -2 \end{bmatrix} \begin{bmatrix} -5 \\ 5 \\ 6 \\ -12 \\ -9 \\ 9 \end{bmatrix} = \begin{bmatrix} -40 \\ 18 \\ -26 \\ 60 \\ 84 \\ -54 \\ 30 \\ -72 \end{bmatrix} \tag{11.16}$$

Adding the last two output sequences, we get the scaled original input $32\{1, 3, 2, 4, 3, -1, 4, 2\}$. ■

11.5 Implementation of the DWT using Factorized Polyphase Matrix

The polyphase matrix $\boldsymbol{P}(z)$ is given by

$$\boldsymbol{P}(z) = \begin{bmatrix} L_e(z) & L_o(z) \\ H_e(z) & H_o(z) \end{bmatrix} \tag{11.17}$$

It is assumed that the determinant of $\boldsymbol{P}(z)$ is equal to 1. Each element of the matrix is a polynomial in z and z^{-1}. The polyphase matrix $\boldsymbol{P}(z)$ is factorized into a product of a set of triangular matrices of the form

$$\begin{bmatrix} 1 & S_N(z) \\ 0 & 1 \end{bmatrix} \quad \text{and} \quad \begin{bmatrix} 1 & 0 \\ T_N(z) & 1 \end{bmatrix} \tag{11.18}$$

and a diagonal matrix of the form

$$\begin{bmatrix} K & 0 \\ 0 & \frac{1}{K} \end{bmatrix} \tag{11.19}$$

This factorization is always possible for FIR filters satisfying the perfect reconstruction conditions. The inverse of a triangular matrix is easy to find.

$$\begin{bmatrix} 1 & S_N(z) \\ 0 & 1 \end{bmatrix} \begin{bmatrix} 1 & -S_N(z) \\ 0 & 1 \end{bmatrix} = \begin{bmatrix} 1 & 0 \\ 0 & 1 \end{bmatrix} \tag{11.20}$$

The inverse of a diagonal matrix is also easy to find.

$$\begin{bmatrix} K & 0 \\ 0 & \frac{1}{K} \end{bmatrix} \begin{bmatrix} \frac{1}{K} & 0 \\ 0 & K \end{bmatrix} = \begin{bmatrix} 1 & 0 \\ 0 & 1 \end{bmatrix} \tag{11.21}$$

In addition, the inverse of a product of invertible matrices is the product of the inverses of the individual matrices in the reverse order. That is,

$$(\boldsymbol{A}(z)\boldsymbol{B}(z))^{-1} = \boldsymbol{B}^{-1}(z)\boldsymbol{A}^{-1}(z)$$

The z-transform of an FIR filter is a Laurent polynomial of the form

$$H(z) = \sum_{n=n_1}^{n_2} h(n)z^{-n}, \quad n_1 \leq n_2$$

where $h(n)$ is the impulse response of the filter. Ordinary polynomials have only nonnegative powers. The degree of a Laurent polynomial is given by $n_2 - n_1$. Let $\boldsymbol{P}(z)$ be a 2×2 matrix of Laurent polynomials, with its determinant equal to 1. Then,

$$\boldsymbol{P}(z) = \begin{bmatrix} K & 0 \\ 0 & \frac{1}{K} \end{bmatrix} \begin{bmatrix} 1 & S_N(z) \\ 0 & 1 \end{bmatrix} \begin{bmatrix} 1 & 0 \\ T_N(z) & 1 \end{bmatrix} \cdots \begin{bmatrix} 1 & S_1(z) \\ 0 & 1 \end{bmatrix} \begin{bmatrix} 1 & 0 \\ T_1(z) & 1 \end{bmatrix} \tag{11.22}$$

where constant $K \neq 0$ and

$$S_1(z), \ldots, S_N(z) \quad \text{and} \quad T_1(z), \ldots, T_N(z)$$

are Laurent polynomials. The factorization is not unique. The filter realization steps are obtained from the factorization. The process is based on the fact that for any two Laurent polynomials $u(z)$ and $v(z) \neq 0$, with the degree of $v(z)$ less than or equal to that of $u(z)$,

$$u(z) = q(z)v(z) + r(z),$$

where the degree of the quotient polynomial $q(z)$ is equal to that of $u(z) - v(z)$ and that of the remainder $r(z)$ is less than that of $v(z)$.

11.5.1 Haar Filter

The transfer function of the Haar analysis filters is

$$L(z) = \frac{1}{\sqrt{2}}(1 + z^{-1}) \qquad H(z) = \frac{1}{\sqrt{2}}(-1 + z^{-1})$$

The first step in the factorization process for a particular filter is to form the polyphase matrix and verify that its determinant is unity. If necessary, we can make the determinant equal to 1 by scaling and shifting by odd number of samples.

$$\boldsymbol{P}(z) = \begin{bmatrix} \frac{1}{\sqrt{2}} & \frac{1}{\sqrt{2}} \\ -\frac{1}{\sqrt{2}} & \frac{1}{\sqrt{2}} \end{bmatrix} = \begin{bmatrix} r1(z) & \frac{1}{\sqrt{2}} \\ r2(z) & \frac{1}{\sqrt{2}} \end{bmatrix} \begin{bmatrix} 1 & 0 \\ q(z) & 1 \end{bmatrix} \tag{11.23}$$

The determinant of $\boldsymbol{P}(z)$ is unity. We have to solve the two equations involving three unknown values

$$r1(z) + \frac{1}{\sqrt{2}}q(z) = \frac{1}{\sqrt{2}}$$

$$r2(z) + \frac{1}{\sqrt{2}}q(z) = -\frac{1}{\sqrt{2}}$$

There is no unique solution. Let $r2(z) = 0$. Then, $q(z) = -1$ and $r1(z) = \sqrt{2}$. The factorization is

$$\begin{bmatrix} \frac{1}{\sqrt{2}} & \frac{1}{\sqrt{2}} \\ -\frac{1}{\sqrt{2}} & \frac{1}{\sqrt{2}} \end{bmatrix} = \begin{bmatrix} \sqrt{2} & \frac{1}{\sqrt{2}} \\ 0 & \frac{1}{\sqrt{2}} \end{bmatrix} \begin{bmatrix} 1 & 0 \\ -1 & 1 \end{bmatrix} \tag{11.24}$$

With the diagonal matrix, the factorization becomes

$$\begin{bmatrix} \frac{1}{\sqrt{2}} & \frac{1}{\sqrt{2}} \\ -\frac{1}{\sqrt{2}} & \frac{1}{\sqrt{2}} \end{bmatrix} = \begin{bmatrix} \sqrt{2} & 0 \\ 0 & \frac{1}{\sqrt{2}} \end{bmatrix} \begin{bmatrix} 1 & \frac{1}{2} \\ 0 & 1 \end{bmatrix} \begin{bmatrix} 1 & 0 \\ -1 & 1 \end{bmatrix} \tag{11.25}$$

The 2-point 1-level DWT of the time-domain signal $\{x(0), x(1)\}$ is defined as

$$\begin{bmatrix} X_\phi(0,0) \\ X_\psi(0,0) \end{bmatrix} = \begin{bmatrix} \frac{1}{\sqrt{2}} & \frac{1}{\sqrt{2}} \\ \frac{1}{\sqrt{2}} & -\frac{1}{\sqrt{2}} \end{bmatrix} \begin{bmatrix} x(0) \\ x(1) \end{bmatrix} \tag{11.26}$$

where the three entities represent, respectively, the coefficient, transform, and input matrices. With factorization, the DWT is defined by

$$\begin{bmatrix} X_\phi(0,0) \\ X_\psi(0,0) \end{bmatrix} = \begin{bmatrix} \sqrt{2} & 0 \\ 0 & \frac{1}{\sqrt{2}} \end{bmatrix} \begin{bmatrix} 1 & \frac{1}{2} \\ 0 & 1 \end{bmatrix} \begin{bmatrix} 1 & 0 \\ -1 & 1 \end{bmatrix} \begin{bmatrix} x(0) \\ x(1) \end{bmatrix} \tag{11.27}$$

Note that the detail coefficients $X_\psi(0,0)$ given by Equation (11.27) *are negatives of those given by Equation* (11.26).

The IDWT, using the rules for finding the inverse of triangular and diagonal matrices and the inverse of a product of matrices, is given by

$$\begin{bmatrix} x(0) \\ x(1) \end{bmatrix} = \begin{bmatrix} 1 & 0 \\ 1 & 1 \end{bmatrix} \begin{bmatrix} 1 & -\frac{1}{2} \\ 0 & 1 \end{bmatrix} \begin{bmatrix} \frac{1}{\sqrt{2}} & 0 \\ 0 & \sqrt{2} \end{bmatrix} \begin{bmatrix} X_\phi(0,0) \\ X_\psi(0,0) \end{bmatrix} \tag{11.28}$$

Example 11.5 Compute the 1-level DWT of the input $x(n)$ using the Haar filters by the matrix factorization approach. Compute the IDWT of the DWT coefficients and verify that the input $x(n)$ is reconstructed.

$$\{x(0) = 4, x(1) = -2\}$$

Solution

Evaluating the factored matrix from right to left, we get

$$\begin{vmatrix} \sqrt{2} \\ -\frac{6}{\sqrt{2}} \end{vmatrix} \begin{vmatrix} 1 \\ -6 \end{vmatrix} \begin{vmatrix} 4 \\ -6 \end{vmatrix} \begin{vmatrix} 4 \\ -2 \end{vmatrix}$$

The IDWT is computed as

$$\begin{vmatrix} 4 \\ -2 \end{vmatrix} \begin{vmatrix} 4 \\ -6 \end{vmatrix} \begin{vmatrix} 1 \\ -6 \end{vmatrix} \begin{vmatrix} \sqrt{2} \\ -\frac{6}{\sqrt{2}} \end{vmatrix}$$

■

11.5.2 D4 Filter

$$L(z) = \frac{1+\sqrt{3}}{4\sqrt{2}} + \frac{3+\sqrt{3}}{4\sqrt{2}}z + \frac{3-\sqrt{3}}{4\sqrt{2}}z^2 + \frac{1-\sqrt{3}}{4\sqrt{2}}z^3$$

$$H(z) = -\frac{(1-\sqrt{3})}{4\sqrt{2}}z^{-2} + \frac{3-\sqrt{3}}{4\sqrt{2}}z^{-1} - \frac{(3+\sqrt{3})}{4\sqrt{2}} + \frac{1+\sqrt{3}}{4\sqrt{2}}z$$

Remember that the determinant of the polyphase matrix must be equal to unity for the matrix factorization process. This requirement, if not already met, can be satisfied by scaling and odd shift in time of the equations defining the filter.

$$\boldsymbol{P}(z) = \begin{bmatrix} \frac{1+\sqrt{3}}{4\sqrt{2}} + \frac{3-\sqrt{3}}{4\sqrt{2}}z & \frac{3+\sqrt{3}}{4\sqrt{2}} + \frac{1-\sqrt{3}}{4\sqrt{2}}z \\ -\frac{(3+\sqrt{3})}{4\sqrt{2}} - \frac{(1-\sqrt{3})}{4\sqrt{2}}z^{-1} & \frac{(1+\sqrt{3})}{4\sqrt{2}} + \frac{(3-\sqrt{3})}{4\sqrt{2}}z^{-1} \end{bmatrix}$$

$$= \begin{bmatrix} \frac{1+\sqrt{3}}{4\sqrt{2}} + \frac{3-\sqrt{3}}{4\sqrt{2}}z & \frac{(1-\sqrt{3})}{\sqrt{2}}z \\ -\frac{(3+\sqrt{3})}{4\sqrt{2}} - \frac{(1-\sqrt{3})}{4\sqrt{2}}z^{-1} & \frac{(1+\sqrt{3})}{\sqrt{2}} \end{bmatrix} \begin{bmatrix} 1 & \sqrt{3} \\ 0 & 1 \end{bmatrix}$$

$$= \begin{bmatrix} \frac{\sqrt{3}-1}{\sqrt{2}} & \frac{(1-\sqrt{3})}{\sqrt{2}}z \\ 0 & \frac{(1+\sqrt{3})}{\sqrt{2}} \end{bmatrix} \begin{bmatrix} 1 & 0 \\ -\frac{\sqrt{3}}{4} - \frac{(\sqrt{3}-2)}{4}z^{-1} & 1 \end{bmatrix} \begin{bmatrix} 1 & \sqrt{3} \\ 0 & 1 \end{bmatrix}$$

$$= \begin{bmatrix} \frac{\sqrt{3}-1}{\sqrt{2}} & 0 \\ 0 & \frac{1+\sqrt{3}}{\sqrt{2}} \end{bmatrix} \begin{bmatrix} 1 & -z \\ 0 & 1 \end{bmatrix} \begin{bmatrix} 1 & 0 \\ -\frac{\sqrt{3}}{4} - \frac{(\sqrt{3}-2)}{4}z^{-1} & 1 \end{bmatrix} \begin{bmatrix} 1 & \sqrt{3} \\ 0 & 1 \end{bmatrix}$$

The IDWT steps are given by

$$\begin{bmatrix} 1 & -\sqrt{3} \\ 0 & 1 \end{bmatrix} \begin{bmatrix} 1 & 0 \\ \frac{\sqrt{3}}{4} + \frac{(\sqrt{3}-2)}{4}z^{-1} & 1 \end{bmatrix} \begin{bmatrix} 1 & z \\ 0 & 1 \end{bmatrix} \begin{bmatrix} \frac{\sqrt{3}+1}{\sqrt{2}} & 0 \\ 0 & \frac{\sqrt{3}-1}{\sqrt{2}} \end{bmatrix}$$

Example 11.6 Compute the 1-level DWT of the input $x(n)$ using the D4 filter by the matrix factorization approach. Assume half-point symmetry of the data. Compute the IDWT of the DWT coefficients and verify that the input $x(n)$ is reconstructed.

$$\boldsymbol{x} = \{1, 3, 2, 4\}$$

Solution

The input, with the half-point symmetry extension, is

$$\boldsymbol{x}_h = \{3, 1, \quad 1, 3, 2, 4, \quad 4, 2\}$$

Consider the first input pair $\{x(-2) = x_e(-1) = 3, x(-1) = x_o(-1) = 1\}$.

$$x_e^1(-1) = x_e(-1) + \sqrt{3}x_o(-1) = 3 + \sqrt{3}$$

$$x_e^1(0) = x_e(0) + \sqrt{3}x_o(0) = 1 + 3\sqrt{3}$$

$$x_o^1(0) = x_o(0) + \left(-\frac{\sqrt{3}}{4}\right) x_e^1(0) - \frac{(\sqrt{3}-2)}{4} x_e^1(-1)$$

$$x_o^1(0) = 3 + \left(-\frac{\sqrt{3}}{4}\right)(1 + 3\sqrt{3}) + \frac{-(\sqrt{3}-2)}{4}(3 + \sqrt{3}) = \frac{(3-\sqrt{3})}{2}$$

$$x_e^2(-1) = x_e^1(-1) - x_o^1(0) = (3 + \sqrt{3}) - \frac{(3-\sqrt{3})}{2} = \frac{3(1+\sqrt{3})}{2}$$

$$X_\phi(-1) = \frac{(\sqrt{3}-1)}{\sqrt{2}} x_e^2(-1) = \frac{(\sqrt{3}-1)}{\sqrt{2}} \frac{3(1+\sqrt{3})}{2} = \frac{3}{\sqrt{2}}$$

$$X_\psi(-1) = \frac{(\sqrt{3}+1)}{\sqrt{2}} x_o^1(0) = \frac{(\sqrt{3}+1)}{\sqrt{2}} \frac{(3-\sqrt{3})}{2} = \sqrt{\frac{3}{2}}$$

$$x_e^1(1) = x_e(1) + \sqrt{3}x_o(1) = 2 + 4\sqrt{3}$$

$$x_o^1(1) = x_o(1) + \left(-\frac{\sqrt{3}}{4}\right) x_e^1(1) - \frac{(\sqrt{3}-2)}{4} x_e^1(0)$$

$$x_o^1(1) = 4 + \left(-\frac{\sqrt{3}}{4}\right)(2 + 4\sqrt{3}) + \frac{-(\sqrt{3}-2)}{4}(1 + 3\sqrt{3}) = \frac{3(\sqrt{3}-1)}{4}$$

$$x_e^2(0) = x_e^1(0) - x_o^1(1) = (1 + 3\sqrt{3}) - \frac{3(\sqrt{3}-1)}{4} = \frac{(7+9\sqrt{3})}{4}$$

$$X_\phi(0) = \frac{(\sqrt{3}-1)}{\sqrt{2}} x_e^2(0) = \frac{(\sqrt{3}-1)}{\sqrt{2}} \frac{(7+9\sqrt{3})}{4} = \frac{10-\sqrt{3}}{2\sqrt{2}}$$

$$X_\psi(0) = \frac{(\sqrt{3}+1)}{\sqrt{2}} x_o^1(1) = \frac{(\sqrt{3}+1)}{\sqrt{2}} \frac{3(\sqrt{3}-1)}{4} = \frac{3}{2\sqrt{2}}$$

$$x_e^1(2) = x_e(2) + \sqrt{3}x_o(2) = 4 + 2\sqrt{3}$$

$$x_o^1(2) = x_o(2) + \left(-\frac{\sqrt{3}}{4}\right) x_e^1(2) - \frac{(\sqrt{3}-2)}{4} x_e^1(1)$$

$$x_o^1(2) = 2 + \left(-\frac{\sqrt{3}}{4}\right)(4+2\sqrt{3}) + \frac{-(\sqrt{3}-2)}{4}(2+4\sqrt{3}) = \frac{\sqrt{3}(1-\sqrt{3})}{2}$$

$$x_e^2(1) = x_e^1(1) - x_o^1(2) = (2+4\sqrt{3}) - \frac{\sqrt{3}(1-\sqrt{3})}{2} = \frac{7(1+\sqrt{3})}{2}$$

$$X_\phi(1) = \frac{(\sqrt{3}-1)}{\sqrt{2}} x_e^2(1) = \frac{(\sqrt{3}-1)}{\sqrt{2}} \frac{7(1+\sqrt{3})}{2} = \frac{7}{\sqrt{2}}$$

$$X_\psi(1) = \frac{(\sqrt{3}+1)}{\sqrt{2}} x_o^1(2) = \frac{(\sqrt{3}+1)}{\sqrt{2}} \frac{\sqrt{3}(1-\sqrt{3})}{2} = -\sqrt{\frac{3}{2}}$$

The IDWT

$$\{X_\phi(1,-1), X_\phi(1,0), X_\phi(1,1)\} = \left\{\frac{3}{\sqrt{2}}, \frac{10-\sqrt{3}}{2\sqrt{2}}, \frac{7}{\sqrt{2}}\right\}$$

$$\{X_\phi(1,-1), X_\phi(1,0), X_\psi(1,1)\} = \left\{\sqrt{\frac{3}{2}}, \frac{3}{2\sqrt{2}}, -\sqrt{\frac{3}{2}}\right\}$$

Multiplying the approximation and detail coefficients, respectively, by

$$\frac{(\sqrt{3}+1)}{\sqrt{2}} \quad \text{and} \quad \frac{(\sqrt{3}-1)}{\sqrt{2}}$$

we get

$$\{x_e^2(-1), x_e^2(0), x_e^2(1)\} = \left\{\frac{3\,(1+\sqrt{3})}{2}, \frac{(7+9\sqrt{3})}{4}, \frac{7(1+\sqrt{3})}{2}\right\}$$

$$\{x_o^1(-1), x_o^1(0), x_o^1(1)\} = \left\{\frac{\sqrt{3}\,(\sqrt{3}-1)}{2}, \frac{3(\sqrt{3}-1)}{4}, \frac{\sqrt{3}(1-\sqrt{3})}{2}\right\}$$

$$x_o^1(0) = x_e^2(-1) + x_o^1(-1) = \frac{3(1+\sqrt{3})}{2} + \frac{\sqrt{3}(\sqrt{3}-1)}{2} = 3+\sqrt{3}$$

$$x_o^1(1) = x_e^2(0) + x_o^1(0) = \frac{(7+9\sqrt{3})}{4} + \frac{3(\sqrt{3}-1)}{4} = 1+3\sqrt{3}$$

$$x_o(0) = x_o^1(-1) + \frac{\sqrt{3}}{4} x_o^1(1) + \frac{(\sqrt{3}-2)}{4} x_o^1(0)$$

$$= \frac{\sqrt{3}(\sqrt{3}-1)}{2} + \frac{\sqrt{3}}{4}(1+3\sqrt{3}) + \frac{(\sqrt{3}-2)}{4}(3+\sqrt{3}) = 3$$

$$x_e(0) = x_o^1(1) - \sqrt{3}x_o(0) = (1+3\sqrt{3}) - 3\sqrt{3} = 1$$

$$x_o^1(2) = x_e^2(1) + x_o^1(1) = \frac{7(1+\sqrt{3})}{2} + \frac{\sqrt{3}(1-\sqrt{3})}{2} = 2(1+2\sqrt{3})$$

$$\begin{aligned}
x_o(1) &= x_o^1(0) + \frac{\sqrt{3}}{4}x_o^1(2) + \frac{(\sqrt{3}-2)}{4}x_o^1(1) \\
&= \frac{3(\sqrt{3}-1)}{4} + \frac{\sqrt{3}}{4}(2(1+2\sqrt{3})) + \frac{(\sqrt{3}-2)}{4}(1+3\sqrt{3}) = 4 \\
x_e(1) &= x_o^1(2) - \sqrt{3}x_o(1) = 2(1+2\sqrt{3}) - 4\sqrt{3} = 2
\end{aligned}$$

■

11.5.3 5/3 Spline Filter

$$\begin{aligned}
L(z) &= -\frac{1}{8}z^2 + \frac{1}{4}z + \frac{3}{4} + \frac{1}{4}z^{-1} - \frac{1}{8}z^{-2} \\
&= \left(-\frac{1}{8}z^2 + \frac{3}{4} - \frac{1}{8}z^{-2}\right) + z\left(\frac{1}{4}z^{-2} + \frac{1}{4}\right) \\
H(z) &= -\frac{1}{4} + \frac{1}{2}z - \frac{1}{4}z^2 \\
&= -\left(\frac{1}{4} + \frac{1}{4}z^2\right) + z\left(\frac{1}{2}\right)
\end{aligned}$$

Note that the highpass transfer function has been multiplied by -1. The constant $\sqrt{2}$ that appears in the numerators is suppressed, for simplicity.

$$\begin{aligned}
\boldsymbol{P}(z) &= \begin{bmatrix} -\frac{1}{8}z + \frac{3}{4} - \frac{1}{8}z^{-1} & \frac{1}{4}z^{-1} + \frac{1}{4} \\ -\frac{1}{4} - \frac{1}{4}z & \frac{1}{2} \end{bmatrix} \\
&= \begin{bmatrix} r1\,(z) & \frac{1}{4}z^{-1} + \frac{1}{4} \\ r2(z) & \frac{1}{2} \end{bmatrix} \begin{bmatrix} 1 & 0 \\ q\,(z) & 1 \end{bmatrix}
\end{aligned}$$

$$\begin{aligned}
-\frac{1}{8}z + \frac{3}{4} - \frac{1}{8}z^{-1} &= q(z)\left(\frac{1}{4}z^{-1} + \frac{1}{4}\right) + r1(z) \\
-\frac{1}{4} - \frac{1}{4}z &= q(z)\left(\frac{1}{2}\right) + r2(z) \\
-\frac{1}{8}z + \frac{3}{4} - \frac{1}{8}z^{-1} &= \left(-\frac{1}{2}z - \frac{1}{2}\right)\left(\frac{1}{4}z^{-1} + \frac{1}{4}\right) + 1
\end{aligned}$$

As the coefficients are symmetric, we select this factorization among three choices.

$$\begin{aligned}
\boldsymbol{P}(z) &= \begin{bmatrix} 1 & \frac{1}{4}z^{-1} + \frac{1}{4} \\ 0 & \frac{1}{2} \end{bmatrix} \begin{bmatrix} 1 & 0 \\ -\frac{1}{2}z - \frac{1}{2} & 1 \end{bmatrix} \\
&= \begin{bmatrix} 1 & 0 \\ 0 & \frac{1}{2} \end{bmatrix} \begin{bmatrix} 1 & \frac{1}{4}\left(1 + z^{-1}\right) \\ 0 & 1 \end{bmatrix} \begin{bmatrix} 1 & 0 \\ -\frac{1}{2}\left(1 + z\right) & 1 \end{bmatrix}
\end{aligned}$$

Inserting the constant $\sqrt{2}$, we get

$$\boldsymbol{P}(z) = \begin{bmatrix} \sqrt{2} & 0 \\ 0 & \frac{1}{\sqrt{2}} \end{bmatrix} \begin{bmatrix} 1 & \frac{1}{4}(1+z^{-1}) \\ 0 & 1 \end{bmatrix} \begin{bmatrix} 1 & 0 \\ -\frac{1}{2}(1+z) & 1 \end{bmatrix} \tag{11.29}$$

For computing the IDWT,

$$\boldsymbol{P}^{-1}(z) = \begin{bmatrix} 1 & 0 \\ \frac{1}{2}(1+z) & 1 \end{bmatrix} \begin{bmatrix} 1 & -\frac{1}{4}(1+z^{-1}) \\ 0 & 1 \end{bmatrix} \begin{bmatrix} \frac{1}{\sqrt{2}} & 0 \\ 0 & \sqrt{2} \end{bmatrix} \tag{11.30}$$

The input sequence $x(n)$ is split into even-indexed sequence $x_e(n)$ and odd-indexed sequence $x_o(n)$. Then

$$x_o^1(n) = x_o(n) - \frac{1}{2}(x_e(n) + x_e(n+1))$$

$$x_e^1(n) = x_e(n) + \frac{1}{4}(x_o^1(n-1) + x_o^1(n))$$

$$x_o^2(n) = \frac{1}{\sqrt{2}} x_o^1(n)$$

$$x_e^2(n) = \sqrt{2} x_e^1(n)$$

The first and third equations implement the highpass section of the filter bank, and the second and fourth equations implement the lowpass section.

Example 11.7 Compute the 1-level DWT of the input $x(n)$ using the 5/3 spline filter by the matrix factorization approach. Assume whole-point symmetry of the data. Compute the IDWT of the DWT coefficients and verify that the input $x(n)$ is reconstructed.

$$\boldsymbol{x} = \{1, 3, 2, 4, 3, -1, 4, 2\}$$

Solution

The input, with the whole-point symmetry extension, is

$$\boldsymbol{x}_w = \{2, 3, \quad 1, 3, 2, 4, 3, -1, 4, 2, \quad 4\}$$

The evaluation of the factored matrices from right to left for the input pairs, starting from the first $\{1, 3\}$ is as follows. Firstly, the initial value is computed as

$$x_o^1(-1) = 3 - \frac{1}{2}(2+1) = \frac{3}{2}$$

$$x_o^1(0) = 3 - \frac{1}{2}(1+2) = \frac{3}{2}, \qquad x_o^2(0) = \frac{1}{\sqrt{2}}\frac{3}{2} = \frac{3}{2\sqrt{2}}$$

$$x_e^1(0) = 1 + \frac{1}{4}\left(\frac{3}{2} + \frac{3}{2}\right) = \frac{7}{4}, \qquad x_e^2(0) = \sqrt{2}\frac{7}{4} = \frac{7}{2\sqrt{2}}$$

$$x_o^1(1) = 4 - \frac{1}{2}(2+3) = \frac{3}{2}, \qquad x_o^2(1) = \frac{1}{\sqrt{2}}\frac{3}{2} = \frac{3}{2\sqrt{2}}$$

$$x_e^1(1) = 2 + \frac{1}{4}\left(\frac{3}{2} + \frac{3}{2}\right) = \frac{11}{4}, \quad x_e^2(1) = \sqrt{2}\frac{11}{4} = \frac{11}{2\sqrt{2}}$$

$$x_o^1(2) = -1 - \frac{1}{2}(3+4) = -\frac{9}{2}, \quad x_o^2(2) = \frac{1}{\sqrt{2}}\left(-\frac{9}{2}\right) = -\frac{9}{2\sqrt{2}}$$

$$x_e^1(2) = 3 + \frac{1}{4}\left(-\frac{9}{2} + \frac{3}{2}\right) = \frac{9}{4}, \quad x_e^2(2) = \sqrt{2}\frac{9}{4} = \frac{9}{2\sqrt{2}}$$

$$x_o^1(3) = 2 - \frac{1}{2}(4+4) = -2, \quad x_o^2(3) = \frac{1}{\sqrt{2}}(-2) = -\sqrt{2}$$

$$x_e^1(3) = 4 + \frac{1}{4}\left(-2 - \frac{9}{2}\right) = \frac{19}{8}, \quad x_e^2(3) = \sqrt{2}\frac{19}{8} = \frac{19}{4\sqrt{2}}$$

The IDWT is computed as

$$x_e^1(n) = \frac{1}{\sqrt{2}}x_e^2(n)z$$

$$x_o^1(n) = \sqrt{2}x_o^2(n)$$

$$x_e(n) = x_e^1(n) - \frac{1}{4}(x_o^1(n-1) + x_o^1(n))$$

$$x_o(n) = x_o^1(n) + \frac{1}{2}(x_e(n) + x_e(n+1))$$

$$x_e^1(0) = \frac{1}{\sqrt{2}}\frac{7}{2\sqrt{2}} = \frac{7}{4}, \quad x_e(0) = \frac{7}{4} - \frac{1}{4}\left(\frac{3}{2} + \frac{3}{2}\right) = 1$$

$$x_o^1(0) = \sqrt{2}\frac{3}{2\sqrt{2}} = \frac{3}{2}, \quad x_o(0) = \frac{3}{2} + \frac{1}{2}\left(1 + \frac{11}{2} - \frac{1}{4}\left(\frac{3}{2} + \frac{3}{2}\right)\right) = 3$$

$$x_e^1(1) = \frac{1}{\sqrt{2}}\frac{11}{2\sqrt{2}} = \frac{11}{4}, \quad x_e(0) = \frac{11}{4} - \frac{1}{4}\left(\frac{3}{2} + \frac{3}{2}\right) = 2$$

$$x_o^1(1) = \sqrt{2}\frac{3}{2\sqrt{2}} = \frac{3}{2}, \quad x_o(0) = \frac{3}{2} + \frac{1}{2}\left(2 + \frac{9}{4} - \frac{1}{4}\left(\frac{3}{2} - \frac{9}{2}\right)\right) = 4$$

$$x_e^1(2) = \frac{1}{\sqrt{2}}\frac{9}{2\sqrt{2}} = \frac{9}{4}, \quad x_e(0) = \frac{9}{4} - \frac{1}{4}\left(\frac{3}{2} - \frac{9}{2}\right) = 3$$

$$x_o^1(2) = \sqrt{2}\frac{-9}{2\sqrt{2}} = -\frac{9}{2}, \quad x_o(0) = -\frac{9}{2} + \frac{1}{2}\left(3 + \frac{19}{8} - \frac{1}{4}\left(-\frac{9}{2} - 2\right)\right) = -1$$

$$x_e^1(3) = \frac{1}{\sqrt{2}}\frac{19}{4\sqrt{2}} = \frac{19}{8}, \quad x_e(0) = \frac{19}{8} - \frac{1}{4}\left(-\frac{9}{2} - 2\right) = 4$$

$$x_o^1(3) = \sqrt{2}(-\sqrt{2}) = -2, \quad x_o(0) = -2 + \frac{1}{2}\left(4 + \frac{19}{8} - \frac{1}{4}\left(-2 - \frac{9}{2}\right)\right) = 2$$

Integer-to-integer version of the DWT is obtained by appropriate rounding of the operands and is presented in Chapter 15 for the $5/3$ spline filter. ■

11.6 Summary

- The DWT is essentially a set of bandpass filters. However, it is efficiently implemented using a set of lowpass and highpass filters recursively.
- The computational complexity of computing the DWT is $O(N)$.
- There are basically two approaches to implement the DWT efficiently. The first approach is to evaluate the required convolutions using the polyphase filter structure. Transform methods are not used to carry out the convolution, as the length of the impulse response of the filters is short.
- The other approach is to factorize the polyphase matrix into a product of a set of sparse matrices.
- The 2-D DWT, with separable filters, is usually computed by the row-column method. That is, 1-D DWT of all the columns is computed, and then the 1-D DWT of all the resulting rows is computed. The order of the computation can also be reversed.
- In the implementation of the DWT, additional memory of about half the size of the data is required. For an in-place computation, data reordering is required.
- In implementing the asymmetric filters, data expansion problem occurs due to the finite length of the data.
- Symmetric filters provide linear phase response and an effective solution to the border problem.

Exercises

11.1 Compute the 2-level DWT of the input $x(n)$ and reconstruct it using the two-channel Haar filter bank.

11.1.1

$$x(n) = \{3, 3, 2, 4\}$$

***11.1.2**

$$x(n) = \{1, 4, -2, 3\}$$

11.1.3

$$x(n) = \{-3, 2, 4, 1\}$$

11.2 Compute the 1-level DWT of the 2-D input $x(n_1, n_2)$ and reconstruct it using a two-channel Haar filter bank.

***11.2.1**

$$\begin{bmatrix} 2 & 1 & 3 & 4 \\ 4 & 1 & -2 & 3 \\ 3 & 1 & -3 & 2 \\ 2 & -2 & 1 & 3 \end{bmatrix}$$

11.2.2

$$\begin{bmatrix} -1 & 3 & 4 & 2 \\ 1 & 4 & 2 & 3 \\ -3 & 2 & 1 & 4 \\ 2 & 4 & 2 & 3 \end{bmatrix}$$

11.2.3

$$\begin{bmatrix} 5 & 4 & 3 & 1 \\ -1 & 4 & 4 & 2 \\ 3 & 4 & 1 & 1 \\ 2 & 1 & 4 & 3 \end{bmatrix}$$

11.3 Show the trace of the 1-D 3-level Haar DWT algorithm for the input $x(n)$. Show the trace for the reconstruction of $x(n)$ from the DWT coefficients.

11.3.1

$$x(n) = \{2, 3, 1, 7, 4, -2, 3, 1\}$$

***11.3.2**

$$x(n) = \{3, 4, 1, 3, -2, 4, 1, 3\}$$

11.3.3

$$x(n) = \{3, 1, 2, 4, -3, -2, 4, 1\}$$

11.4 Compute the 1-level DWT of the input $x(n)$ using the D4 filter by the convolution approach. Assume half-point symmetry of the data. Compute the IDWT of the DWT coefficients and verify that the input $x(n)$ is reconstructed.

11.4.1

$$x(n) = \{2, 1, 3, 4\}$$

11.4.2

$$x(n) = \{5, -2, 1, 3\}$$

***11.4.3**

$$x(n) = \{3, 4, -2, -3\}$$

11.5 Compute the 1-level DWT of the input $x(n)$ using the $5/3$ spline filter by the convolution approach. Assume whole-point symmetry of the data. Compute the IDWT of the DWT coefficients and verify that the input $x(n)$ is reconstructed.

***11.5.1**

$$x(n) = \{3, 1, 1, 3, 2, 4, 4, 2\}$$

11.5.2

$$x(n) = \{1, 2, 2, 1, 3, 4, 4, 3\}$$

11.5.3

$$x(n) = \{-2, 5, 5, -2, 1, 3, 3, 1\}$$

11.6 Compute the 1-level DWT of the input $x(n)$ using the CDF $9/7$ filter by the convolution approach. Assume whole-point symmetry of the data. Compute the IDWT of the DWT coefficients and verify that the input $x(n)$ is reconstructed.

11.6.1

$$x(n) = \{3, 1, 1, 3, 2, 4, 4, 2, 2, 1, 1, -2, 2, 3, 4, 2\}$$

***11.6.2**

$$x(n) = \{1, 2, 2, 1, 3, 4, 4, 3, 1, 3, 2, -1, 3, 1, 4, -3\}$$

11.6.3

$$x(n) = \{-2, 5, 5, -2, 1, 3, 3, 1, -2, 4, 3, -1, 2, 3, -2, 1\}$$

11.7 Compute the 1-level DWT of the input $x(n)$ using the $4/4$ spline filter by the convolution approach. Assume half-point symmetry of the data. Compute the IDWT of the DWT coefficients and verify that the input $x(n)$ is reconstructed.

11.7.1

$$x(n) = \{2, 1, 1, -2, 2, 3, 4, 2\}$$

11.7.2

$$x(n) = \{1, 2, 2, 1, 3, 4, 4, 3\}$$

***11.7.3**

$$x(n) = \{-2, 4, 3, -1, 2, 3, -2, 1\}$$

11.8 Compute the DWT of the input $x(n)$ using the Haar filter by the matrix factorization approach. Compute the IDWT of the DWT coefficients and verify that the input $x(n)$ is reconstructed.

11.8.1

$$x(n) = \{2, 3\}$$

***11.8.2**

$$x(n) = \{7, 3\}$$

11.9 Compute the DWT of the input $x(n)$ using the D4 filter by the matrix factorization approach. Assume half-point symmetry of the data. Compute the IDWT of the DWT coefficients and verify that the input $x(n)$ is reconstructed.

11.9.1

$$x(n) = \{2, 1, 3, 4\}$$

11.9.2

$$x(n) = \{5, -2, 1, 3\}$$

11.9.3

$$x(n) = \{3, 4, -2, -3\}$$

11.10 Compute the DWT of the input $x(n)$ using the $5/3$ spline filter by the matrix factorization approach. Assume whole-point symmetry of the data. Compute the IDWT of the DWT coefficients and verify that the input $x(n)$ is reconstructed.

11.10.1

$$x(n) = \{3, 1, 1, 3, 2, 4, 4, 2\}$$

11.10.2

$$x(n) = \{1, 2, 2, 1, 3, 4, 4, 3\}$$

11.10.3

$$x(n) = \{-2, 5, 5, -2, 1, 3, 3, 1\}$$

12

The Discrete Wavelet Packet Transform

In the DWT, at any level of decomposition, only the approximation part (the low-frequency part of the spectrum) is further decomposed. In the discrete wavelet packet transform (DWPT), the detail part (the high-frequency part of the spectrum) is also further decomposed. In contrast to the unique DWT representation, this decomposition results in many representations of the signal. The advantage is that the most optimum representation of the signal with respect to some criterion can be selected, using the redundancy. This approach provides a better solution to some applications than that is possible with the DWT. The order of the computational complexity of the DWPT is, however, $O(N\log_2 N)$ compared to $O(N)$ of the DWT.

12.1 The Discrete Wavelet Packet Transform

The 2-level N-point DWPT is obtained by taking the 1-level DWT of both the lower- and upper-halves of the N 1-level DWT coefficients. This process continues until we reach the lowest scale of decomposition. For example, with $N = 4$, the 2-level Haar DWPT transform matrix is

$$\boldsymbol{W}p_{4,0} = \begin{bmatrix} \boldsymbol{W}_{2,0} & \boldsymbol{0}_2 \\ \boldsymbol{0}_2 & \boldsymbol{W}_{2,0} \end{bmatrix} \boldsymbol{W}_{4,1}$$

$$\boldsymbol{W}p_{4,0} = \frac{1}{2}\begin{bmatrix} 1 & 1 & 1 & 1 \\ 1 & 1 & -1 & -1 \\ 1 & -1 & 1 & -1 \\ 1 & -1 & -1 & 1 \end{bmatrix} = \begin{bmatrix} \frac{1}{\sqrt{2}} & \frac{1}{\sqrt{2}} & 0 & 0 \\ \frac{1}{\sqrt{2}} & -\frac{1}{\sqrt{2}} & 0 & 0 \\ 0 & 0 & \frac{1}{\sqrt{2}} & \frac{1}{\sqrt{2}} \\ 0 & 0 & \frac{1}{\sqrt{2}} & -\frac{1}{\sqrt{2}} \end{bmatrix} \begin{bmatrix} \frac{1}{\sqrt{2}} & \frac{1}{\sqrt{2}} & 0 & 0 \\ 0 & 0 & \frac{1}{\sqrt{2}} & \frac{1}{\sqrt{2}} \\ \frac{1}{\sqrt{2}} & -\frac{1}{\sqrt{2}} & 0 & 0 \\ 0 & 0 & \frac{1}{\sqrt{2}} & -\frac{1}{\sqrt{2}} \end{bmatrix} \quad (12.1)$$

The basis functions are shown in Figure 12.1. The basis functions are the same as those of the discrete Walsh transform, except for a scale factor.

Discrete Wavelet Transform: A Signal Processing Approach, First Edition. D. Sundararajan.

Companion Website: www.wiley.com/go/sundararajan/wavelet

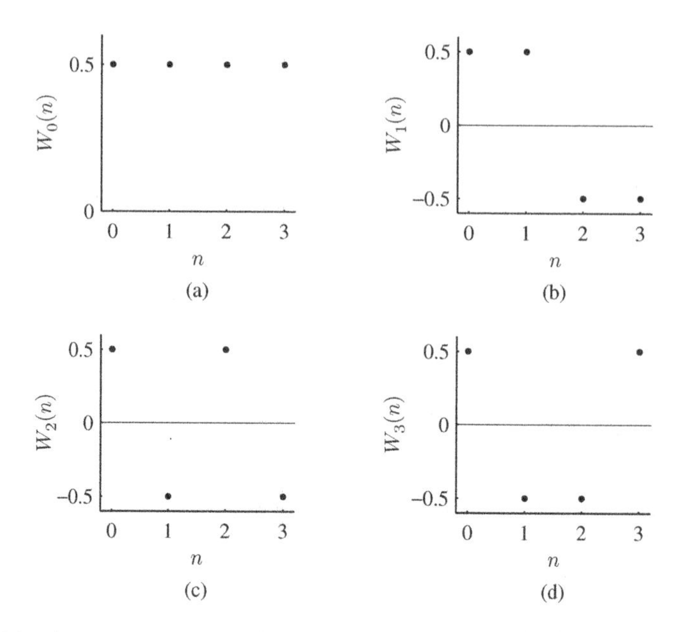

Figure 12.1 2-level Haar DWPT basis functions with $N = 4$. (a) $W_0(n)$; (b) $W_1(n)$; (c) $W_2(n)$; (d) $W_3(n)$

Example 12.1 Find the DWPT of

$$\{x(0) = 2, x(1) = 3, x(2) = 4, x(3) = 1\}$$

Solution

Using $\boldsymbol{W}p_{4,0}$ defined earlier, is given by

$$\begin{bmatrix} X(0) \\ X(1) \\ X(2) \\ X(3) \end{bmatrix} = \frac{1}{2} \begin{bmatrix} 1 & 1 & 1 & 1 \\ 1 & 1 & -1 & -1 \\ 1 & -1 & 1 & -1 \\ 1 & -1 & -1 & 1 \end{bmatrix} \begin{bmatrix} 2 \\ 3 \\ 4 \\ 1 \end{bmatrix} = \begin{bmatrix} 5 \\ 0 \\ 1 \\ -2 \end{bmatrix}$$

Multiplying each basis function by the corresponding DWPT coefficient and summing the products give the time-domain signal.

$$\begin{array}{llllll} \frac{1}{2}(1 & 1 & 1 & 1) & 5 & + \\ \frac{1}{2}(1 & 1 & -1 & -1) & 0 & + \\ \frac{1}{2}(1 & -1 & 1 & -1) & 1 & + \\ \frac{1}{2}(1 & -1 & -1 & 1) & -2 & = \\ ---- & ---- & ---- & ---- & & \\ 2 & 3 & 4 & 1 & & \end{array}$$

Formally, the IDWPT gets back the original input samples.

$$\begin{bmatrix} x(0) \\ x(1) \\ x(2) \\ x(3) \end{bmatrix} = \frac{1}{2} \begin{bmatrix} 1 & 1 & 1 & 1 \\ 1 & 1 & -1 & -1 \\ 1 & -1 & 1 & -1 \\ 1 & -1 & -1 & 1 \end{bmatrix} \begin{bmatrix} 5 \\ 0 \\ 1 \\ -2 \end{bmatrix} = \begin{bmatrix} 2 \\ 3 \\ 4 \\ 1 \end{bmatrix}$$

Applying Parseval's theorem,

$$(2^2 + 3^2 + 4^2 + 1^2) = 30 = (5^2 + 0^2 + 1^2 + (-2)^2)$$

Figure 12.2(a) shows the complete wavelet packet decomposition of the signal $\{2, 3, 4, 1\}$. Figure 12.2(b)–(e) show different sets of DWPT coefficients; each set is a different representation of the same signal. There are two conditions for a valid representation of a signal. The first one is that it should be possible to reconstruct the original signal. The second condition is that there should be no overlap between the selected coefficients. Subject to these conditions, the concatenation of each set of the selected coefficients is a representation of the signal. Including the signal itself, there are five representations of the signal $\{2, 3, 4, 1\}$. The coefficient sets $\{5, 5, -1, 3\}/\sqrt{2}$ and $\{5, 0, -1/\sqrt{2}, 3/\sqrt{2}\}$ are, respectively, the 1-level and 2-level DWT representations. The DWPT coefficient set $\{5, 0, 1, -2\}$ has been presented in the example. The DWPT coefficient set $\{5/\sqrt{2}, 5/\sqrt{2}, 1, -2\}$ corresponds to the representation

$$\begin{aligned} &(5\{1, 1, 0, 0\}/2) + (5\{0, 0, 1, 1\}/2) + (1\{1, -1, 1, -1\}/2) - (2\{1, -1, -1, 1\}/2) \\ &= \{2, 3, 4, 1\} \end{aligned}$$

The DWPT coefficient set $\{5, 0, -1/\sqrt{2}, 3/\sqrt{2}\}$ corresponds to the representation

$$\begin{aligned} &(5\{1, 1, 1, 1\}/2) + (0\{1, 1, -1, -1\}/2) - (1\{1, -1, 0, 0\}/2) + (3\{0, 0, 1, -1\}/2) \\ &= \{2, 3, 4, 1\} \end{aligned}$$

The spectral decomposition for a 3-level DWPT is shown in Table 12.1. The spectrum of the signal has been decomposed into eight equal subbands. The subbands are not in natural order. The ordering is shown in the last row. They are in gray-code order. In a gray-code order of a sequence of numbers, there is only one bit that is different in the binary representation of two consecutive numbers. The spectrum of a real aperiodic discrete signal is continuous, periodic of period 2π, and symmetric. The magnitude profile of the spectrum in the interval from π to 2π is the frequency reversal about $\omega = \pi$ (mirror image) of that in the interval 0 to π. In the DWT analysis, we decompose only the low-frequency component of the spectrum. The application of lowpass and highpass filters results in the expected range of frequencies of the decomposed

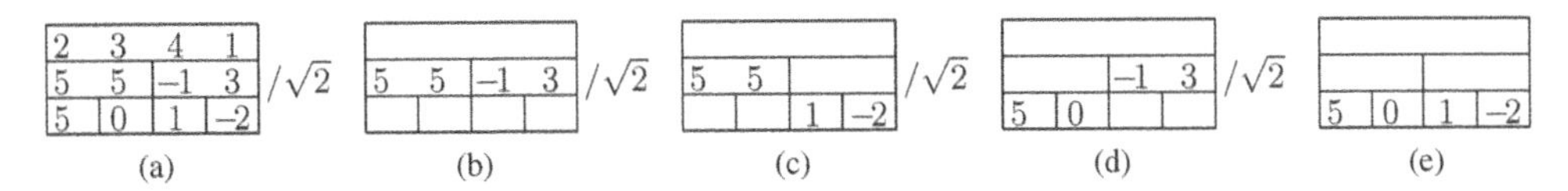

Figure 12.2 (a) The complete wavelet packet decomposition of the signal $\{2, 3, 4, 1\}$; (b)–(e) different representations of the signal by the DWPT coefficients

Table 12.1 Subband decomposition in a 3-level DWPT

$0-\pi$ radians							
$l,\ 0-\frac{\pi}{2}$				$h,\ \pi-\frac{\pi}{2}$			
$l,\ 0-\frac{\pi}{4}$		$h,\ \frac{\pi}{2}-\frac{\pi}{4}$		$l,\ \pi-3\frac{\pi}{4}$		$h,\ \frac{\pi}{2}-3\frac{\pi}{4}$	
$l,\ 0-\frac{\pi}{8}$	$h,\ \frac{\pi}{4}-\frac{\pi}{8}$	$l,\ 3\frac{\pi}{8}-\frac{\pi}{2}$	$h,\ 3\frac{\pi}{8}-\frac{\pi}{4}$	$l,\ 3\frac{\pi}{4}-7\frac{\pi}{8}$	$h,\ \pi-7\frac{\pi}{8}$	$l,\ 5\frac{\pi}{8}-3\frac{\pi}{4}$	$h,\ 5\frac{\pi}{8}-\frac{\pi}{2}$
0 (000)	1 (001)	3 (011)	2 (010)	6 (110)	7 (111)	5 (101)	4 (100)

Table 12.2 Subband decomposition in a 3-level DWPT, natural order

$0-\pi$ radians							
$l,\ 0-\frac{\pi}{2}$				$h,\ \pi-\frac{\pi}{2}$			
$l,\ 0-\frac{\pi}{4}$		$h,\ \frac{\pi}{2}-\frac{\pi}{4}$		$h,\ \frac{\pi}{2}-3\frac{\pi}{4}$		$l,\ \pi-3\frac{\pi}{4}$	
$l,\ 0-\frac{\pi}{8}$	$h,\ \frac{\pi}{4}-\frac{\pi}{8}$	$h,\ \frac{\pi}{4}-3\frac{\pi}{8}$	$l,\ 3\frac{\pi}{8}-\frac{\pi}{2}$	$l,\ \frac{\pi}{2}-5\frac{\pi}{8}$	$h,\ 3\frac{\pi}{4}-5\frac{\pi}{8}$	$h,\ 7\frac{\pi}{8}-3\frac{\pi}{4}$	$l,\ 7\frac{\pi}{8}-\pi$
0	1	2	3	4	5	6	7

components. In the DWPT analysis, we decompose the high-frequency component as well. The application of lowpass and highpass filters and downsampling results in the interchange of components corresponding to low and high frequencies, due to aliasing. In order to get the decompositions in the natural order, the decomposed components of each high-frequency component must be swapped. Consider the entries in Table 12.1. The entry $0-\pi$ in the first row indicates the spectral content of a discrete signal. When the signal is decomposed into low- and high-frequency components, the range of the frequencies of the subband components is $0-\frac{\pi}{2}$ and $\pi-\frac{\pi}{2}$. The alias of a high frequency near $\omega=\pi$ is a low frequency near $\omega=0$. The naturally ordered spectral decomposition for a 3-level DWPT is shown in Table 12.2. ■

12.1.1 Number of Representations

For a signal with only one element $N=1$, there is only one representation. That is the signal itself. With $N=2$, there are two choices. One is the signal itself, and the other is the approximation and the detail coefficients. For example, the Haar DWT of $\{2,3\}$ is $\{5,-1\}/\sqrt{2}$. The two representations of the signal are

$$[\{2,3\}] \quad \text{and} \quad [\{5,-1\}/\sqrt{2}]$$

With $N=4$, there are two possible (1-level and 2-level) DWT decompositions of $\{2,3,4,1\}$ resulting in the two sets of coefficients

$$[\{5,5,-1,3\}/\sqrt{2}], [\{5,0,-1/\sqrt{2},3/\sqrt{2}\}]$$

Either set of coefficients is a complete representation of the corresponding signal. We choose the one that yields a better representation for a particular application. In the DWPT, there are four (two new) possible decompositions resulting in the four sets of coefficients

$$[\{5,5,-1,3\}/\sqrt{2}], [\{5,0,-1/\sqrt{2},3/\sqrt{2}\}], [\{5/\sqrt{2},5/\sqrt{2},1,-2\}], [\{5,0,1,-2\}]$$

Each of the representations is a complete representation of the signal. Each representation spans the complete spectrum of the signal. We choose the one that is most advantageous for a particular application with respect to some criterion. For each of the k choices in the left side of the tree, there are k choices in the right. Therefore, there are k^2 representations. Including the signal itself, there are $k^2 + 1$ choices. The number of decompositions $N(k)$ for a k-level decomposition is given by

$$N(k) = (N(k-1))^2 + 1, \quad N(0) = 1$$

For a data length 4 and two levels of decomposition, there are $2^2 + 1 = 5$ possible representations as shown in Figure 12.2. For a data length 8 and three levels of decomposition, there are $5^2 + 1 = 26$ possible representations. The number of representations is a function of the signal length and the number of levels of decomposition. A signal of any length with one level of decomposition has only two representations.

The advantage of DWPT is that it is possible to select the optimal representation of the signal with respect to some criterion. However, there are two problems to be solved. One is that the possible number of representations increases rapidly as the data length and levels of decomposition increase. The second problem is choosing the optimal representation from a large number of representations. This selection is called the choice of representation.

12.2 Best Representation

Example 12.2 Find the 3-level DWPT of

$$\{3, 4, 2, 4, 4, 5, 1, 7\}$$

Let the cost function be the number of elements with magnitude greater than 1. Find the best DWPT representation that minimizes the cost function.

Solution

The 3-level Haar DWPT of the signal

$$\{3, 4, 2, 4, 4, 5, 1, 7\}$$

is shown in Table 12.3. Using the criterion that the magnitude of the coefficient is greater than 1, the cost values are shown in Table 12.4. For example, there are seven coefficients with magnitude greater than 1 in the first row of Table 12.3, and the cost function, shown in the first row of Table 12.4, is 7. The search for the best representation starts from the bottom row of Table 12.4. The cost values are shown in boldface. Now, we compare the first cost value (2) in the third row with the sum of those $(1 + 1 = 2)$ under it in the fourth row. As the values are equal, we select the cost value (2) in the third row. The last cost value (1) is selected as its value is less than the sum of those under it in the fourth row. All of the values in the third row are selected (shown in boldface), and all of the values in the fourth row are discarded, as shown in Table 12.5. In a similar way, we compare and select the cost values in the second and third rows. As the first cost value (4) in the second row is greater than the sum $(2 + 0 = 2)$ of those under it, the third row cost values are optimum and remain selected. The cost value (4) is replaced by (2). As the second cost value (2) in the second row is less than the sum $(2 + 1 = 3)$ of those under it, the second row cost value is optimum and remains selected, as shown in Table 12.6. Now, we compare and select the cost values in the first and second rows.

Table 12.3 The 3-level Haar DWPT of the signal $\{3, 4, 2, 4, 4, 5, 1, 7\}$

3 4 2 4 4 5 1 7							
4.9497	4.2426	6.3640	5.6569	−0.7071	−1.4142	−0.7071	−4.2426
6.5000	8.5000	0.5000	0.5000	−1.5000	−3.5000	0.5000	2.5000
10.6066	−1.4142	0.7071	0.0000	−3.5355	1.4142	2.1213	−1.4142

Table 12.4 The cost function is the number of elements with magnitude greater than 1

7							
4				2			
2		0		2		1	
1	**1**	**0**	**0**	**1**	**1**	**1**	**1**

Table 12.5 Second stage in finding the best representation

7							
4				2			
2		**0**		**2**		**1**	
1	1	0	0	1	1	1	1

Table 12.6 Third stage in finding the best representation

7							
2				**2**			
2		**0**		2		1	
1	1	0	0	1	1	1	1

As the cost value (7) in the first row is greater than the sum $(2+2=4)$ of those under it, the second row cost values are optimum and remain selected. The cost value (7) is replaced by (4), as shown in Table 12.7. The best representation includes four values greater than 1, and the corresponding DWPT coefficients are

6.5000	8.5000	0.5000	0.5000	−0.7071	−1.4142	−0.7071	−4.2426

Table 12.7 Fourth stage in finding the best representation

4							
2				**2**			
2		**0**		2		1	
1	1	0	0	1	1	1	1

In the case of the availability of more than one selection, the one with minimum levels of decomposition is to be chosen. ■

While, in practice, we have to use a computer to find the best representation, for understanding the algorithms, it is good to consider all possible representations. Then, it is obvious to find the best one. This can be done only for small amount of data. For 1-D signals, a signal with four elements is feasible.

Example 12.3 The 2-level Haar DWPT of the signal

$$\{2, 3, 4, 1\}$$

is shown in Figure 12.2(a). Let the cost function be the number of elements with magnitude greater than 2. Find the best DWPT representation that minimizes the cost function: (i) by writing all possible representations and (ii) using the algorithm described in Example 12.2.

Solution

Writing out all the five representations, we get

$$\begin{array}{rrrrcr} 2 & 3 & 4 & 1 & \rightarrow & 2 \\ 5/\sqrt{2} & 5/\sqrt{2} & -1/\sqrt{2} & 3/\sqrt{2} & \rightarrow & 3 \\ 5 & 0 & -1/\sqrt{2} & 3/\sqrt{2} & \rightarrow & 2 \\ 5/\sqrt{2} & 5/\sqrt{2} & 1 & -2 & \rightarrow & 2 \\ 5 & 0 & 1 & -2 & \rightarrow & 1 \end{array}$$

It is obvious that the minimum cost value is 1 with one possible best representation. Using the algorithm, we get the same results as shown in Tables 12.8–12.10.

Table 12.8 The cost function is the number of elements with magnitude greater than 2

<table>
<tr><td colspan="4">2</td></tr>
<tr><td colspan="2">2</td><td colspan="2">1</td></tr>
<tr><td>1</td><td>0</td><td>0</td><td>0</td></tr>
</table>

Table 12.9 Second stage in finding the best representation

<table>
<tr><td colspan="4">2</td></tr>
<tr><td colspan="2">1</td><td colspan="2">0</td></tr>
<tr><td>1</td><td>0</td><td>0</td><td>0</td></tr>
</table>

Table 12.10 Third stage in finding the best representation

<table>
<tr><td colspan="4">1</td></tr>
<tr><td colspan="2">1</td><td colspan="2">0</td></tr>
<tr><td>1</td><td>0</td><td>0</td><td>0</td></tr>
</table>

■

12.2.1 Cost Functions

A cost function, which is positive and real valued, is used to compare several representations of the same signal and assign a number that indicates the quality of representation with respect to some criterion. For example, the representation that uses fewer coefficients is the best in image compression. While the cost function based on threshold is simple, more useful cost functions are based on signal energy and other measures. A cost function is a measure of concentration, expressed as a real value, of some quality, such as the energy, of a finite-length signal. The set of large numbers plays a dominant role in the representation of a signal. The cost of the best representation should be optimum. For ease of implementation, certain properties are required for a cost function. One property is that it is additive. Let the cost function of vectors $\boldsymbol{a}$ and $\boldsymbol{b}$ be $C(\boldsymbol{a})$ and $C(\boldsymbol{b})$, respectively. Then, the cost function of the concatenation of $\boldsymbol{a}$ and $\boldsymbol{b}$ is the sum of the individual cost functions of those vectors.

$$C([\boldsymbol{a},\boldsymbol{b}]) = C(\boldsymbol{a}) + C(\boldsymbol{b})$$

The second property is that the cost function of a null element vector is zero.

Consider the signal $\{2,2\}$. The scaled Haar DWT coefficients are $\{4,0\}$. The transform values have higher energy concentration than the input. While the example values are ideal, typical practical signals exhibit advantageous properties such as higher energy concentration, when the transform is taken. In this section, we present some examples of best representation. For each application, the cost function has to be suitably designed.

The Cost Function Based on Signal Energy

The cost function of a vector $\boldsymbol{a}$ based on signal energy is defined as

$$C(\boldsymbol{a}) = \sum_n |a(n)|^p$$

where $0 < p < \infty$. Signal energy is defined with $p = 2$. If the DWPT is energy preserving, then, with $p = 2$, the best representation algorithm will always return the given signal. For $0 < p < 2$, this cost function is useful. If $p > 1$, $|a(n)|^p < |a(n)|, |a(n)| < 1$. In this case, coefficients with smaller magnitudes form the best representation.

Example 12.4 The 2-level Haar DWPT of the signal

$$\{2, 3, 4, 1\}$$

is shown in Figure 12.2(a). Let the cost function be the energy of the signal using the norm $\sum_n |a(n)|^{1.5}$. Find the best 2-level Haar DWPT representation that maximizes the cost function: (i) by writing all possible representations and (ii) using the algorithm.

Solution

The 2-level Haar DWPT of the signal

$$\{2, 3, 4, 1\}$$

is shown in Figure 12.2(a). The energy of the signal is

$$(2^2 + 3^2 + 4^2 + 1^2) = 30$$

Let $p = 1.5$. Then, writing all the five representations using energy with $p = 1.5$, we get

$$\begin{array}{rrrrcr}
2.8284 & 5.1962 & 8.0000 & 1.0000 & = & 17.0246 \\
6.6479 & 6.6479 & 0.5946 & 3.0897 & = & 16.9800 \\
6.6479 & 6.6479 & 1.0000 & 2.8284 & = & 17.1242 \\
11.1803 & 0 & 0.5946 & 3.0897 & = & 14.8646 \\
11.1803 & 0 & 1.0000 & 2.8284 & = & 15.0088
\end{array}$$

The representation by the set of coefficients

$$\{5/\sqrt{2}, 5/\sqrt{2}, 1, -2\}$$

is the best and gives the highest energy.

Table 12.11 shows the energy of the 2-level Haar DWPT representation of the signal with $p = 1.5$. Tables 12.12–12.14 show the stages in finding the best representation.

Table 12.11 The energy of the 2-level Haar DWPT representation of the signal with $p = 1.5$

2.8284	5.1962	8.0000	1.0000
6.6479	6.6479	0.5946	3.0897
11.1803	0	1.0000	2.8284

Table 12.12 The cost function is $\sum_n |a(n)|^{1.5}$

17.0246			
13.2957		3.6843	
11.1803	**0**	**1.0000**	**2.8284**

Table 12.13 Second stage in finding the best representation

17.0246			
13.2957		3.8284	
11.1803	0	**1.0000**	**2.8284**

Table 12.14 Third stage in finding the best representation

17.1242			
13.2957		3.8284	
11.1803	0	**1.0000**	**2.8284**

■

Example 12.5 The 3-level Haar DWPT of the signal

$$\{3, 4, 2, 4, 4, 5, 1, 7\}$$

is shown in Table 12.3. Let the cost function be the energy of the signal using the norm $|a(n)|^{1.9}$. Find the best DWPT representation that maximizes the cost function.

Solution

The energy of the signal is

$$(3^2 + 4^2 + 2^2 + 4^2 + 4^2 + 5^2 + 1^2 + 7^2) = 136$$

With $p = 1.9$, the signal energy is 116.2011. Table 12.15 shows the energy of the 3-level Haar DWPT representation of the signal with $p = 1.9$. Tables 12.16–12.18 show the stages in finding the best representation. In this case, the best representation is the signal itself.

Table 12.15 The energy of the 3-level Haar DWPT representation of the signal with $p = 1.9$

8.0636	13.9288	3.7321	13.9288	13.9288	21.2835	1.0000	40.3354
20.8789	15.5779	33.6575	26.9087	0.5176	1.9319	0.5176	15.5779
35.0377	58.3305	0.2679	0.2679	2.1606	10.8076	0.2679	5.7028
88.8372	1.9319	0.5176	0.0000	11.0170	1.9319	4.1740	1.9319

Table 12.16 The cost function is $\sum_n |a(n)|^{1.9}$

116.2011							
97.0230				18.5450			
93.3682		0.5359		12.9682		5.9707	
88.8372	**1.9319**	**0.5176**	**0.0000**	**11.0170**	**1.9319**	**4.1740**	**1.9319**

Table 12.17 Second stage in finding the best representation

116.2011							
97.0230				18.5450			
93.3682		**0.5359**		**12.9682**		6.1059	
88.8372	1.9319	0.5176	0.0000	11.0170	1.9319	**4.1740**	**1.9319**

Table 12.18 Third stage in finding the best representation

116.2011							
97.0230				19.0741			
93.3682		0.5359		**12.9682**		6.1059	
88.8372	1.9319	0.5176	0.0000	11.0170	1.9319	**4.1740**	**1.9319**

■

The DWPT best representation of a signal is of three types: (i) it may be the DWT representation; (ii) it may be the input signal itself; or (iii) it may be a representation that is better than that of the input or DWT. Even if it is a better representation, the advantage should be significant to justify the increased cost of computation. One critical factor that has to be taken into account in deciding to use the DWPT is the characteristics of the class of signals of the application. This point is true to any processing by any transform.

12.3 Summary

- In the DWT, only the approximation part is further decomposed.
- In the DWPT, both the approximation part and the detail part are further decomposed.
- The DWPT provides the best representation of a signal in some applications.
- Decomposing both the approximation part and the detail part results in a large number of representations of a signal.
- Using a cost function such as signal energy and threshold, the best representation is determined.

Exercises

12.1 Find the Haar DWPT transform matrix for $N = 8$.

12.2 Find the 2-level Haar DWPT of $x(n)$. Write down all the possible representations of the signal. Let the cost function be the number of elements with magnitude greater than 1. Find the best DWPT representation that minimizes the cost function.

***12.2.1** $x(n) = \{2, 2, -3, 4\}$.
12.2.2 $x(n) = \{3, 1, 2, 4\}$.

12.3 Find the 3-level DWPT of $x(n)$. Let the cost function be the number of elements greater than 2. Find the best DWPT representation that minimizes the cost function.

12.3.1

$$\{3, 2, 2, 4, 4, 2, 1, 7\}$$

***12.3.2**

$$\{1, 7, 2, 3, 1, 1, 4, 3\}$$

13

The Discrete Stationary Wavelet Transform

The DWT is a time-variant transform. That is, the DWT of a signal and that of its shifted version (even with periodic extension) are, in general, not time-shifted versions of each other. It is expected, as the computation of the DWT involves the downsampling operation, which is time variant. Consider the signal

$$x(n) = \ldots, x(-2), x(-1), x(0), x(1), x(2), \ldots$$

The signal after downsampling by a factor 2 is

$$x_d(n) = \ldots, x(-2), x(0), x(2), \ldots$$

The downsampled signal consists of the even-indexed samples of $x(n)$. The signal delayed by one sample interval and downsampled by a factor 2 is

$$x_d(n) = \ldots, x(-3), x(-1), x(1), \ldots$$

The downsampled signal consists of the odd-indexed samples of $x(n)$. The two signals are different. To be time invariant, a delayed input should result in a delayed output. It is assumed that in this chapter, the signal is periodically extended at the borders.

13.1 The Discrete Stationary Wavelet Transform

13.1.1 The SWT

The 1-level Haar DWT of

$$x(n) = \{2, 3, -4, 5, 1, 5, 3, -2\}$$

is

$$\{5, 1, 6, 1, -1, -9, -4, 5\}/\sqrt{2}$$

Discrete Wavelet Transform: A Signal Processing Approach, First Edition. D. Sundararajan.

Companion Website: www.wiley.com/go/sundararajan/wavelet

The 1-level Haar DWT of $x_s(n)$ (the periodically right-shifted version of $x(n)$ by one position)

$$\{-2, 2, 3, -4, 5, 1, 5, 3\}$$

is

$$\{0, -1, 6, 8, -4, 7, 4, 2\}/\sqrt{2}$$

The pairings of the values for the DWT computation are totally different in the two cases. Therefore, the two sets of DWT values are different. In particular, the DWT coefficients at discontinuities of a signal are greatly perturbed. The DFT of $\{x(0) = 2, x(1) = 3\}$ is $\{X(0) = 5, X(1) = -1\}$, and the DFT of $\{x(0) = 3, x(1) = 2\}$ is $\{X(0) = 5, X(1) = 1\}$. The DFT of a data and that of its time-shifted version are related by the time-shift property. The SWT is essentially a DWT without downsampling, with assumed periodicity of the data. The analysis and detail coefficients of the 1-level SWT using Haar filters of $x(n)$ are

$$\{5, -1, 1, 6, 6, 8, 1, 0\}/\sqrt{2}$$

$$\{-1, 7, -9, 4, -4, 2, 5, -4\}/\sqrt{2}$$

The analysis and detail coefficients of the SWT of $x_s(n)$ are

$$\{0, 5, -1, 1, 6, 6, 8, 1\}/\sqrt{2}$$

$$\{-4, -1, 7, -9, 4, -4, 2, 5\}/\sqrt{2}$$

The SWT values for the two sets of the data are shifted versions. Of course, the SWT values are of double the length. The cost of restoring time invariance is that the data are decomposed into 2^j different representations, where j is the number of levels of decomposition. The SWT can be computed with other DWT filters as well.

13.1.2 The ISWT

In computing the DWT, the even-indexed coefficients are retained in the downsampling stage. However, the odd-indexed coefficients are equally capable of representing the data. Therefore, we can reconstruct the signal from either set of coefficients. Let us reconstruct $x(n)$ from its SWT coefficients. Using the even-indexed and odd-indexed coefficients, we form two sets. Using the even-indexed approximation and detail coefficients, we get

$$\{5, 1, 6, 1, \quad -1, -9, -4, 5\}/\sqrt{2}$$

This is the standard DWT of the input. The other set is formed using the odd-indexed coefficients of the SWT output.

$$\{-1, 6, 8, 0, \quad 7, 4, 2, -4\}/\sqrt{2}$$

The Haar IDWT of either set (taking the right circular shift, by one position, in the case of the odd-indexed coefficients) gives

$$\{2, 3, -4, 5, 1, 5, 3, -2\}$$

Normally, the average of the two sets yields the output signal, as the coefficients may be modified in processing the signal. Let us reconstruct the signal $x(n)$ from its SWT approximation coefficients alone. Assume that some operation on the detail coefficients resulted in making all of them zero. Using the even-indexed approximation and detail coefficients, we get

$$\{5, 1, 6, 1, \quad 0, 0, 0, 0\}/\sqrt{2}$$

The inverse DWT of this set is

$$re(n) = \{10, 10, 2, 2, 12, 12, 2, 2\}/4$$

Using the odd-indexed approximation and detail coefficients, we get

$$\{-1, 6, 8, 0, \quad 0, 0, 0, 0\}/\sqrt{2}$$

The inverse DWT of this set, circularly right shifted by one position, is

$$ros(n) = \{0, -2, -2, 12, 12, 16, 16, 0\}/4$$

The average of $re(n)$ and $ros(n)$ is the reconstructed signal

$$xr(n) = \{5, 4, 0, 7, 12, 14, 9, 1\}/4$$

The average of the inverse DWTs of each of the two different representations of the signal is the inverse discrete stationary wavelet transform (ISWT).

For the shifted input $xs(n)$, using the even-indexed approximation and detail coefficients, we get

$$\{0, -1, 6, 8, \quad 0, 0, 0, 0\}/\sqrt{2}$$

The inverse DWT of this set is

$$re(n) = \{0, 0, -2, -2, 12, 12, 16, 16\}/4$$

Using the odd-indexed approximation and detail coefficients, we get

$$\{5, 1, 6, 1, \quad 0, 0, 0, 0\}/\sqrt{2}$$

The inverse DWT of this set, circularly right shifted by one position, is

$$ros(n) = \{2, 10, 10, 2, 2, 12, 12, 2\}/4$$

The average of $re(n)$ and $ros(n)$ is the reconstructed signal

$$xr(n) = \{1, 5, 4, 0, 7, 12, 14, 9\}/4$$

At each level of decomposition, the choice of selecting the odd- or even-indexed coefficients is there. For j levels of decomposition, there are 2^j different sets of coefficients representing the signal. The average of the inverses of each of the 2^j decompositions is the inverse SWT.

13.1.3 Algorithms for Computing the SWT and the ISWT

Let us find the 2-level SWT of $x(n) = \{2, 3, -4, 5\}$. The 1-level Haar SWT decomposition yields the coefficients shown in Table 13.1. The 2-level Haar SWT decomposition, by decomposing the 1-level approximation coefficients, yields the coefficients shown in Table 13.2. The 1-level detail coefficients, in the first row of Table 13.1, are left alone. The even-indexed output of the 1-level SWT yields $\{3.5355, 0.7071\}$. Assuming periodic extension, we get $\{3.5355, 0.7071, 3.5355\}$. The SWT of these values is $\{3, 3, 2, -2\}$, shown in the first and third columns of the fourth and sixth rows of Table 13.2. The odd-indexed output of the 1-level SWT yields $\{-0.7071, 4.9497\}$. Assuming periodic extension, we get $\{-0.7071, 4.9497, -0.7071\}$. The SWT of these values is $\{3, 3, -4, 4\}$, shown in the second and fourth columns of the fourth and sixth rows of Table 13.2. Computation of the 2-level 4-point SWT using a two-stage two-channel analysis filter bank is shown in Figure 13.1. Note that the upsampled filter coefficients are used in the second stage.

As periodicity is assumed, it is tempting to use the DFT for the SWT computation, and it also helps to understand the SWT. Let us redo the example using the DFT. The input is

$$x(n) = \{2, 3, -4, 5\}$$

Table 13.1 1-level Haar SWT of $\{2, 3, -4, 5\}$

$X_{\psi(1,0)}$	$X_{\psi(1,1)}$	$X_{\psi(1,0)}$	$X_{\psi(1,1)}$
−0.7071	4.9497	−6.3640	2.1213
$X_{\phi(1,0)}$	$X_{\phi(1,1)}$	$X_{\phi(1,0)}$	$X_{\phi(1,1)}$
3.5355	−0.7071	0.7071	4.9497

Table 13.2 2-level Haar SWT of $\{2, 3, -4, 5\}$

$X_{\psi(1,0)}$	$X_{\psi(1,1)}$	$X_{\psi(1,0)}$	$X_{\psi(1,1)}$
−0.7071	4.9497	−6.3640	2.1213
$X_{\psi(0,0,0)}$	$X_{\psi(0,1,0)}$	$X_{\psi(0,0,1)}$	$X_{\psi(0,1,1)}$
2	−4	−2	4
$X_{\phi(0,0,0)}$	$X_{\phi(0,1,0)}$	$X_{\phi(0,0,1)}$	$X_{\phi(0,1,1)}$
3	3	3	3

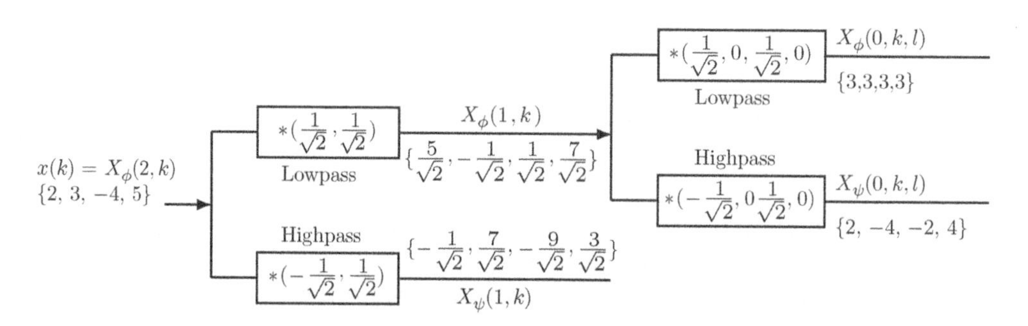

Figure 13.1 Computation of a 2-level 4-point SWT using a two-stage two-channel analysis filter bank

The filters are

$$l(n) = \{1, 1\}/\sqrt{2}, \qquad h(n) = \{-1, 1\}/\sqrt{2}$$

The DFT of the zero-appended filters

$$l(n) = \{1, 1, 0, 0\}/\sqrt{2}, \qquad h(n) = \{-1, 1, 0, 0\}/\sqrt{2}$$

(to make the filter length the same as that of the data) is

$$L(k) = \{2, 1 - j, 0, 1 + j\}/\sqrt{2}$$
$$H(k) = \{0, -1 - j, -2, -1 + j\}/\sqrt{2}$$

The DFT of $x(n)$ is

$$X(k) = \{6, 6 + j2, -10, 6 - j2\}$$

The pointwise product of $L(k)$ and $X(k)$ is

$$L(k)X(k) = \{12, 8 - j4, 0, 8 + j4\}/\sqrt{2}$$

The IDFT of $L(k)X(k)$ is

$$\{7, 5, -1, 1\}/\sqrt{2}$$

Circularly left shifting by one position, we get the approximation coefficients

$$\{5, -1, 1, 7\}/\sqrt{2}$$

The pointwise product of $H(k)$ and $X(k)$ is

$$H(k)X(k) = \{0, -4 - j8, 20, -4 + j8\}/\sqrt{2}$$

The IDFT of $H(k)X(k)$ is

$$\{3, -1, 7, -9\}/\sqrt{2}$$

Circularly left shifting by one position, we get the detail coefficients

$$\{-1, 7, -9, 3\}/\sqrt{2}$$

For the second level decomposition, we get

$$x(n) = \{5, -1, 1, 7\}/\sqrt{2}$$

The upsampled filters are

$$l(n) = \{1, 0, 1, 0\}/\sqrt{2}, \qquad h(n) = \{-1, 0, 1, 0\}/\sqrt{2}$$

The DFT of the filters is

$$L(k) = \{2, 0, 2, 0\}/\sqrt{2}, \qquad H(k) = \{0, -2, 0, -2\}/\sqrt{2}$$

The DFT of $x(n)$ is

$$X(k) = \{12, 4 + j8, 0, 4 - j8\}/\sqrt{2}$$

The pointwise product of $L(k)$ and $X(k)$ is

$$L(k)X(k) = \{12, 0, 0, 0\}$$

The IDFT of $L(k)X(k)$ is

$$\{3, 3, 3, 3\}$$

Circularly left shifting by two positions, we get the approximation coefficients

$$\{3, 3, 3, 3\}$$

Although, in this case, it makes no difference, the number of shifts required is 2^{s-1}, assuming s is the number of the levels of decomposition starting from 1. The pointwise product of $H(k)$ and $X(k)$ is

$$H(k)X(k) = \{0, -4 - j8, 0, -4 + j8\}$$

The IDFT of $H(k)X(k)$ is

$$\{-2, 4, 2, -4\}$$

Circularly left shifting by two positions, we get the detail coefficients

$$\{2, -4, -2, 4\}$$

ISWT

Computation of a partial result of a 2-level 4-point ISWT using a two-stage two-channel synthesis filter bank is shown in Figure 13.2. This computation is the same as that carried out in the IDWT. Computation of a partial result of a 2-level 4-point ISWT using a two-stage two-channel synthesis filter bank for different set of SWT coefficients that involves shifting is shown in Figures 13.3–13.5. In these figures, the letter C indicates right circular shift by one position. The average of the four partial results is the ISWT of the input.

Let us use the DFT for the ISWT computation. The upsampled synthesis filters are

$$\tilde{l}(n) = \{1, 0, 1, 0\}/\sqrt{2}, \qquad \tilde{h}(n) = \{1, 0, -1, 0\}/\sqrt{2}$$

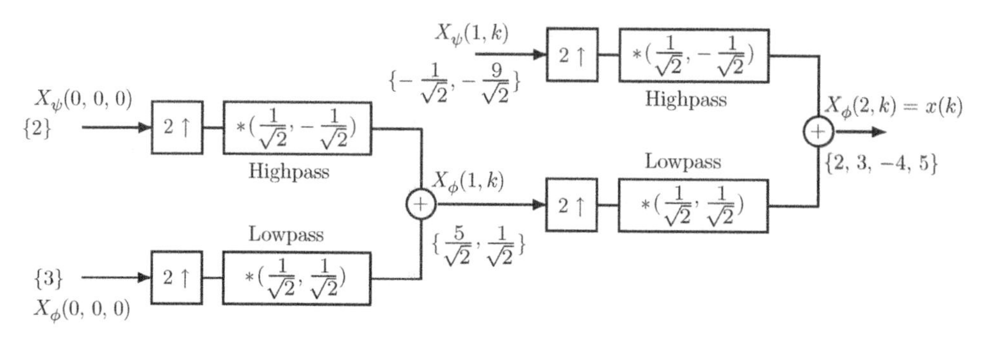

Figure 13.2 Computation of a partial result of a 2-level 4-point ISWT using a two-stage two-channel synthesis filter bank

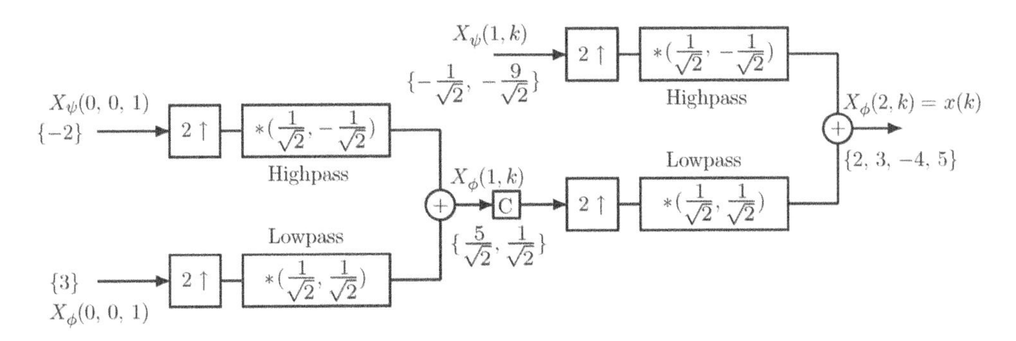

Figure 13.3 Computation of a partial result of a 2-level 4-point ISWT using a two-stage two-channel synthesis filter bank. Letter C indicates right circular shift by one position

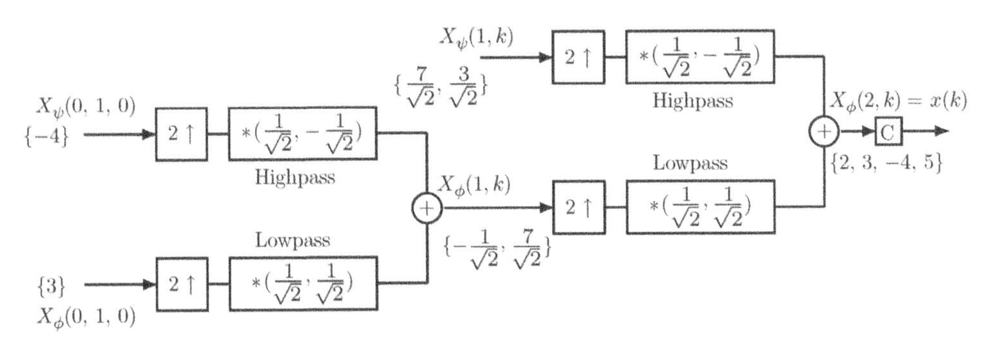

Figure 13.4 Computation of a partial result of a 2-level 4-point ISWT using a two-stage two-channel synthesis filter bank. Letter C indicates right circular shift by one position

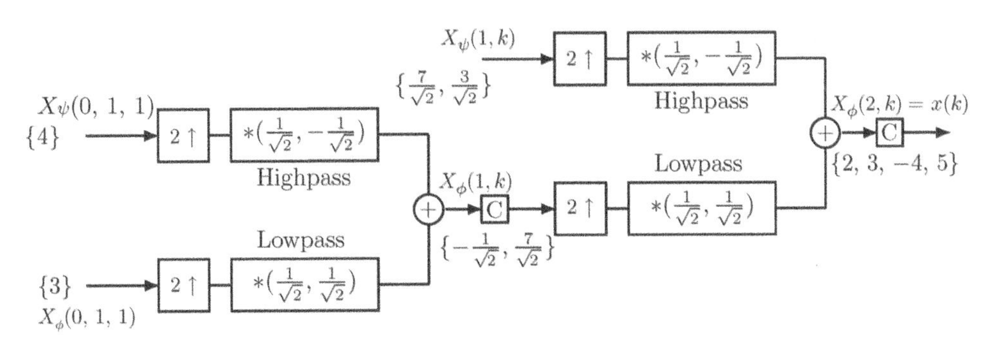

Figure 13.5 Computation of a partial result of a 2-level 4-point ISWT using a two-stage two-channel synthesis filter bank. Letter C indicates right circular shift by one position

The DFT of the filters is

$$\tilde{L}(k) = \{2, 0, 2, 0\}/\sqrt{2}, \qquad \tilde{H}(k) = \{0, 2, 0, 2\}/\sqrt{2}$$

The DFT of approximation coefficients $\{3, 3, 3, 3\}$ in the last row of Table 13.2 is

$$Xa(k) = \{12, 0, 0, 0\}$$

The DFT of detail coefficients $\{2, -4, -2, 4\}$ in the fourth row of Table 13.2 is

$$Xd(k) = \{0, 4 + j8, 0, 4 - j8\}$$

The pointwise product of $\tilde{L}(k)$ and $Xa(k)$

$$\tilde{L}(k)Xa(k) = \{24, 0, 0, 0\}/\sqrt{2}$$

The pointwise product of $\tilde{H}(k)$ and $Xd(k)$

$$\tilde{H}(k)Xd(k) = \{0, 8 + j16, 0, 8 - j16\}/\sqrt{2}$$

The sum of $\tilde{L}(k)Xa(k)$ and $\tilde{H}(k)Xd(k)$ divided by 2 is

$$\{12, 4 + j8, 0, 4 - j8\}/\sqrt{2}$$

The IDFT of these values is

$$\{5, -1, 1, 7\}/\sqrt{2}$$

For the second stage, the filters are

$$\tilde{l}(n) = \{1, 1\}/\sqrt{2}, \qquad \tilde{h}(n) = \{1, -1\}/\sqrt{2}$$

The DFT of the zero-appended filters (to make the filter length the same as that of the data) is

$$\tilde{L}(k) = \{2, 1 - j, 0, 1 + j\}/\sqrt{2}$$

$$\tilde{H}(k) = \{0, 1 + j, 2, 1 - j\}/\sqrt{2}$$

The DFT of approximation coefficients

$$\{5, -1, 1, 7\}/\sqrt{2}$$

is

$$Xa(k) = \{12, 4 + j8, 0, 4 - j8\}/\sqrt{2}$$

The DFT of detail coefficients

$$\{-0.7071, 4.9497, -6.3640, 2.1213\}$$

is

$$Xd(k) = \{-0.0000, 8 - j4, -20, 8 + j4\}/\sqrt{2}$$

The pointwise product of $\tilde{L}(k)$ and $Xa(k)$

$$\tilde{L}(k)Xa(k) = \{12, 6 + j2, 0, 6 - j2\}$$

The pointwise product of $\tilde{H}(k)$ and $Xd(k)$

$$H(k)Xd(k) = \{0, 6 + j2, -20, 6 - j2\}$$

The sum of $\tilde{L}(k)Xa(k)$ and $\tilde{H}(k)Xd(k)$ divided by 2 is

$$\{6, 6 + j2, -10, 6 - j2\}$$

The IDFT of these values is

$$\{2, 3, -4, 5\}$$

For using different filters, shifting of the data has to be adjusted.

13.1.4 2-D SWT

Let us compute the 1-level 2-D Haar SWT of the input

$$x(n_1, n_2) = \begin{bmatrix} 2 & -3 \\ 1 & 2 \end{bmatrix}$$

In computing the 2-D SWT, two downsampling operations are left out compared to the row-column computation of the 2-D DWT. Therefore, while the DWT is not expansive, the output of the 2-D SWT is 4 times expansive. That is, we compute convolution only with assumed periodic extension at the borders. Applying the lowpass filter to each row, we get the partial result

$$\frac{1}{\sqrt{2}} \begin{bmatrix} -1 & -1 \\ 3 & 3 \end{bmatrix}$$

The application of lowpass filter to the columns of the partial result yields the approximation coefficients

$$X_\phi^0(n_1, n_2) = \begin{bmatrix} 1 & 1 \\ 1 & 1 \end{bmatrix}$$

The application of highpass filter to the columns of the partial result yields the V detail coefficients

$$X_V^0(n_1, n_2) = \begin{bmatrix} -2 & -2 \\ 2 & 2 \end{bmatrix}$$

Applying the highpass filter to each row of $x(n_1, n_2)$, we get the partial result

$$\frac{1}{\sqrt{2}} \begin{bmatrix} 5 & -5 \\ -1 & 1 \end{bmatrix}$$

The application of lowpass filter to the columns of the partial result yields the H detail coefficients

$$X_H^0(n_1, n_2) = \begin{bmatrix} 2 & -2 \\ 2 & -2 \end{bmatrix}$$

The application of highpass filter to the columns of the partial result yields the D detail coefficients

$$X_D^0(n_1, n_2) = \begin{bmatrix} 3 & -3 \\ -3 & 3 \end{bmatrix}$$

ISWT

To find the inverse, four different combinations of the coefficients are formed, and the 2-D IDWT is computed and averaged. Combining the four top-left coefficients of the four coefficient matrices, we get

$$\begin{bmatrix} 1 & 2 \\ -2 & 3 \end{bmatrix}$$

This is the standard DWT of the input. The 2-D IDWT of this set yields the input

$$\begin{bmatrix} 2 & -3 \\ 1 & 2 \end{bmatrix}$$

Combining the four top-right coefficients of the four coefficient matrices, we get

$$\begin{bmatrix} 1 & -2 \\ -2 & -3 \end{bmatrix}$$

The 2-D IDWT of this set yields

$$\begin{bmatrix} -3 & 2 \\ 2 & 1 \end{bmatrix}$$

Circularly right shifting the rows by one position, we get the input. Combining the four bottom-left coefficients of the four coefficient matrices, we get

$$\begin{bmatrix} 1 & 2 \\ 2 & -3 \end{bmatrix}$$

The 2-D IDWT of this set yields

$$\begin{bmatrix} 1 & 2 \\ 2 & -3 \end{bmatrix}$$

Circularly shifting the columns up by one position, we get the input. Combining the four bottom-right coefficients of the four coefficient matrices, we get

$$\begin{bmatrix} 1 & -2 \\ 2 & 3 \end{bmatrix}$$

The 2-D IDWT of this set yields

$$\begin{bmatrix} 2 & 1 \\ -3 & 2 \end{bmatrix}$$

Circularly right shifting the rows by one position and the columns up by one position, we get the input.

13.2 Summary

- The DWT is a shift-variant transform.
- The SWT, which is essentially a DWT without downsampling, is a shift-invariant transform.
- In the DWT, the even-indexed coefficients are retained after downsampling. However, the odd-indexed coefficients also represent the signal. Therefore, all the possible DWTs are computed in each stage of decomposition. This results in a set of coefficients to represent the signal.
- The SWT is useful in applications such as denoising and detecting discontinuities.

Exercises

13.1 Compute the 1-level Haar SWT of $x(n)$. Compute the ISWT of the resulting coefficients and verify that the input is reconstructed. Let the input data be circularly left shifted by two positions. Compute the 1-level SWT of the shifted data and verify that they are the same as that of $x(n)$, but shifted by two positions.

13.1.1

$$x(n) = \{2, 1, 2, 4, -1, 2, 1, 4\}$$

13.1.2

$$x(n) = \{1, 1, 3, -4, 1, 2, 1, -1\}$$

13.1.3

$$x(n) = \{4, 1, -3, 4, 1, 2, 2, 3\}$$

13.1.4

$$x(n) = \{2, 1, 3, 2, 1, 4, 1, -3\}$$

13.1.5

$$x(n) = \{-2, 1, -1, 4, 3, 2, 1, 2\}$$

13.2 Compute the 1-level Haar SWT of $x(n)$. Replace all the detail coefficients by zero. Compute the ISWT of the resulting SWT coefficients.

13.2.1

$$x(n) = \{3, 2, 1, 2, -1, 2, 1, 4\}$$

***13.2.2**

$$x(n) = \{1, 4, 1, -3, 1, 2, 1, -1\}$$

13.2.3

$$x(n) = \{4, 1, -3, 4, 1, 1, 3, -4\}$$

13.2.4

$$x(n) = \{1, 2, 1, -1, 1, 4, 1, -3\}$$

13.2.5

$$x(n) = \{2, 1, -1, 4, 3, 2, 1, 2\}$$

13.3 Compute the 2-level Haar SWT of $x(n)$. Compute the ISWT of the resulting coefficients and verify that the input is reconstructed. Let the input data be circularly right shifted by one position. Compute the 1-level SWT of the shifted data and verify that they are the same as that of $x(n)$, but shifted by one position. Use DFT and IDFT in computing the SWT and the ISWT.

13.3.1

$$x(n) = \{2, 1, 2, 4\}$$

13.3.2

$$x(n) = \{1, 1, 3, -4\}$$

13.3.3

$$x(n) = \{4, 1, -3, 4\}$$

13.4 Compute the 1-level Haar 2-D SWT of $x(n_1, n_2)$. Compute the ISWT of the resulting coefficients and verify that the input is reconstructed.

13.4.1

$$x(n_1, n_2) = \begin{bmatrix} 1 & 2 \\ 3 & 4 \end{bmatrix}$$

13.4.2

$$x(n_1, n_2) = \begin{bmatrix} 2 & 1 \\ 3 & 4 \end{bmatrix}$$

13.4.3

$$x(n_1, n_2) = \begin{bmatrix} 3 & -1 \\ -2 & 4 \end{bmatrix}$$

13.5 Compute the 1-level Haar 2-D SWT of $x(n_1, n_2)$. Let the rows be circularly left-shifted by one position. Compute the 1-level 2-D SWT of the shifted data and verify that they are the same as that of $x(n_1, n_2)$, but the rows circularly left-shifted by one position.

13.5.1

$$x(n_1, n_2) = \begin{bmatrix} 1 & 2 \\ 3 & 4 \end{bmatrix}$$

13.5.2

$$x(n_1, n_2) = \begin{bmatrix} 2 & 1 \\ 3 & 4 \end{bmatrix}$$

13.5.3

$$x(n_1, n_2) = \begin{bmatrix} 3 & -1 \\ -2 & 4 \end{bmatrix}$$

14

The Dual-Tree Discrete Wavelet Transform

A major drawback of the DWT is that it is shift variant. It is due to the fact that the phase information of the signal is lost in its DWT representation. Another problem is that the directional selectivity of multidimensional DWT is not good. Both of the drawbacks can be alleviated by increasing the redundancy of the transform coefficients. In Chapter 13, we described the SWT, which is one of the versions of the DWT that is time invariant. In this chapter, another version of the DWT is introduced that is nearly shift invariant but with much less redundancy. By using a pair of real DWTs, both the shift invariance and directional selectivity can be obtained to a good degree with a redundancy of 2^d, where d is the dimension of the signal. For 1-D signals, the redundancy is by a factor of 2. The output of the first real DWT is the real part of the complex transform, while the second real DWT yields the imaginary part. This version of the DWT is called the DTDWT. As the Fourier transform does not have these drawbacks, the idea is to design the DWT similar to the Fourier transform, but without giving up the local nature of the DWT. The basis functions are complex wavelets with the imaginary part being the Hilbert transform of the other (approximately). The advantages of the DTDWT include (i) nearly shift invariant, (ii) good directional selectivity, (iii) no oscillations in the transform magnitude near discontinuities of a signal, and (iv) no aliasing. The PR conditions of the DWT hold only if no processing of the transform coefficients is carried out. However, processing of the transform coefficients (which is inevitable), such as thresholding and quantization, makes the PR conditions not perfect. This leads to artifacts in the reconstructed signal. Instead of real-valued scaling and wavelet functions in the DWT, the scaling and wavelet functions of the DTDWT are complex valued.

$$\phi(t) = \phi_r(t) + j\phi_i(t) \quad \text{and} \quad \psi(t) = \psi_r(t) + j\psi_i(t)$$

Similarly, the transform coefficients are represented as

$$X_\phi(j,k) = Xr_\phi(j,k) + jXi_\phi(j,k) \quad \text{and} \quad X_\psi(j,k) = Xr_\psi(j,k) + jXi_\psi(j,k)$$

Discrete Wavelet Transform: A Signal Processing Approach, First Edition. D. Sundararajan.

Companion Website: www.wiley.com/go/sundararajan/wavelet

This approach, which is more advantageous, in designing complex filters is to make the real and imaginary parts individually orthogonal or biorthogonal. It should be mentioned that there are other complex DWTs.

14.1 The Dual-Tree Discrete Wavelet Transform

Let $\boldsymbol{W}_0$ and $\boldsymbol{W}_1$ be the orthogonal square transform matrices of the pair of real DWTs forming the DTDWT transform matrix $\boldsymbol{W}$. Then,

$$\boldsymbol{W} = \begin{bmatrix} \boldsymbol{W}_0 \\ \boldsymbol{W}_1 \end{bmatrix} \tag{14.1}$$

which is a rectangular matrix. Note that

$$\boldsymbol{W}_0^T \boldsymbol{W}_0 = \boldsymbol{I} \quad \text{and} \quad \boldsymbol{W}_1^T \boldsymbol{W}_1 = \boldsymbol{I}$$

The real and imaginary parts of the DTDWT of a real signal x are given, respectively, by

$$\boldsymbol{X}_r = \boldsymbol{W}_0 \boldsymbol{x} \qquad \text{and} \qquad \boldsymbol{X}_i = \boldsymbol{W}_1 \boldsymbol{x}$$

The complex coefficients are represented as

$$\boldsymbol{X}_r + j\boldsymbol{X}_i$$

The left inverse of $\boldsymbol{W}$ is given by

$$\boldsymbol{W}^{-1} = \frac{1}{2} \begin{bmatrix} \boldsymbol{W}_0^{-1} & \boldsymbol{W}_1^{-1} \end{bmatrix} \tag{14.2}$$

It can be verified that

$$\boldsymbol{W}^{-1} \boldsymbol{W} = \frac{1}{2} \begin{bmatrix} \boldsymbol{W}_0^{-1} & \boldsymbol{W}_1^{-1} \end{bmatrix} \begin{bmatrix} \boldsymbol{W}_0 \\ \boldsymbol{W}1 \end{bmatrix} = \frac{1}{2}(\boldsymbol{I} + \ \boldsymbol{I}) = \boldsymbol{I}$$

Splitting the constant factor $1/2$, we get the forward and inverse DTDWT transform matrices as

$$\boldsymbol{W} = \frac{1}{\sqrt{2}} \begin{bmatrix} \boldsymbol{W}_0 \\ \boldsymbol{W}_1 \end{bmatrix} \qquad \text{and} \qquad \boldsymbol{W}^{-1} = \frac{1}{\sqrt{2}} \begin{bmatrix} \boldsymbol{W}_0^{-1} & \boldsymbol{W}_1^{-1} \end{bmatrix} \tag{14.3}$$

14.1.1 Parseval's Theorem

For orthogonal transforms, Parseval's theorem states that the sum of the squared-magnitude of a time-domain sequence (energy) equals the sum of the squared-magnitude of the corresponding transform coefficients. In the case of DTDWT, the sum of the squared-magnitude of two sets of coefficients has to be taken. That is,

$$\sum_n |x(n)|^2 = \sum_{j,k} (|X_r(j,k)|^2 + |X_i(j,k)|^2)$$

Two stages of the DTDWT analysis filter bank are shown in Figure 14.1. It consists of a pair of the two stages of the DWT analysis filter bank. The top cascade is called Tree R (tree that

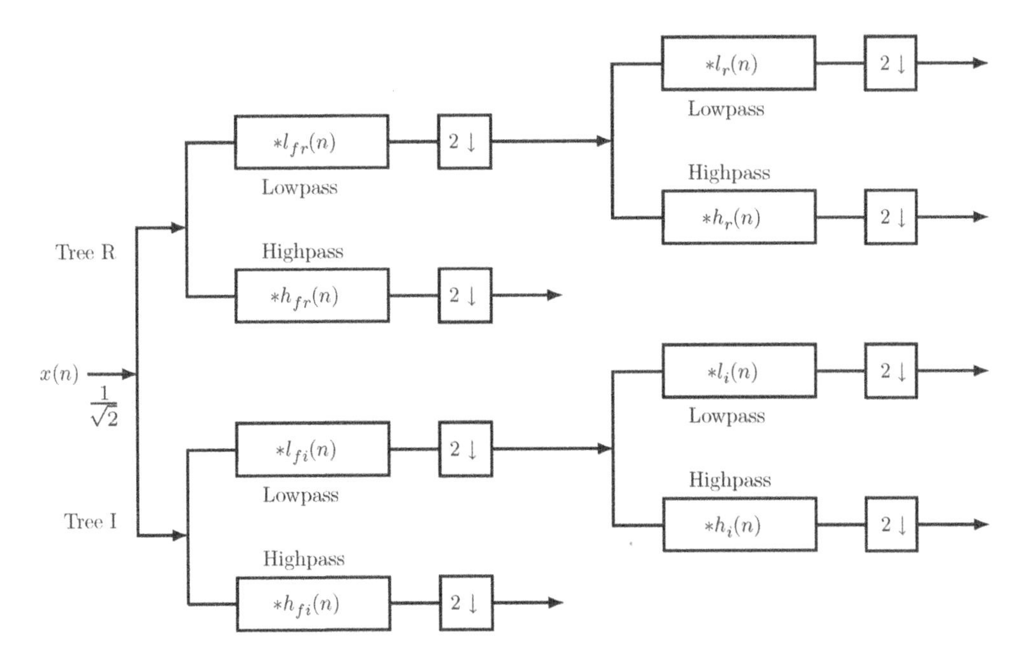

Figure 14.1 DTDWT analysis filter bank

produces the real part of the complex DTDWT coefficient), and the bottom is called Tree I (tree that produces the imaginary part of the complex DTDWT coefficient). Therefore, the outputs of the Tree R and Tree I are the real and imaginary parts, respectively, of the complex DTDWT coefficients. Both of the trees decompose the input $x(n)$. The set of filters of Tree R is different from that of Tree I. Each set satisfies the PR conditions

$$\tilde{L}(e^{j\omega})L(e^{j\omega}) + \tilde{H}(e^{j\omega})H(e^{j\omega}) = 2e^{-jK\omega}$$

$$\tilde{L}(e^{j\omega})L(e^{j(\omega+\pi)}) + \tilde{H}(e^{j\omega})H(e^{j(\omega+\pi)}) = 0$$

Example 14.1 The impulse response of the first-stage analysis filters of Tree R is given in Table 14.1. The synthesis filters are the time-reversed and shifted versions of the corresponding analysis filters. Verify that they satisfy the perfect reconstruction conditions.

Solution

The solution is similar to that presented in Chapter 4.

$$l_{fr}(n) * \tilde{l}_{fr}(n) = \{0, 0, -0.0010, 0, 0.0166, 0, -0.1064, 0, 0.5908, 1, 0.5908, 0,$$
$$-0.1064, 0, 0.0166, 0, -0.0010, 0, 0\}$$

$$h_{fr}(n) * \tilde{h}_{fr}(n) = \{0, 0, 0.0010, 0, -0.0166, 0, 0.1064, 0, -0.5908, 1, -0.5908, 0,$$
$$0.1064, 0, -0.0166, 0, 0.0010, 0, 0\}$$

The sum of the two convolutions is

$$l_{fr}(n) * \tilde{l}_{fr}(n) + h_{fr}(n) * \tilde{h}_{fr}(n) = \{0, 0, 0, 0, 0, 0, 0, 0, 0, 2, 0, 0, 0, 0, 0, 0, 0, 0, 0\}$$

The DTFT of this sequence is of the form $2e^{-jK\omega}$. The first of the two conditions is met.

$$((-1)^n l_{fr}(n)) * \tilde{l}_{fr}(n) = \{0, 0, 0.0010, 0.0020, -0.0146, 0, 0.1376, -0.0020, -0.3467, 0, 0.3467, -0.0020, -0.1376, 0, 0.0146, 0.0020, -0.0010, 0, 0\}$$

$$((-1)^n h_{fr}(n)) * \tilde{h}_{fr}(n) = \{0, 0, -0.0010, -0.0020, 0.0146, 0, -0.1376, 0.0020, 0.3467, 0, -0.3467, 0.0020, 0.1376, 0, -0.0146, -0.0020, 0.0010, 0, 0\}$$

The sum of the two convolutions is

$$((-1)^n l_{fr}(n)) * \tilde{l}_{fr}(n) + ((-1)^n h_{fr}(n)) * \tilde{h}_{fr}(n) = \{0, 0, 0, 0, 0, 0, 0, 0, 0, 0, 0, 0, 0, 0, 0, 0, 0, 0, 0\}$$

The second of the two conditions is also met. ■

In addition to the PR conditions, the complex wavelets satisfy the Hilbert transform condition approximately. The Hilbert transform is used in applications such as sampling of bandpass signals and single-sideband amplitude modulation. Given a real signal, its Hilbert transform is obtained by adding $-\pi/2$ radians to every frequency component of the signal. For example, the Hilbert transform of $\cos(\omega n)$ is

$$\cos\left(\omega n - \frac{\pi}{2}\right) = \sin(\omega n)$$

Obviously, the signal and its transform belong to the same domain, in contrast to two domains in the case of other transforms. The complex signal constructed using a signal as the real part and its Hilbert transform as the imaginary part has only one-sided spectrum. The complex signal $\cos\left(\frac{2\pi}{4}n\right) + j\sin\left(\frac{2\pi}{4}n\right) = e^{j\frac{2\pi}{4}n}$ has one-sided spectrum. That is, $X(e^{j\omega}) = \delta\left(\omega - \frac{2\pi}{4}\right)$.

The synthesis filters are the time-reversed and shifted versions of the respective analysis filters. That is, the order of the coefficients is reversed. The first-stage filters are different from those of the subsequent stages for each tree. Therefore, four sets of filters are required. For example, one set of DTDWT filter coefficients is given in Tables 14.1–14.4. Other sets of filters are also available. For good accuracy in computation, they should be used with double

Table 14.1 First-stage filter coefficients – Tree R

n	Lowpass, $l_{fr}(n)$	Highpass, $h_{fr}(n)$
0	0	0
1	−0.08838834764832	−0.01122679215254
2	0.08838834764832	0.01122679215254
3	0.69587998903400	0.08838834764832
4	0.69587998903400	0.08838834764832
5	0.08838834764832	−0.69587998903400
6	−0.08838834764832	0.69587998903400
7	0.01122679215254	−0.08838834764832
8	0.01122679215254	−0.08838834764832
9	0	0

Table 14.2 First-stage filter coefficients – Tree I

n	Lowpass, $l_{fi}(n)$	Highpass, $h_{fi}(n)$
0	0.01122679215254	0
1	0.01122679215254	0
2	−0.08838834764832	−0.08838834764832
3	0.08838834764832	−0.08838834764832
4	0.69587998903400	0.69587998903400
5	0.69587998903400	−0.69587998903400
6	0.08838834764832	0.08838834764832
7	−0.08838834764832	0.08838834764832
8	0	0.01122679215254
9	0	−0.01122679215254

Table 14.3 Second and subsequent stages filter coefficients – Tree R

n	Lowpass, $l_r(n)$	Highpass, $h_r(n)$
0	0.03516384	0
1	0	0
2	−0.08832942	−0.11430184
3	0.23389032	0
4	0.76027237	0.58751830
5	0.58751830	−0.76027237
6	0	0.23389032
7	−0.11430184	0.08832942
8	0	0
9	0	−0.03516384

Table 14.4 Second and subsequent stages filter coefficients – Tree I

n	Lowpass, $l_i(n)$	Highpass, $h_i(n)$
0	0	−0.03516384
1	0	0
2	−0.11430184	0.08832942
3	0	0.23389032
4	0.58751830	−0.76027237
5	0.76027237	0.58751830
6	0.23389032	0
7	−0.08832942	−0.11430184
8	0	0
9	0.03516384	0

precision. In the examples, four digits are used to represent numbers for compactness. In other aspects, the filter banks are similar to those of the DWT.

The magnitude of the frequency responses of the analysis filters of stage 2 is shown in Figure 14.2, which is the same for both the trees. Two stages of the DTDWT synthesis filter bank are shown in Figure 14.3.

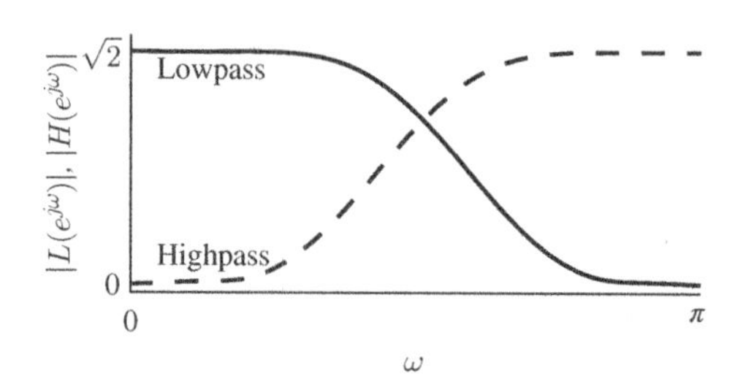

Figure 14.2 The magnitude of the frequency response of the analysis filters

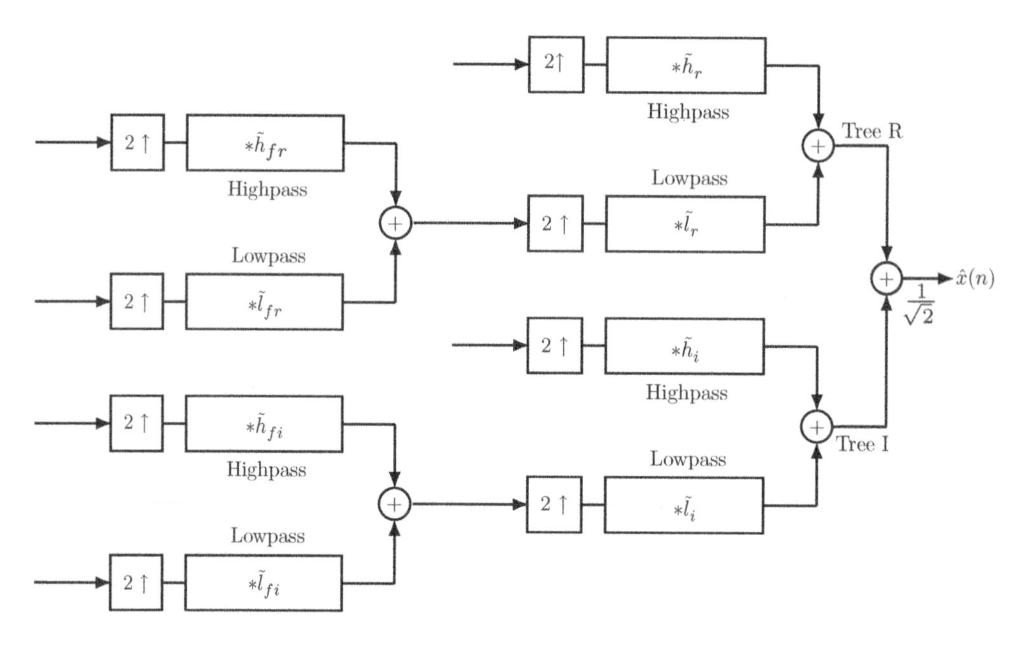

Figure 14.3 DTDWT synthesis filter bank

14.2 The Scaling and Wavelet Functions

Approximations of the scaling function derived from the second-stage synthesis lowpass filter $\tilde{l}_r(n)$ of Tree R are shown in Figure 14.4. This function is derived using Equation (8.13). Approximations of the wavelet function derived from the second-stage synthesis highpass filter $\tilde{h}_r(n)$ of Tree R are shown in Figure 14.5. This function is derived using Equation (8.15). Approximations of the scaling function derived from the second-stage synthesis lowpass filter $\tilde{l}_i(n)$ of Tree I are shown in Figure 14.6. This function is derived using Equation (8.13). Approximations of the wavelet function derived from the second-stage synthesis highpass filter $\tilde{l}_i(n)$ of Tree I are shown in Figure 14.7. This function is derived using Equation (8.15). The magnitude of the DFT of the approximation of the complex signal $\psi_r(t) + j\psi_i(t)$, the real part being the Tree R wavelet function (Figure 14.5(f)) and the imaginary part being the Tree I wavelet function (Figure 14.7(f)), is shown in Figure 14.8. The spectrum is approximately one sided. The magnitude of the spectrum for the negative frequencies is almost zero.

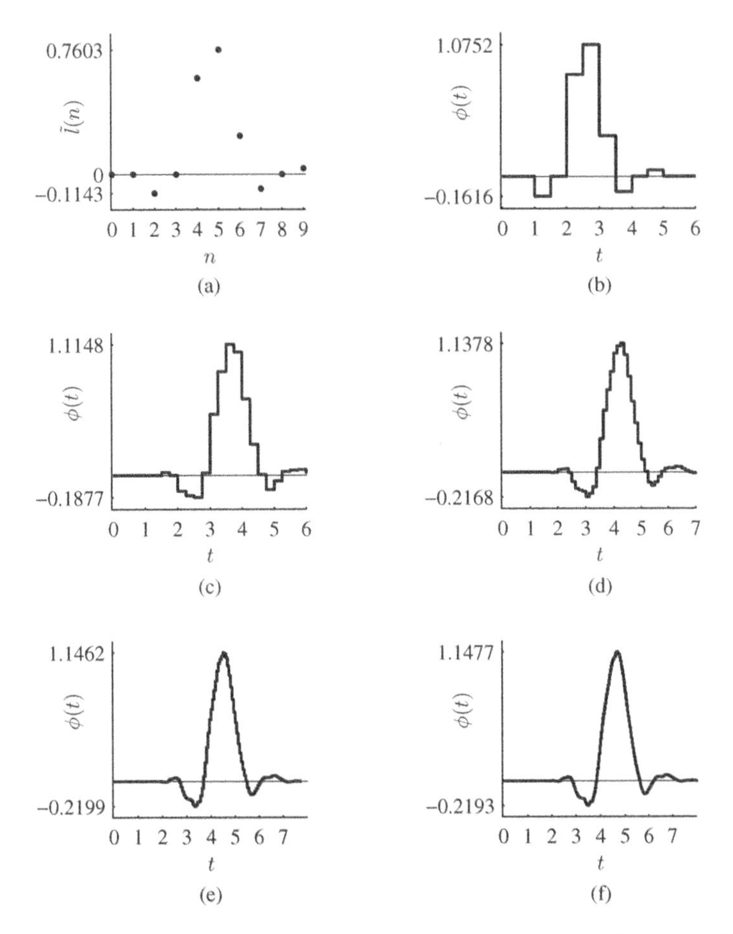

Figure 14.4 Approximations of the scaling function derived from the second-stage synthesis lowpass filter of Tree R. (a) Filter coefficients; (b) after one stage; (c) after two cascade stages; (d) after three cascade stages; (e) after four cascade stages; (f) after five cascade stages

14.3 Computation of the DTDWT

Example 14.2 Find the 1-level DTDWT of $x(n)$. Verify that $x(n)$ is reconstructed by computing the inverse DTDWT (IDTDWT). Verify Parseval's theorem.

$$x(n) = \{2, 3, 1, 4, 3, 1, 2, 2, 4, 2, 1, 3, 1, 1, 2, 2, 2, 3, 4, 1, 2, 3, 4, 1, 1, 2, 2, 4, 4, 3, 1, -2\}$$

Solution

The length of $x(n)$ is 32. Firstly, we give three examples of individual coefficient computation. The lowpass filter coefficients of the first stage of Tree R are

$$l_{fr}(n) = \{0, -0.0884, 0.0884, 0.6959, 0.6959, 0.0884, -0.0884, 0.0112, 0.0112, 0\}$$

In computing the DTDWT, we use periodic convolution. The time-reversed filter coefficients are

$$\{0, 0.0112, 0.0112, -0.0884, 0.0884, 0.6959, 0.6959, 0.0884, -0.0884, 0\}$$

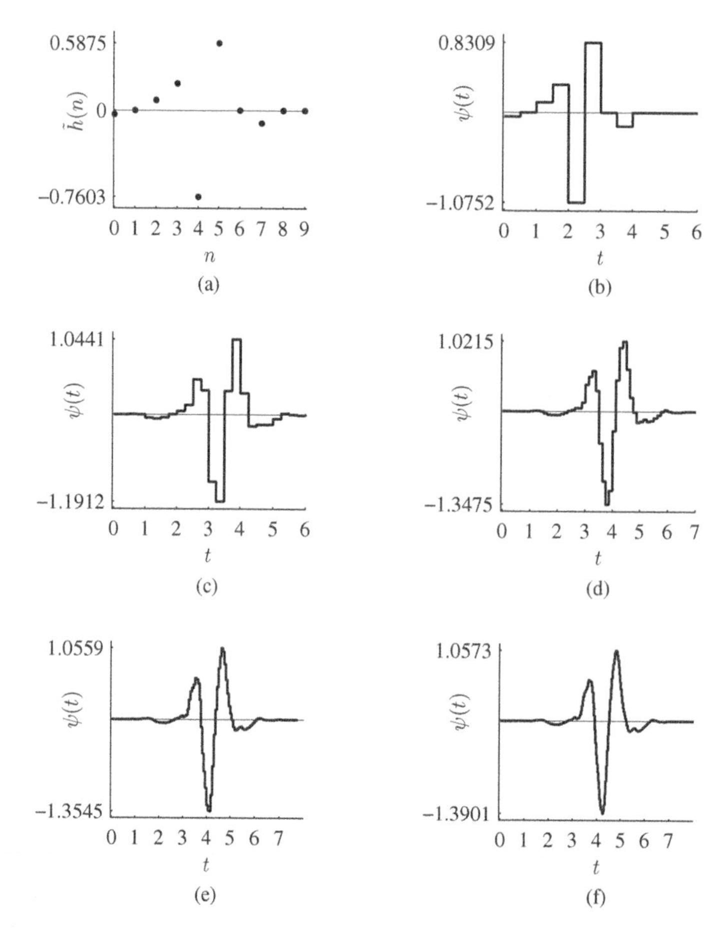

Figure 14.5 Approximations of the wavelet function derived from the second-stage synthesis highpass filter of Tree R. (a) Filter coefficients; (b) after one stage; (c) after two cascade stages; (d) after three cascade stages; (e) after four cascade stages; (f) after five cascade stages

The first coefficient is computed using the following 10 data values. The last six values are the first six values of the given input $x(n)$. The first four values are the last four values of $x(n)$. In this chapter, periodicity of the input $x(n)$ is assumed.

$$x_1(n) = \{4, 3, 1, -2, \quad 2, 3, 1, 4, 3, 1\}$$

The sum of the pointwise product of these values with the time-reversed filter coefficients is the first output value, 3.2704. All of the values have to be divided by the constant $\sqrt{2}$. Although the computation looks not as simple as that with Haar filter due to the length and the precision of the filter, this computation can also be verified easily by hand calculation to a good approximation. By rounding the coefficients, we get

$$\{0, 0, 0, -0.1, 0.1, 0.7, 0.7, 0.1, -0.1, 0\}$$

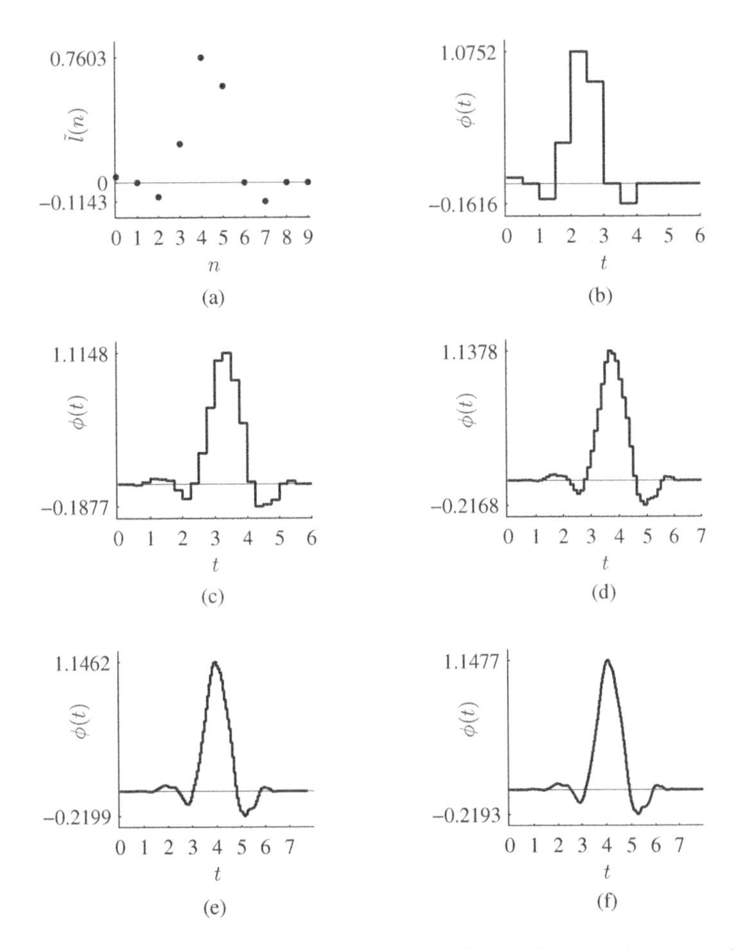

Figure 14.6 Approximations of the scaling function derived from the second-stage synthesis lowpass filter of Tree I. (a) Filter coefficients; (b) after one stage; (c) after two cascade stages; (d) after three cascade stages; (e) after four cascade stages; (f) after five cascade stages

Now, the pointwise product of the coefficients and $x_1(n)$ is

$$= 0 + 0 + 0 - 0.1(-2) + 0.1(2) + 0.7(3) + 0.7(1) + 0.1(4) - 0.1(3) + 0$$
$$= -0.1(1) + 0.1(6) + 0.7(4) = 0.1(5) + 0.7(4) = 3.3 \approx 3.2704$$

This procedure helps our understanding and gives confidence in programming any algorithm.

The second set of double-shifted data values is

$$x_2(n) = \{1, -2, \ 2, 3, 1, 4, 3, 1, 2, 2\}$$

and the corresponding output is 4.6061. The third set of double-shifted data values is

$$x_3(n) = \{2, 3, 1, 4, 3, 1, 2, 2, 4, 2\}$$

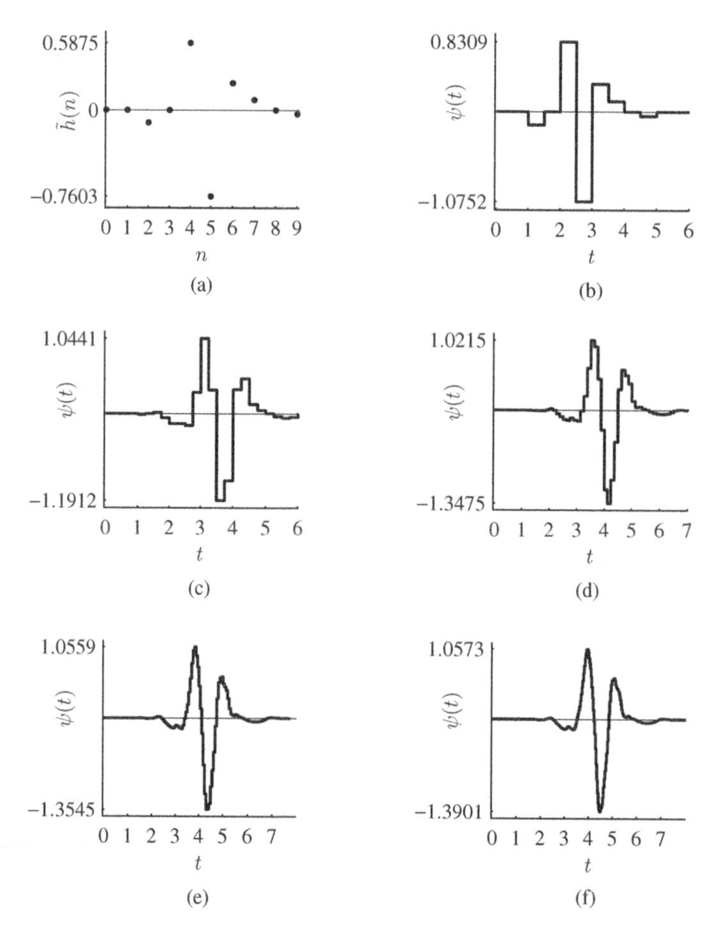

Figure 14.7 Approximations of the wavelet function derived from the second-stage synthesis highpass filter of Tree I. (a) Filter coefficients; (b) after one stage; (c) after two cascade stages; (d) after three cascade stages; (e) after four cascade stages; (f) after five cascade stages

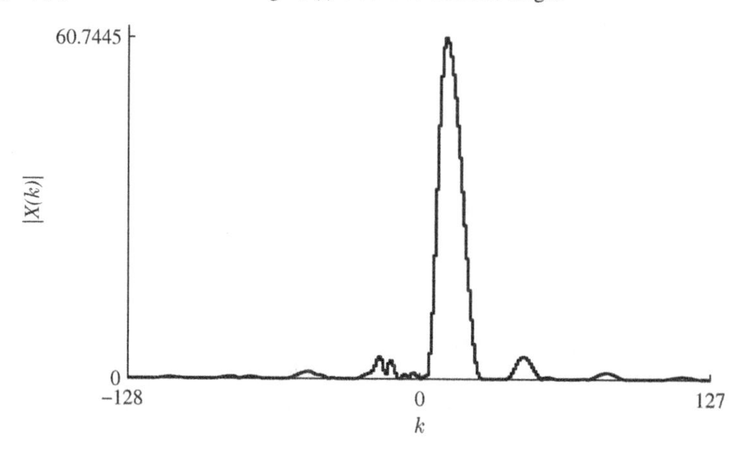

Figure 14.8 The magnitude of the DFT of the approximation of the complex signal $\psi_r(t) + j\psi_i(t)$, the real part being the Tree R wavelet function and the imaginary part being the Tree I wavelet function

and the corresponding output is 1.8673. Similarly, the other 13 approximation coefficients are computed. This procedure is repeated for the other three filters. The 16 approximation coefficients corresponding to the lowpass filter of Tree R are

$$\{3.2704, 4.6061, 1.8673, 4.4306, 2.4749, 2.6740, 1.9445, 2.8284,$$
$$4.8165, 2.1325, 5.0381, 1.5138, 2.8619, 5.7665, 2.4747, 0.0897\}/\sqrt{2}$$

The 16 detail coefficients corresponding to the highpass filter of Tree R are

$$\{-2.7724, 1.9994, 0.5851, -0.7732, -1.3694, 0.5078, 1.3919, -0.7071,$$
$$0.3424, -0.7955, -0.6959, -0.7843, -0.2652, 0.5528, -0.0448, 0.7070\}/\sqrt{2}$$

The 16 approximation coefficients corresponding to the lowpass filter of Tree I are

$$\{2.9939, 3.7895, 3.1159, 2.8284, 4.0210, 2.6517, 1.6246, 2.7512,$$
$$3.8007, 3.5355, 3.5130, 3.5467, 1.7241, 4.3411, 5.3693, -0.8167\}/\sqrt{2}$$

The 16 detail coefficients corresponding to the highpass filter of Tree I are

$$\{0.1543, 2.1425, -1.2922, -0.1433, -1.4142, 1.7454, -0.0223, -0.2428,$$
$$0.6075, -2.0877, 0.7071, -1.9445, 0.6187, 1.0047, -0.0659, -1.8885\}/\sqrt{2}$$

Parseval's theorem: It can be verified that the sum of the squared-magnitude of the input $x(n)$ is 203 and that of the four sets of the coefficients is also the same.

The inverse:

The upsampled convolutions are carried out by convolving the coefficients with the even- and odd-indexed synthesis filter coefficients. The synthesis filters are the time-reversed and shifted versions of the analysis filters. The time-reversed lowpass synthesis filter coefficients of the first stage of Tree R are

$$h_o(n) = \{0, 0.0884, 0.6959, -0.0884, 0.0112\}$$
$$h_e(n) = \{-0.0884, 0.6959, 0.0884, 0.0112, 0\}$$

The last two and the first three approximation coefficients of Tree R are

$$X(k) = \{2.4747, 0.0897, 3.2704, 4.6061, 1.8673\}/\sqrt{2}$$

The sum of pointwise product of $X(k)$ and $h_e(n)$ is $0.1844/\sqrt{2}$. The sum of pointwise product of $X(k)$ and $h_o(n)$ is $1.8975/\sqrt{2}$. The last and the first four detail coefficients of Tree R are

$$X(k) = \{0.0897, 3.2704, 4.6061, 1.8673, 4.4306\}/\sqrt{2}$$

The sum of pointwise product of $X(k)$ and $h_e(n)$ is $2.6959/\sqrt{2}$. The sum of pointwise product of $X(k)$ and $h_o(n)$ is $3.3790/\sqrt{2}$. The four partial results are

$$\{0.1844, 1.8975, 2.6959, 3.3790\}\sqrt{2}$$

For Tree R, the time-reversed odd- and even-indexed synthesis highpass filter coefficients are

$$h_o(n) = \{0, 0.0112, 0.0884, 0.6959, -0.0884\}$$

$$h_e(n) = \{-0.0112, 0.0884, -0.6959, -0.0884, 0\}$$

The last two and the first three detail coefficients of Tree R are

$$X(k) = \{-0.0448, 0.7070, -2.7724, 1.9994, 0.5851\}/\sqrt{2}$$

The sum of pointwise product of $X(k)$ and $h_e(n)$ is $1.8156/\sqrt{2}$. The sum of pointwise product of $X(k)$ and $h_o(n)$ is $1.1025/\sqrt{2}$. The last and the first four detail coefficients of Tree R are

$$X(k) = \{0.7070, -2.7724, 1.9994, 0.5851, -0.7732\}/\sqrt{2}$$

The sum of pointwise product of $X(k)$ and $h_e(n)$ is $-1.6961/\sqrt{2}$. The sum of pointwise product of $X(k)$ and $h_o(n)$ is $0.6212/\sqrt{2}$. The four partial results are

$$\{1.8156, 1.1025, -1.6961, 0.6212\}\sqrt{2}$$

The sum of pairwise Tree R partial results, divided by $\sqrt{2}$, yields

$$\{1, 1.5, 0.5, 2\}$$

Similar computation with Tree I also yields the same values, resulting in the first four values of the reconstructed input

$$\{2, 1, 3, 4\}$$

■

In principle, the computation of the 2-level DTDWT and IDTDWT is the same as that of 1-level, except that we use a different set of filters for the second and subsequent stages. One set of filters is shown in Tables 14.3 and 14.4. The input to the 2-level computation is the two approximation coefficients of the 1-level DTDWT of Tree R and Tree I.

Example 14.3 Find the 2-level DTDWT of $x(n)$. Verify Parseval's theorem.

$$x(n) = \{2, 3, 1, 4, 3, 1, 2, 2, 4, 2, 1, 3, 1, 1, 2, 2, 2, 3, 4, 1, 2, 3, 4, 1, 1, 2, 2, 4, 4, 3, 1, -2\}$$

Solution

The 1-level approximation coefficients of Tree R, from Example 14.2, are

$$\{2.3125, 3.2570, 1.3204, 3.1329, 1.7500, 1.8908, 1.3750, 2.0000,$$
$$3.4058, 1.5079, 3.5625, 1.0704, 2.0237, 4.0775, 1.7499, 0.0634\}$$

First, we give three examples of individual coefficient computation. The lowpass filter coefficients of the second stage of Tree R are

$$l(n) = \{0.0352, 0, -0.0883, 0.2339, 0.7603, 0.5875, 0, -0.1143, 0, 0\}$$

In computing the DTDWT, we use periodic convolution. The time-reversed filter coefficients are

$$\{0, 0, -0.1143, 0, 0.5875, 0.7603, 0.2339, -0.0883, 0, 0.0352\}$$

The first coefficient is computed using the following 10 approximation coefficients. The last six values are the first six values of approximation coefficients. The first four values are the last four.

$$X_\phi(n) = \{2.0237, 4.0775, 1.7499, 0.0634, 2.3125, 3.2570, 1.3204, 3.1329, 1.7500, 1.8908\}$$

The sum of the pointwise product of these values with the time-reversed filter coefficients is the first output value, 3.7334. The second set of double-shifted data approximation coefficients is

$$X_\phi(n) = \{1.7499, 0.0634, 2.3125, 3.2570, 1.3204, 3.1329, 1.7500, 1.8908, 1.3750, 2.0000\}$$

and the corresponding output is 3.2060. The third set of double-shifted approximation coefficients is

$$X_\phi(n) = \{2.3125, 3.2570, 1.3204, 3.1329, 1.7500, 1.8908, 1.3750, 2.0000, 3.4058, 1.5079\}$$

and the corresponding output is 2.5128. Similarly, the other five 2-level approximation coefficients are computed. This procedure is repeated for other filters. The 2-level approximation coefficients corresponding to the lowpass filter of Tree R are

$$\{3.7334, 3.2060, 2.5128, 2.8294, 3.8723, 2.6330, 4.4000, 1.2085\}$$

The 2-level detail coefficients corresponding to the highpass filter of Tree R are

$$\{-0.1045, 1.5251, 0.3199, 0.5077, -1.2980, -1.9404, 1.2949, -0.6583\}$$

The 2-level approximation coefficients corresponding to the lowpass filter of Tree I are

$$\{2.5277, 3.2091, 3.3885, 1.9955, 3.6789, 3.3990, 3.2580, 2.9385\}$$

The 2-level detail coefficients corresponding to the highpass filter of Tree I are

$$\{-0.6015, 0.2940, 0.3455, -0.3681, 0.2413, -0.1780, -1.1588, 3.1916\}$$

Parseval's theorem: It can be verified that the sum of the squared-magnitude of the input $x(n)$ is 203 and that of the six sets of the coefficients is also the same. ■

Example 14.4 Find the 2-level DTDWT of $x(n) = \delta(n-6)$ and $x(n) = \delta(n-9)$. Compare the energy of the 1-level detail coefficients to that of the 1-level DWT coefficients computed using Daubechies orthogonal filter of length 12.

Solution

The input is

$$\delta(n-6) = \{0, 0, 0, 0, 0, 0, 1, 0, 0, 0, 0, 0, 0, 0, 0, 0, 0, 0, 0, 0, 0, 0, 0, 0, 0, 0, 0, 0, 0, 0, 0, 0\}$$

The input is shown in Figure 14.9(a), and the magnitude of its 1-level complex DTDWT coefficients is shown in Figure 14.9(b). The coefficients are computed as they were computed in

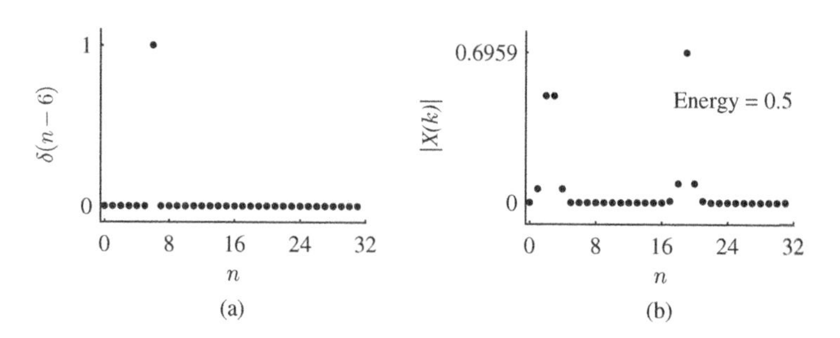

Figure 14.9 (a) The delayed unit-impulse $\delta(n-6)$; (b) the magnitude of the 1-level complex DTDWT coefficients and the energy of the detail coefficients

Example 14.3. In addition to demonstrating the shift invariance, this example serves as another example of DTDWT computation. The 16 approximation coefficients corresponding to the lowpass filter of Tree R are

$$\{-0, -0.0625, 0.4921, 0.0625, 0.0079, 0, 0, 0, -0, 0, 0, 0, -0, 0, 0, -0\}$$

The 16 detail coefficients corresponding to the highpass filter of Tree R are

$$\{0, -0.0079, 0.0625, -0.4921, -0.0625, 0, 0, -0, -0, 0, 0, 00, 0, -0, 0\}$$

The eight approximation coefficients of the second stage of Tree R are

$$\{0.0621, 0.3385, -0.0516, -0.0009, 0, 0, -0.0022, 0.0077\}$$

The eight detail coefficients of the second stage of Tree R are

$$\{-0.0439, -0.3520, 0.0520, -0.0166, -0.0003, -0, 0, 0.0071\}$$

The 16 approximation coefficients corresponding to the lowpass filter of Tree I are

$$\{-0, 0.0079, 0.0625, 0.4921, -0.0625, -0, -0, 0, 0, -0, 0, 0, -0, -0, 0, 0\}$$

The 16 detail coefficients corresponding to the highpass filter of Tree I are

$$\{0, -0, -0.0625, -0.4921, 0.0625, -0.0079, 0, -0, -0, 0, -0, 0, 0, 0, -0, -0\}$$

The 8 approximation coefficients corresponding to the second stage of Tree I are

$$\{-0.0516, 0.3385, 0.0621, 0.0077, -0.0022, 0, -0, -0.0009\}$$

The 8 detail coefficients of the second stage of Tree I are

$$\{0.0520, -0.3520, -0.0439, 0.0071, 0, 0, -0.0003, -0.0166\}$$

The complex DTDWT coefficients are formed with the Tree R coefficients as the real part and the corresponding Tree I coefficients as the imaginary part. The magnitude of the complex coefficients is

$$\{0.0807, 0.4787, 0.0807, 0.0078, 0.0022, 0, 0.0022, 0.0078, 0.0681, 0.4978, 0.0681, 0.0181,$$
$$0.0003, 0, 0.0003, 0.0181, 0, 0.0079, 0.0884, 0.6959, 0.0884, 0.0079, 0, 0, 0, 0, 0, 0, 0, 0, 0, 0\}$$

Few coefficients contribute to the energy significantly. This is the advantage of the DWT, localization. It is well-known that the DFT spectrum of an impulse has equal amplitude for all of the coefficients in the spectrum (Figure 4.2(b)).

Now, consider the input

$$\delta(n-9) = \{0,0,0,0,0,0,0,0,0,1,0,0,0,0,0,0,0,0,0,0,0,0,0,0,0,0,0,0,0,0,0,0\}$$

The input is shown in Figure 14.10(a), and the magnitude of its 1-level complex DTDWT coefficients is shown in Figure 14.10(b). The 16 approximation coefficients corresponding to the lowpass filter of Tree R are

$$\{0,0,-0,0.0625,0.4921,-0.0625,0.0079,0,00,0-0,0-0,0,0\}$$

The 16 detail coefficients corresponding to the highpass filter of Tree R are

$$\{-0,-0,0,0.0079,0.0625,0.4921,-0.0625,-0,-0,0,0,0,0,-0,-0,-0\}$$

The eight approximation coefficients of the second stage of Tree R are

$$\{-0.0077,0.1681,0.2434,-0.0516,-0.0009,0,0,0.0022\}$$

The eight detail coefficients of the second stage of Tree R are

$$\{-0.0071,0.0439,-0.3962,0.0228,-0.0166,-0.0003,0,0\}$$

The 16 approximation coefficients corresponding to the lowpass filter of Tree I are

$$\{0,-0,0.0079,-0.0625,0.4921,0.0625,0,0,0,0,0,-0,-0,0,-0,0\}$$

The 16 detail coefficients corresponding to the highpass filter of Tree I are

$$\{0,-0,0,-0.0625,0.4921,0.0625,0.0079,0,00,0,-0,-0,-0,-0,0\}$$

The eight approximation coefficients corresponding to the second stage of Tree I are

$$\{0.0071,-0.0378,0.3955,-0.0286,0.0173,0,-0,0\}$$

The eight detail coefficients of the second stage of Tree I are

$$\{-0.0059,0.1728,0.2407,-0.0562,-0,0,-0,0.0022\}$$

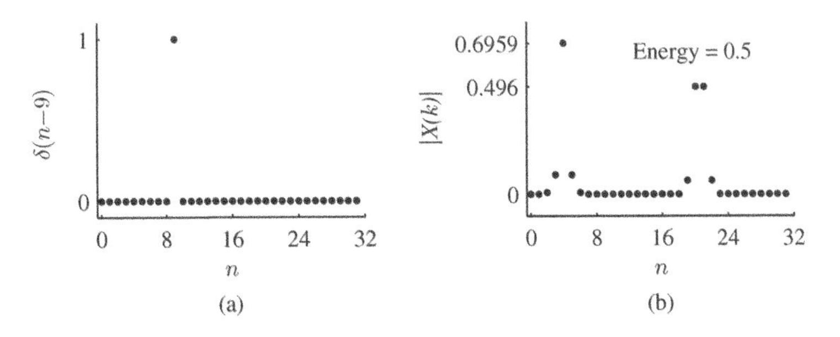

Figure 14.10 (a) The delayed unit-impulse $\delta(n-9)$; (b) the magnitude of the 1-level DTDWT complex coefficients and the energy of the detail coefficients

The complex DTDWT coefficients are formed with the Tree R coefficients as the real part and the corresponding Tree I coefficients as the imaginary part. The magnitude of the complex coefficients are

$$\{0.0105, 0.1723, 0.4644, 0.0590, 0.0173, 0, 0, 0.0022, 0.0092, 0.1783, 0.4636, 0.0607,$$
$$0.0166, 0.0003, 0, 0.0022, 0, 0, 0, 0.0630, 0.4960, 0.4960, 0.0630, 0, 0, 0, 0, 0, 0, 0, 0, 0\}$$

Only few coefficients contribute to the energy significantly. The shift of the signal has not affected the energy. This characteristic of the DTDWT is similar to that of the DFT.

The 1-level DWT coefficients of the delayed unit-impulse $\delta(n-6)$ with Daubechies orthogonal filter of length 12 and the energy of the detail coefficients and the DWT coefficients of $\delta(n-9)$ and the energy of the detail coefficients are shown, respectively, in Figure 14.11(a) and (b). There is a significant change in the energy due to the shift of the input.

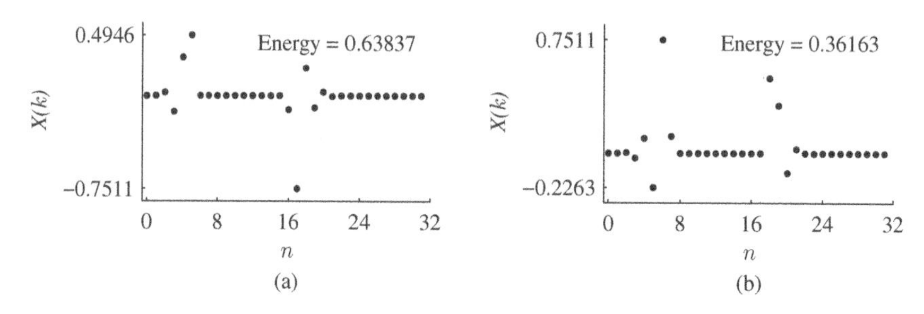

Figure 14.11 (a) The 1-level DWT coefficients of the delayed unit-impulse $\delta(n-6)$ with Daubechies orthogonal filter of length 12 and the energy of the detail coefficients; (b) the DWT coefficients of $\delta(n-9)$ and the energy of the detail coefficients ■

14.4 Summary

- The DTDWT is designed similar to the Fourier transform, but with the local nature of the DWT retained.
- The DTDWT is essentially a pair of standard DWTs. It has two cascades of filter banks.
- Different sets of filters are used for the first and the rest of the filter bank stages.
- The wavelet functions of the real and imaginary trees approximately form an Hilbert transform pair.
- Each set of lowpass and highpass filters is orthogonal or biorthogonal and satisfies the PR property.
- Synthesis filters are the time-reversed and shifted versions of analysis filters.
- The DTDWT is nearly shift invariant.
- The DTDWT has good directional selectivity in multidimensions.
- The DTDWT is 2 times expansive for 1-D signals.
- The DTDWT is a good alternative to the DWT in applications in which the DWT has shortcomings.

Exercises

14.1 The impulse response of the first-stage analysis filters of Tree I is given in Table 14.2. The synthesis filters are the time-reversed and shifted versions of the corresponding analysis filters. Verify that they satisfy the perfect reconstruction conditions.

14.2 The impulse response of the second-stage analysis filters of Tree R is given in Table 14.3. The synthesis filters are the time-reversed and shifted versions of the corresponding analysis filters. Verify that they satisfy the perfect reconstruction conditions.

14.3 The impulse response of the second-stage analysis filters of Tree I is given in Table 14.4. The synthesis filters are the time-reversed and shifted versions of the corresponding analysis filters. Verify that they satisfy the perfect reconstruction conditions.

15

Image Compression

A discrete image $x(n_1, n_2)$ is a 2-D signal represented by an $M \times N$ matrix of elements, called pixels (picture elements). The value of a pixel is a three-element vector in the case of a color image. In a black and white image, each value, called gray level value, is an integer from a specified range. Usually, gray level value 0 is colored black, and the highest value is colored white. In this chapter, we assume that the image matrix is square and the dimension is a power of 2. Compression algorithms are applicable to 1-D signals as well, as a 1-D signal is a $1 \times N$ matrix. A picture is worth more than 10,000 words. At the same time, a picture requires more amount of storage. For example, a 256×256 image with 8 bits per pixel (gray level values varying from 0 to 255) requires $256 \times 256 \times 8 = 524,288$ bits of storage. The problem of image compression is to reduce this storage requirement as much as possible, without any loss of information (lossless compression) or with some acceptable loss of information (lossy compression).

Data compression is finding the minimum amount of data required to represent a certain amount of information, and it is a necessity for efficient storage and transmission of data. The basic property of signals that enables compression and reconstruction is that the spectrum of practical signals tends to fall off to insignificant levels at high frequencies. Therefore, the higher-frequency components of a signal can be coded with less number of bits or discarded altogether. There are redundancies in most of the signals. The redundancies in an image can be classified into three major types: (i) coding redundancy, (ii) spatial redundancy, and (iii) irrelevant information. A gray level value in an image is usually represented using 8 bits. However, the probability of occurrence of these values varies. Therefore, variable- length codes can be used to reduce the amount of storage required. The values with a higher probability of occurrence can be coded with less number of bits and vice versa. Furthermore, the values of neighboring pixels are correlated spatially. The application of a transform, such as the DWT, decorrelates and, thus, enables the representation of an image with reduced storage. In addition, certain types of information are ignored by the human visual system or not required for the intended application of an image. Consider the 512 samples of an arbitrary 1-D signal, shown in Figure 15.1(a). The 3-level Haar DWT decomposition of the signal is shown in Figure 15.1(b). The amplitude of the signal level, in each of the last three components, is much closer to zero than that of the preceding lower-frequency component.

Discrete Wavelet Transform: A Signal Processing Approach, First Edition. D. Sundararajan.

Companion Website: www.wiley.com/go/sundararajan/wavelet

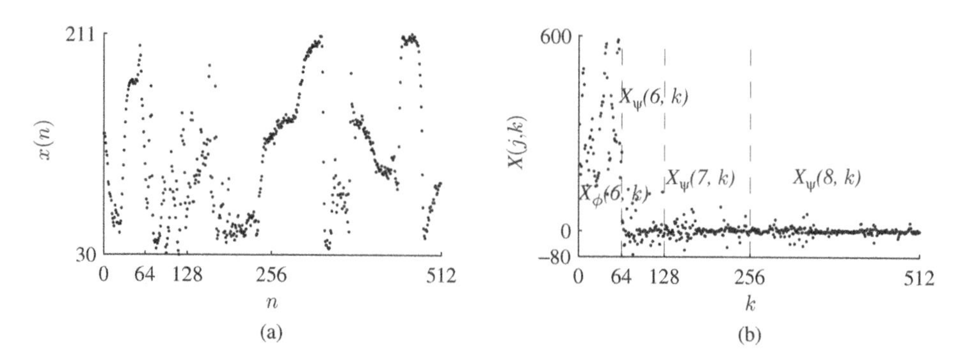

Figure 15.1 (a) An arbitrary signal; (b) the 3-level Haar DWT decomposition of the signal

15.1 Lossy Image Compression

There are three basic steps, shown in Figure 15.2(a), in lossy transform image compression: (i) transformation, (ii) quantization, and (iii) coding. There are two basic steps, shown in Figure 15.2(b), in image reconstruction: (i) decoding and (ii) inverse transformation. There is no inverse quantization, as quantization is an irreversible process.

15.1.1 Transformation

In general, the transformation of a signal into another equivalent representation makes it easier to interpret or process. The DWT of a signal decomposes it into subband components. Figure 15.3 shows the 3-level DWT decomposition of a 16×16 image. With three levels, we get $3(3) + 1 = 10$ subband components. The given image of size 16×16 is considered as the approximation component $\boldsymbol{X}_{\phi}^{4}$ with no detail components. In 1-level of decomposition, we get the approximation component $\boldsymbol{X}_{\phi}^{3}$ (the top left 8×8 subimage) and three detail components $\boldsymbol{X}_{\mathrm{V}}^{3}$, $\boldsymbol{X}_{\mathrm{H}}^{3}$, and $\boldsymbol{X}_{\mathrm{D}}^{3}$, each of size 8×8. In 2-level of decomposition, the approximation component $\boldsymbol{X}_{\phi}^{3}$ is further decomposed into $\boldsymbol{X}_{\phi}^{2}\boldsymbol{X}_{\mathrm{V}}^{2}$, $\boldsymbol{X}_{\mathrm{H}}^{2}$ and $\boldsymbol{X}_{\mathrm{D}}^{2}$, each of size 4×4. In 3-level of decomposition, the approximation component $\boldsymbol{X}_{\phi}^{2}$ is further decomposed into $\boldsymbol{X}_{\phi}^{1}\boldsymbol{X}_{\mathrm{V}}^{1}$, $\boldsymbol{X}_{\mathrm{H}}^{1}$ and

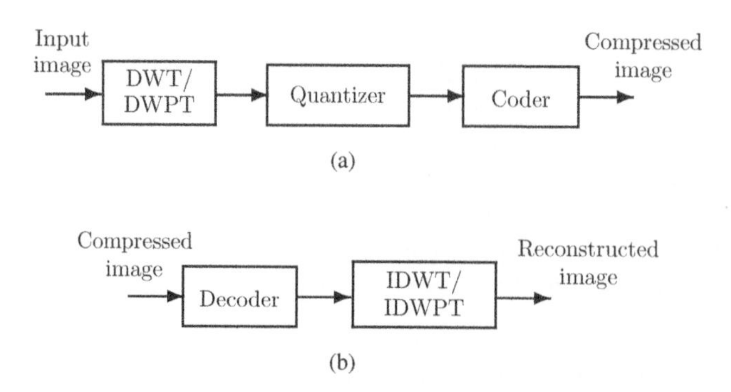

Figure 15.2 (a) Image compression using DWT/DWPT; (b) reconstructing a compressed image

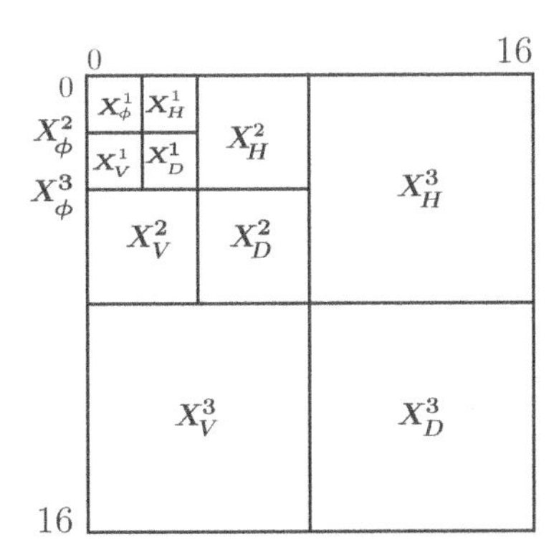

Figure 15.3 Labeling of the DWT coefficients of 3-level decomposition of a 16×16 image

X_{D}^1, each of size 2×2. There is one more level of decomposition possible with each component having one value. Usually, three or four levels of decompositions are carried out for a 512×512 image. The magnitudes of the values of the components decrease with increasing label values. This step is invertible, except for the numerical errors in the computation of the transform, and considered as lossless part of the compression procedure. After the transformation, we are able to discard or use less number of bits to code the signal components with smaller amplitudes in the quantization and thresholding step.

The DWT is more often used for image compression. The major advantage is the structure of the DWT transformed image. Exploiting this structure, the use of the DWT gives better reconstructed images and compression ratios, particularly at lower bit rates. Note that the compressed image must also store the information where the nonzero values are located. The alternative of using the DWPT has the advantage of optimized compression with respect to some criterion. However, the disadvantages are the computation and storage of the best basis. This overhead is reduced in the case of compressing a class of images (such as fingerprint images) that can use the same best basis. As the DWT basis functions are local and the computational complexity of DWT is $O(N)$, subdivision of the original image is unnecessary. This has the advantage of reducing the blocking artifact of other transforms. As the number of decomposition levels is increased, the number of resolutions used in the representation of an image increases, as well as the processing time.

The most suitable DWT filters are of biorthogonal type. The advantage is that longer symmetric filters are available. A symmetric filter handles the border problem more effectively. Furthermore, symmetric filters provide linear phase response, which is essential in the analysis of images. Shifting of the image contents between the subbands does not occur with linear phase filters. The location of features such as a ridge remains the same within each subband. An image is less tolerant to phase distortion compared to amplitude distortion.

A smoother filter yields minimum number of coefficients for the smooth parts of an image. That is, the frequency response of the filter must have more number of zeros at $\omega = \pi$. This feature is also desirable for reconstruction filters, as it results in fewer artifacts in the reconstructed image. Longer filters provide more zeros. However, they require more computation time and result in oscillations in the reconstructed image. The CDF 9/7 filter is often used for lossy image compression. For lossless image compression, the 5/3 spline filter is often used.

15.1.2 Quantization

Even if the pixel values are integers, the output of most of the transforms is a set of real values. These values have to be mapped to approximations so that the conflicting requirements of reducing the loss of information and reducing the bits to store the transformed image are satisfied to the stated specifications. This step is irreversible. Quantization is simply sampling of the amplitude. Each value is expressed using finite number of digits. Let Δ be the quantization step size. A smaller step size provides more number of quantization levels with reduced error in the representation of signal samples. The uniform scalar quantization function is given by

$$q(x) = \begin{cases} \left\lfloor \dfrac{|x|}{\Delta} \right\rfloor & \text{if } x \geq 0 \\ -\left\lfloor \dfrac{|x|}{\Delta} \right\rfloor & \text{if } x < 0 \end{cases}$$

where x is the input value and q maps x to an integer. All the values with magnitude less than Δ are quantized to zero. The quantization function truncates the quotient of dividing the magnitude of the input value by the quantization step size Δ to an integer and then attaches the corresponding sign. Figure 15.4 shows the quantization scheme with input values ranging from -10 to 10. The signal being quantized is shown by a continuous line. The input values are

$$\{10, 9.2388, 7.0711, 3.8268, 0, -3.8268, -7.0711, -9.2388, -10\}$$

With $\Delta = 2.5$, the quantized representations are

$$\{4, 3, 2, 1, 0, -1, -2, -3, -4\}$$

and the approximate values are

$$\{10, 7.5, 5, 2.5, 0, -2.5, -5, -7.5, -10\}$$

The number 9.2388 is approximated to $(3)(2.5) = 7.5$, and the integer 3 represents this value in the quantized set. With a smaller $\Delta = 1.25$, the quantized representations are

$$\{8, 7, 5, 3, 0, -3, -5, -7, -8\}$$

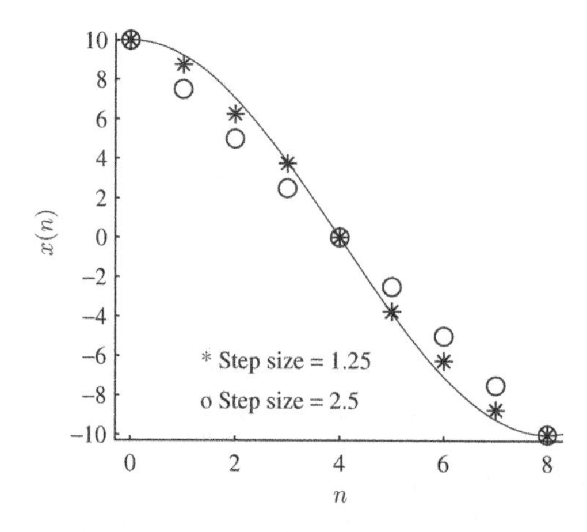

Figure 15.4 Uniform scalar quantization

and the approximate values are

$$\{10, 8.75, 6.25, 3.75, 0, -3.75, -6.25, -8.75, -10\}$$

The number 9.2388 is approximated to $(7)(1.25) = 8.75$, and the integer 7 represents this value in the quantized set. A smaller quantization step size provides more levels and, therefore, results in a better approximation. Several other quantization schemes are used. The objective is to minimize the quantization error.

Usually, a threshold $T > 0$ is specified, and all the values with magnitude less than T are quantized to zero. Hard thresholding method, shown in Figure 15.5(a), is defined as

$$g_h(x) = \begin{cases} 0, & \text{if } |x| \le T \\ x, & \text{if } |x| > T \end{cases}$$

In hard thresholding, the value of the function is retained, if its magnitude is greater than a chosen threshold value. Otherwise, the value of the function is set to zero. Soft thresholding method, shown in Figure 15.5(b), is defined as

$$g_s(x) = \begin{cases} 0, & \text{if } |x| \le T \\ x - T, & \text{if } x > T \\ x + T, & \text{if } x < -T \end{cases}$$

The difference in soft thresholding is that the value of the function is made closer to zero by adding or subtracting the chosen threshold value from it, if its magnitude is greater than the threshold. With thresholding, the quantization map becomes $q(x)g(x)$. For image compression, hard thresholding is usually preferred. Let the input be

$$x(n) = \{116.5000, 132.7813, -155.5625, 60.3438\}$$

and the threshold values be 70 and 120. Hard thresholding with threshold 70 yields

$$x(n) = \{116.5000, 132.7813, -155.5625, 0\}$$

Hard thresholding with threshold 120 yields

$$x(n) = \{0, 132.7813, -155.5625, 0\}$$

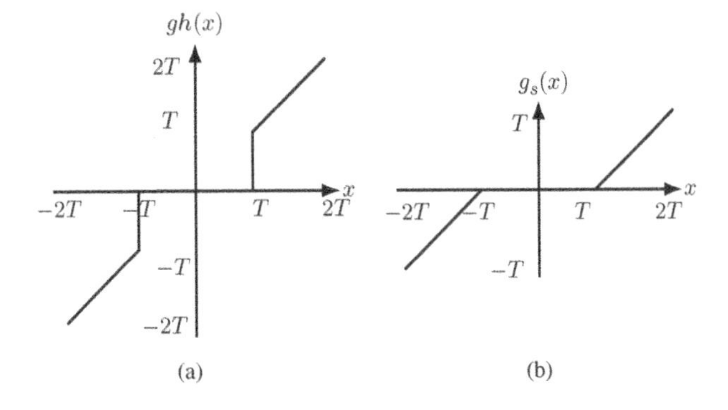

Figure 15.5 (a) Hard thresholding; (b) soft thresholding

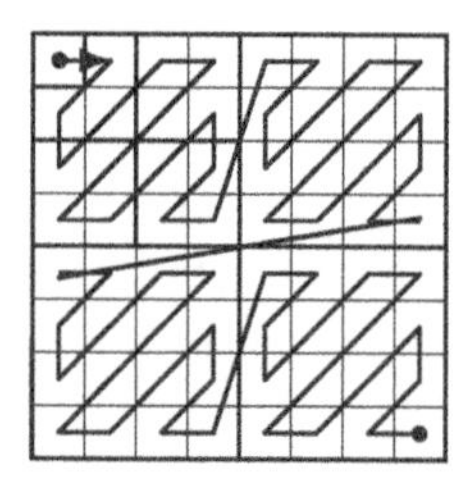

Figure 15.6 Zigzag scan order of the coefficients of the 1-level DWT decomposition of a 2-D 8×8 image

Soft thresholding with threshold 70 yields

$$x(n) = \{46.5000, 62.7813, -85.5625, 0\}$$

Soft thresholding with threshold 120 yields

$$x(n) = \{0, 12.7813, -35.5625, 0\}$$

The coefficients in the higher-indexed subbands must be scanned after the coefficients in the lower-indexed subbands are scanned. One of the possible orders of the coefficients of the 1-level DWT decomposition of a 2-D 8×8 image is shown in Figure 15.6. The $N \times N$ matrix of coefficients is mapped to a $1 \times N^2$ vector. This puts the low-frequency coefficients, which have, in general, high magnitudes, in the top of the vector and the high-frequency coefficients in the bottom. Zigzag scanning helps in getting a high runlength of zeros.

15.1.3 Coding

The image is represented, after transformation and quantization, by a set of integers. An image consists of $N \times N$ pixels. Each pixel is coded using p bits. Then, the total number of bits required to store the image is $N \times N \times p$. In the transformed image, as the energy of an image is concentrated in a small set of coefficients, using a shorter bit-sequence representation for more probable ones and a longer bit-sequence representation for others reduces the total number of bits required. Various types of coding methods have been developed to take advantage of this characteristic of the transformed and quantized image.

Huffman Coding

Huffman coding is a popular method to code images. When each character is coded individually, Huffman code is optimum and yields the smallest number of bits to code an image. Consider coding the 4×4 image

23	44	23	44
32	44	51	23
23	23	23	32
32	44	23	44

Table 15.1 Assignment of Huffman codes

Character	Frequency	Relative frequency	Code	Total Bits
23	7	$\frac{7}{16}$	0	7
44	5	$\frac{5}{16}$	10	10
32	3	$\frac{3}{16}$	110	9
51	1	$\frac{1}{16}$	111	3

A list of unique values of the image and the relative frequencies is created, as shown in Table 15.1. Then, a binary tree is constructed with the unique values as its leaves. The higher the frequency of a value, the more closer is its location to the root. A right leaf is assigned the bit 1, and the left leaf is assigned the bit 0. This choice is arbitrary and could be reversed. The position of a value in the tree indicates its code. We simply read the bits from the root to the location of the value. The higher the frequency of a value, the fewer the bits used to encode it. Consider the binary tree shown in Figure 15.7. A two-leaf tree is constructed from the two lowest frequencies (1/16 and 3/16). The sum 4/16 is placed at the node connecting the two leaves. Now, a left node is created with frequency 5/16. A two-leaf tree is constructed from the two frequencies (5/16 and 4/16). The sum 9/16 is placed at the node connecting the two leaves. One more left node with frequency 7/16 completes the tree. The most frequently occurring number 23 is placed in the leaf that is located to the left of the root. Therefore, its code is 0. The number 44 is placed in the first left node to the right side of the root. Therefore, its code is 10. The code for the example image is

0 10 0 10 110 10 111 0 0 0 0 110 110 10 0 10

The total number of bits to code the image is $7 + 10 + 9 + 3 = 29$ bits compared to $(16)(8) = 128$ bits without coding. Therefore, the average number of bits required to code a character is reduced to $29/16 = 1.8125$ bits from 8. For coding a large number of characters, the Huffman coding algorithm is not trivial. If precomputed tables are available, near optimal coding can be achieved for a class of images.

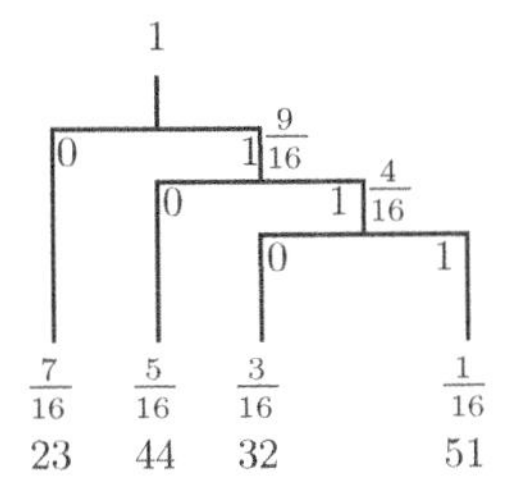

Figure 15.7 Huffman coding tree

Decoding

The decoding is carried out just by looking at the first few characters until a code corresponding to a value is found. The first bit is 0. Therefore, the first value from the tree is 23. The next bit is 1, and it is not a code. We look at 10; it corresponds to 44, and it goes on until all the bits are decoded.

Entropy

Entropy is a measure to find the effectiveness of compression algorithms. It gives the minimum number of bits per pixel (bpp) for compressing an image by a lossless compression algorithm. Bpp is defined as the number of bits used to represent the image divided by the number of pixels. Let the number of distinct values in an image be N and the frequency of their occurrence be

$$s_1, s_2, \dots, s_N$$

Entropy is defined as

$$E = \sum_{k=1}^{N} p(s_k) \log_2 \left(\frac{1}{p(s_k)} \right)$$

where $p(s_k)$ is the probability of occurrence of s_k. The term $\log_2(1/p(s_k))$ gives the number of bits required to represent $1/p(s_k)$. By multiplying this factor with $p(s_k)$, we get the bpp required to code s_k. The sum of bpp for all the distinct values yields the bpp to code the image. This equation can be equivalently written as

$$E = -\sum_{k=1}^{N} p(s_k) \log_2(p(s_k))$$

Let $N = 8$ be the number of values in a set with each value being distinct. The probability of their occurrence is

$$\{1, 1, 1, 1, 1, 1, 1, 1\}/8$$

Then,

$$E = -(8)\frac{1}{8}\log_2\left(\frac{1}{8}\right) = (8)\frac{1}{8}\log_2(8) = 3$$

That is, when each value is distinct, the number of bits required is 3. When all the values are the same, $E = 0$. For all other cases, entropy varies between 0 and $\log_2(N)$. In typical images, considerable amount of repetition of values occurs. For the example 4×4 image presented in the previous subsection,

$$E = -\left(\frac{7}{16}\log_2\left(\frac{7}{16}\right) + \frac{5}{16}\log_2\left(\frac{5}{16}\right) + \frac{3}{16}\log_2\left(\frac{3}{16}\right) + \frac{1}{16}\log_2\left(\frac{1}{16}\right) \right) = 1.7490$$

The actual value of 1.8125 bpp is quite close with the ideal value 1.7490.

Signal-to-Noise Ratio

Let the $N \times N$ input image be $x(n_1, n_2)$ and that reconstructed after compression and decompression be $\hat{x}(n_1, n_2)$. Then, a quality measure of the reconstructed image is the signal-to-noise ratio, expressed in decibels, defined as

$$\text{SNR} = 10\log_{10}\left(\frac{\sum_{n_1=0}^{N-1}\sum_{n_2=0}^{N-1}\hat{x}^2(n_1, n_2)}{\sum_{n_1=0}^{N-1}\sum_{n_2=0}^{N-1}(x(n_1, n_2) - \hat{x}(n_1, n_2))^2}\right)$$

Compute the signal-to-noise ratio in decibels of the input $x(n)$ and its approximation $\hat{x}(n)$.

$$x(n) = \{178, 187, 189, 192\}, \quad \hat{x}(n) = \{177.4643, 186.2579, 188.4563, 190.6548\}$$

SNR = 46.7499 dB

15.1.4 Compression Algorithm

The dimension N of the $N \times N$ image, the filters used for the DWT computation, the number of levels of decomposition, the quantization method and the threshold value, and the coding algorithm all have to be fixed appropriately.

1. Level shift the image. Let the gray levels be represented using p bits. Then, each level-shifted pixel value of the image is obtained by subtracting $2^{(p-1)}$. This ensures that the DWT coefficients are more evenly distributed around zero and quantization will be more effective.
2. Compute the 2-D DWT of the level-shifted image, resulting in $N \times N$ coefficients, usually with real values, over some range.
3. Quantize the coefficients to q quantization levels, so that the fidelity of the reconstructed image is adequate. Each value in a range is mapped to an integer value.
4. Threshold the coefficients, if necessary, so that coefficients with value less than a chosen threshold are replaced by zero.
5. Code the resulting sequence of symbols using a suitable coder, so that the redundancy is exploited to reduce the number of bits required to store the compressed image.

Firstly, we present a 4×4 example using Haar filters, so that similar examples can be worked out by hand calculation and the basic compression process be understood. Consider the 4×4 image matrix, shown in Table 15.2, with the pixels represented by 8 bits. The image is level shifted by subtracting 128 from each pixel value. The resulting image is shown in Table 15.3. The 2-D 1-level Haar DWT of the level-shifted image is shown in Table 15.4. The transform coefficients have to be quantized. The quantized levels are assumed to be -63 to 63. The maximum value of the coefficients is $X(1, 1) = 138.5$. Therefore, the quantization step is

Table 15.2 Example image

172	188	189	186
178	187	189	192
188	190	196	197
191	193	197	199

Table 15.3 Level-shifted image

44	60	61	58
50	59	61	64
60	62	68	69
63	65	69	71

Table 15.4 The row DWT on the left and the 2-D Haar DWT of the level-shifted image on the right

$\frac{1}{\sqrt{2}}$

104	119	−16	3
109	125	−9	−3
122	137	−2	−1
128	140	−2	−2

106.5	122.0	−12.5	0
125.0	138.5	−2.0	−1.5
−2.5	−3.0	−3.5	3.0
−3.0	−1.5	0	0.5

Table 15.5 Quantized image

48	55	−5	0
56	63	0	0
−1	−1	−1	1
−1	0	0	0

138.5/63 $= 2.1984$. The resulting quantized coefficient matrix is shown in Table 15.5. The quantized 4×4 matrix is converted to a 1×16 vector by arranging the values in the zigzag order. The approximation coefficients are followed by the H, V, and D detail coefficients. We get the vector

$$\{48, 55, 56, 63, \quad -5, 0, 0, 0, \quad -1, -1, -1, 0, \quad -1, 1, 0, 0\}$$

The Huffman code of the image is

$$\{0010\ 00001\ 00000\ 00011\ 00010\ 1\ 1\ 1\ 01\ 01\ 01\ 1\ 01\ 0011\ 1\ 1\}$$

A small space between the codes is given for easy readability. It can be verified that the code corresponds to the 1-D vector. The compressed image requires 42 bits and that of the input image is $16 \times 8 = 128$. The compression ratio (cr) is defined as the ratio of the bpp of the given image and that of its compressed version. Therefore, the compression ratio is $cr = 128/42 = 3.0476$.

The bits per pixel is bpp $= 8/3.0476 = 2.6250$.

Table 15.6 2-D Haar DWT of the reconstructed level-shifted image

105.5238	120.9127	−10.9921	0
123.1111	138.5000	0	0
−2.1984	−2.1984	−2.1984	2.1984
−2.1984	0	0	0

Table 15.7 Level-shifted reconstructed image

45.0675	58.2579	60.4563	58.2579
49.4643	58.2579	60.4563	62.6548
60.4563	60.4563	69.2500	69.2500
62.6548	62.6548	69.2500	69.2500

Table 15.8 Reconstructed image

173.0675	186.2579	188.4563	186.2579
177.4643	186.2579	188.4563	190.6548
188.4563	188.4563	197.2500	197.2500
190.6548	190.6548	197.2500	197.2500

For reconstructing the image, the code is decoded using the dictionary, and the resulting 1-D vector is converted to the 4×4 matrix. The values are multiplied by the quantization step, 2.1984, to get the values shown in Table 15.6. The 2-D Haar IDWT yields the values shown in Table 15.7. These values are level shifted by adding 128 to get the reconstructed image shown in Table 15.8. SNR is 44.4712 dB.

Consider the 8×8 image matrix, shown in Table 15.9, with the pixels represented by 8 bits. The level-shifted image, shown in Table 15.10, is obtained by subtracting $2^{(8-1)} = 2^7 = 128$ from each value. The 1-level 2-D DWT is computed using 5/3 spline filter, assuming whole-point symmetry extension at the borders. The resulting transform representation of the image is shown in Table 15.11. The maximum value of the coefficients is $X(2,1) = 160.6563$. Let us say, the quantized coefficient range is -63 to 63. Therefore, the quantization step is $160.6563/63 = 2.5501$. The resulting quantized coefficient matrix is shown in Table 15.12. No thresholding is done in this example. If thresholding is applied, the number of independent

Table 15.9 Example image

172	188	189	186	198	195	195	192
178	187	189	192	197	195	189	177
188	190	196	197	199	193	171	124
191	193	197	199	192	158	111	110
196	199	199	189	149	108	111	113
200	200	182	130	100	98	108	113
204	178	117	85	100	96	104	108
173	100	84	85	95	98	96	100

Table 15.10 Level-shifted image

44	60	61	58	70	67	67	64
50	59	61	64	69	67	61	49
60	62	68	69	71	65	43	−4
63	65	69	71	64	30	−17	−18
68	71	71	61	21	−20	−17	−15
72	72	54	2	−28	−30	−20	−15
76	50	−11	−43	−28	−32	−24	−20
45	−28	−44	−43	−33	−30	−32	−28

Table 15.11 2-D DWT of the image using 5/3 spline filter, assuming whole-point symmetry extension

93.8750	119.4375	134.4375	140.6875	−7.8750	6.0000	2.1250	−3.5000
116.5000	132.7813	155.5625	60.3438	2.0000	0.4375	−11.0625	38.3750
136.6875	160.6563	36.7500	−53.1250	−1.1875	−12.6250	16.6250	−7.8750
142.3750	−33.7813	−81.9063	−50.4375	−5.8750	20.4375	1.8750	−4.5000
1.6250	2.5625	1.0625	−8.9375	0.3750	1.5000	−0.6250	6.5000
1.3750	1.3750	−20.6875	21.2500	−0.3750	−1.3750	6.7500	10.7500
0.2500	−22.1875	24.1875	−3.0000	−0.2500	−3.3750	4.0000	1.0000
54.0000	39.7500	−1.8750	5.8750	−23.0000	9.5000	4.2500	0

Table 15.12 Quantized image

36	46	52	55	−3	2	0	−1
45	52	61	23	0	0	−4	15
53	63	14	−20	0	−4	6	−3
55	−13	−32	−19	−2	8	0	−1
0	1	0	−3	0	0	0	2
0	0	−8	8	0	0	2	4
0	−8	9	−1	0	−1	1	0
21	15	0	2	−9	3	1	0

values reduces and the compression ratio will increase at the cost of more degradation of the reconstructed image. The quantized values are reordered using the zigzag pattern, shown in Table 15.13. The resulting 1-D data are

$$\{36, 46, 45, 53, 52, 52, 55, 61, 63, 55, -13, 14, 23, -20, -32, -19,$$
$$-3, 2, 0, 0, 0, 0, -1, -4, -4, -2, 8, 6, 15, -3, 0, -1,$$
$$0, 1, 0, 0, 0, 0, -3, -8, -8, 21, 15, 9, 8, -1, 0, 2,$$
$$0, 0, 0, 0, 0, 0, 2, 2, -1, -9, 3, 1, 4, 0, 1, 0\}$$

Table 15.13 Zigzag pattern

0	1	5	6	16	17	21	22
2	4	7	12	18	20	23	28
3	8	11	13	19	24	27	29
9	10	14	15	25	26	30	31
32	33	37	38	48	49	53	54
34	36	39	44	50	52	55	60
35	40	43	45	51	56	59	61
41	42	46	47	57	58	62	63

This reordering results in long runs of zeros that can be taken as an advantage in coding. The Huffman coded image is given by

{010011 111101 010010 101100 10001 10001 10011 101101 111100 10011
010111 101001 010000 010101 101010 010100 1101 0111 00 00 00 00
0110 11000 11000 101011 11001 010001 10000 1101 00 0110 00 1110
00 00 00 00 1101 10010 10010 101111 10000 101110 11001 0110 00
0111 00 00 00 00 00 00 0111 0111 0110 11111 010110 1110 101000 00 1110 00}

Any signal represents some information. The information is carried in a signal with some amount of redundant data. Compression refers to the process of representing a signal with a smaller amount of data by reducing the redundant data and retaining the information content subject to the constraints such as the fidelity and the amount of storage available. In the example, the number of pixels is 64 and the number of bits used to represent the image is 267. Therefore,

$$\text{bpp} = \frac{267}{64} = 4.1719$$

rather than 8 bits used without compression. In practice, bpp as low as 0.1 is achieved with acceptable image degradation. For the example image, $cr = 8/4.1719 = 1.9176$.

15.1.5 Image Reconstruction

The compressed image is first decoded. The resulting values are multiplied by the quantization step. Then, the 2-D IDWT is computed. Finally, the level-shifting is undone to obtain the reconstructed image. The reconstructed image is shown in Table 15.14. SNR is 43.7517 dB.

In compression, there is always a compromise between a high compression ratio and the image quality. Taking the DWT of an image puts most of the information in the approximation part, as a typical image is composed of stronger low-frequency components than the high-frequency components. The effect of varying the parameters such as the quantization step, the threshold and the resulting compression ratio, and the energy of the reconstructed image is presented, and the images are shown in Figures 15.8–15.17. The parameters have to

Table 15.14 Reconstructed image

171.0103	187.1392	188.0880	187.1392	196.3105	195.3618	194.4130	191.8830
178.6004	187.4555	188.7205	190.4599	197.2593	195.9943	187.1392	173.2241
186.1905	190.3017	194.4130	196.3105	198.2080	191.5667	169.7453	124.2050
191.2505	194.1758	197.1012	199.7102	190.9342	156.2255	112.6618	111.3967
196.3105	198.0499	199.7893	185.3998	148.2402	110.7642	111.2386	113.7686
200.1056	198.3662	180.1817	129.8975	101.1185	99.4582	109.1830	111.7130
203.9006	177.1773	117.5637	85.7803	99.5373	99.5373	104.5973	107.1273
173.5404	101.2767	87.2034	83.2503	94.4772	98.9048	98.2723	100.8023

Figure 15.8 A reconstructed 512 × 512 image after a 1-level DWT decomposition using 5/3 spline filters with the quantization step 4.2068 and the threshold value 0

be chosen to suit the requirements of the particular application. Figure 15.8 shows a reconstructed 512 × 512 image with 256 gray levels after a 1-level DWT decomposition using 5/3 spline filters assuming whole-point symmetry. The input image is shown in Figure 8.12. The quantization step is 4.2068 quantizing the DWT coefficients in the range from −63 to 63. No thresholding has been carried out. The bpp of the compressed image is 3.1069. The energy of the compressed image is 99.47% of that of the original image.

The quantization step has been changed to 8.5494 quantizing the DWT coefficients in the range from −32 to 32. No thresholding has been carried out. Figure 15.9 shows the reconstructed image. The bpp of the compressed image is 2.4152. The energy of the compressed image is 98.96% of that of the original image.

The quantization step has been changed to 17.6688 quantizing the DWT coefficients in the range from −16 to 16. No thresholding has been carried out. Figure 15.10 shows the

Figure 15.9 A reconstructed 512 × 512 image after a 1-level DWT decomposition using 5/3 spline filters with the quantization step 8.5494 and the threshold value 0

Figure 15.10 A reconstructed 512 × 512 image after a 1-level DWT decomposition using 5/3 spline filters with the quantization step 17.6688 and the threshold value 0

Figure 15.11 A reconstructed 512×512 image after a 1-level DWT decomposition using 5/3 spline filters with the quantization step 37.8616 and the threshold value 0

reconstructed image. The bpp of the compressed image is 1.9513. The energy of the compressed image is 97.96% of that of the original image.

The quantization step has been changed to 37.8616 quantizing the DWT coefficients in the range from -8 to 8. No thresholding has been carried out. Figure 15.11 shows the reconstructed image. The bpp of the compressed image is 1.5514. The energy of the compressed image is 95.98% of that of the original image.

Now, the quantization step is 4.2068 quantizing the DWT coefficients in the range from -64 to 64. Thresholding has been set at 8.4137, which is 2 times the quantization step. Figure 15.12 shows the reconstructed image. The bpp of the compressed image is 2.1080. The energy of the compressed image is 98.48% of that of the original image.

The quantization step is 4.2068 quantizing the DWT coefficients in the range from -64 to 64. Thresholding has been set at 12.6205, which is 3 times the quantization step. Figure 15.13 shows the reconstructed image. The bpp of the compressed image is 1.8796. The energy of the compressed image is 97.28% of that of the original image.

The quantization step is 4.2068 quantizing the DWT coefficients in the range from -64 to 64. Thresholding has been set at 16.8274, which is 4 times the quantization step. Figure 15.14 shows the reconstructed image. The bpp of the compressed image is 1.6780. The energy of the compressed image is 95.99% of that of the original image.

The quantization step is 4.2068 quantizing the DWT coefficients in the range from -64 to 64. Thresholding has been set at 21.0342, which is 5 times the quantization step. Figure 15.15 shows the reconstructed image. The bpp of the compressed image is 1.5181. The energy of the compressed image is 94.58% of that of the original image.

In the next two cases, 2-level DWT decomposition has been used. The quantization step is 7.9856 quantizing the DWT coefficients in the range from -64 to 64. Thresholding has been

Figure 15.12 A reconstructed 512×512 image after a 1-level DWT decomposition using 5/3 spline filters with the quantization step 4.2068 and the threshold value 8.4137

Figure 15.13 A reconstructed 512×512 image after a 1-level DWT decomposition using 5/3 spline filters with the quantization step 4.2068 and the threshold value 12.6205

Figure 15.14 A reconstructed 512×512 image after a 1-level DWT decomposition using 5/3 spline filters with the quantization step 4.2068 and the threshold value 16.8274

Figure 15.15 A reconstructed 512×512 image after a 1-level DWT decomposition using 5/3 spline filters with the quantization step 4.2068 and the threshold value 21.0342

Figure 15.16 A reconstructed 512 × 512 image after a 2-level DWT decomposition using 5/3 spline filters with the quantization step 7.9856 and the threshold value 0

set at 0. Figure 15.16 shows the reconstructed image. The bpp of the compressed image is 1.7950. The energy of the compressed image is 99.41% of that of the original image.

The quantization step is 16.2287 quantizing the DWT coefficients in the range from −32 to 32. Thresholding has also been set at 16.2287. Figure 15.17 shows the reconstructed image.

Figure 15.17 A reconstructed 512 × 512 image after a 2-level DWT decomposition using 5/3 spline filters with the quantization step 16.2287 and the threshold value 16.2287

The bpp of the compressed image is 1.0546. The energy of the compressed image is 91.62% of that of the original image. It is observed that thresholding distorts an image more than quantization does.

15.2 Lossless Image Compression

For lossless image compression, the quantization and thresholding steps are not applied. Furthermore, integer-to-integer mapping version of the DWT enables exact reconstruction of the image possible. For integer-to-integer transform, a modified form of 5/3 spline filter is often used. Other filters can also be used. The 5/3 spline filter is easily adopted to map integers to rational numbers. The modified filter with polyphase factorization yields integer-to-integer DWT. The 5/3 spline analysis filters for this DWT are given by

$$l = \frac{\sqrt{2}}{8}(-1, 2, 6, 2, -1) \quad \text{and} \quad h = \frac{\sqrt{2}}{8}(-2, 4, -2)$$

By multiplying the first filter by $1/\sqrt{2}$ and the second filter by $\sqrt{2}$, we get

$$l = \frac{1}{8}(-1, 2, 6, 2, -1) \quad \text{and} \quad h = \frac{1}{2}(-1, 2, -1)$$

The 8×8 forward transform matrix is given by

$$\boldsymbol{W}_8 = \begin{bmatrix} \frac{3}{4} & \frac{1}{4} & -\frac{1}{8} & 0 & 0 & 0 & -\frac{1}{8} & \frac{1}{4} \\ -\frac{1}{8} & \frac{1}{4} & \frac{3}{4} & \frac{1}{4} & -\frac{1}{8} & 0 & 0 & 0 \\ 0 & 0 & -\frac{1}{8} & \frac{1}{4} & \frac{3}{4} & \frac{1}{4} & -\frac{1}{8} & 0 \\ -\frac{1}{8} & 0 & 0 & 0 & -\frac{1}{8} & \frac{1}{4} & \frac{3}{4} & \frac{1}{4} \\ -\frac{1}{2} & 1 & -\frac{1}{2} & 0 & 0 & 0 & 0 & 0 \\ 0 & 0 & -\frac{1}{2} & 1 & -\frac{1}{2} & 0 & 0 & 0 \\ 0 & 0 & 0 & 0 & -\frac{1}{2} & 1 & -\frac{1}{2} & 0 \\ -\frac{1}{2} & 0 & 0 & 0 & 0 & 0 & -\frac{1}{2} & 1 \end{bmatrix}$$

The 8×8 inverse transform matrix is given by

$$\widetilde{\boldsymbol{W}}_8 = \begin{bmatrix} 1 & \frac{1}{2} & 0 & 0 & 0 & 0 & 0 & \frac{1}{2} \\ 0 & \frac{1}{2} & 1 & \frac{1}{2} & 0 & 0 & 0 & 0 \\ 0 & 0 & 0 & \frac{1}{2} & 1 & \frac{1}{2} & 0 & 0 \\ 0 & 0 & 0 & 0 & 0 & \frac{1}{2} & 1 & \frac{1}{2} \\ -\frac{1}{4} & \frac{3}{4} & -\frac{1}{4} & -\frac{1}{8} & 0 & 0 & 0 & -\frac{1}{8} \\ 0 & -\frac{1}{8} & -\frac{1}{4} & \frac{3}{4} & -\frac{1}{4} & -\frac{1}{8} & 0 & 0 \\ 0 & 0 & 0 & -\frac{1}{8} & -\frac{1}{4} & \frac{3}{4} & -\frac{1}{4} & -\frac{1}{8} \\ -\frac{1}{4} & -\frac{1}{8} & 0 & 0 & 0 & -\frac{1}{8} & -\frac{1}{4} & \frac{3}{4} \end{bmatrix}$$

Consider the computation of the 1-level detail coefficients with the input

$$\{x(0), x(1), \ldots, x(7)\}$$

Let us partition the input into even- and odd-indexed values $e(n)$ and $o(n)$.

$$\{e(0) = x(0), e(1) = x(2), e(2) = x(4), e(3) = x(6)\}$$
$$\{o(0) = x(1), o(1) = x(3), o(2) = x(5), o(3) = x(7)\}$$

From the transform matrix and the input, we get

$$X_{\psi}(2,0) = o(0) - \frac{1}{2}(e(0) + e(1))$$
$$X_{\psi}(2,1) = o(1) - \frac{1}{2}(e(1) + e(2))$$
$$X_{\psi}(2,2) = o(2) - \frac{1}{2}(e(2) + e(3))$$
$$X_{\psi}(2,3) = o(3) - \frac{1}{2}(e(3) + e(4))$$

where $e(4)$ is a border extension value that is to be specified. These equations can be written as

$$X_{\psi}(2,k) = o(k) - \frac{1}{2}(e(k) + e(k+1)), k = 0, 1, 2, 3 \quad (15.1)$$

Consider the computation of $X_{\phi}(2,1)$.

$$\begin{aligned}
X_{\phi}(2,1) &= -\frac{1}{8}x(0) + \frac{1}{4}x(1) + \frac{3}{4}x(2) + \frac{1}{4}x(3) - \frac{1}{8}x(4) \\
&= x(2) - \frac{1}{8}x(0) + \frac{1}{4}x(1) + \frac{3}{4}x(2) - x(2) + \frac{1}{4}x(3) - \frac{1}{8}x(4) \\
&= x(2) - \frac{1}{8}x(0) + \frac{1}{4}x(1) - \frac{1}{4}x(2) + \frac{1}{4}x(3) - \frac{1}{8}x(4) \\
&= x(2) - \frac{1}{8}(x(0) - 2x(1) + 2x(2) - 2x(3) + x(4)) \\
&= e(1) + \frac{1}{8}(2x(3) - x(2) - x(4) + 2x(1) - x(0) - x(2)) \\
&= e(1) + \frac{1}{4}\left(o(1) - \frac{1}{2}(e(1) + e(2)) + o(0) - \frac{1}{2}(e(0) + e(1))\right) \\
&= e(1) + \frac{1}{4}(X_{\psi}(2,1) + X_{\psi}(2,0))
\end{aligned}$$

Similarly,

$$X_{\phi}(2,0) = e(0) + \frac{1}{4}(X_{\psi}(2,0) + X_{\psi}(2,-1))$$
$$X_{\phi}(2,1) = e(1) + \frac{1}{4}(X_{\psi}(2,1) + X_{\psi}(2,0))$$
$$X_{\phi}(2,2) = e(2) + \frac{1}{4}(X_{\psi}(2,2) + X_{\psi}(2,1))$$
$$X_{\phi}(2,3) = e(3) + \frac{1}{4}(X_{\psi}(2,3) + X_{\psi}(2,2))$$

where $X_\psi(2,-1)$ is to be computed using a border extension value that is to be specified. These equations can be written as

$$X_\phi(2,k) = e(k) + \frac{1}{4}(X_\psi(2,k) + X_\psi(2,k-1)), k = 0,1,2,3 \tag{15.2}$$

The inverse transform equations are readily obtained by first solving Equation (15.2) for $e(k)$ and then solving Equation (15.1) for $o(k)$

$$e(k) = X_\phi(2,k) - \frac{1}{4}(X_\psi(2,k) + X_\psi(2,k-1)), k = 0,1,2,3$$

$$o(k) = X_\psi(2,k) + \frac{1}{2}(e(k) + e(k+1)), k = 0,1,2,3$$

Let the input be

$$\boldsymbol{x} = \{1,3,2,4,3,-1,4,2\}$$

The input, with the whole-point symmetry extension, is

$$\boldsymbol{xe}_w = \{3, \quad 1,3,2,4,3,-1,4,2, \quad 4\}$$

Let us partition the input into even- and odd-indexed values $e(n)$ and $o(n)$.

$$\{e(0) = x(0) = 1, e(1) = x(2) = 2, e(2) = x(4) = 3, e(3) = x(6) = 4\}$$

and

$$\{o(0) = x(1) = 3, 0(1) = x(3) = 4, o(2) = x(5) = -1, o(3) = x(7) = 2\}$$

The DWT is computed as

$$X_\psi(2,0) = 3 - \frac{1}{2}(1+2) = 1.5$$

$$X_\psi(2,1) = 4 - \frac{1}{2}(2+3) = 1.5$$

$$X_\psi(2,2) = -1 - \frac{1}{2}(3+4) = -4.5$$

$$X_\psi(2,3) = 2 - \frac{1}{2}(4+4) = -2$$

$$X_\phi(2,0) = 1 + \frac{1}{4}(1.5+1.5) = 1.75$$

$$X_\phi(2,1) = 2 + \frac{1}{4}(1.5+1.5) = 2.75$$

$$X_\phi(2,2) = 3 + \frac{1}{4}(-4.5+1.5) = 2.25$$

$$X_\phi(2,3) = 4 + \frac{1}{4}(-2-4.5) = 2.375$$

The IDWT is computed as

$$e(0) = 1.75 - \frac{1}{4}(1.5 + 1.5) = 1$$

$$e(1) = 2.75 - \frac{1}{4}(1.5 + 1.5) = 2$$

$$e(2) = 2.25 - \frac{1}{4}(-4.5 + 1.5) = 3$$

$$e(3) = 2.375 - \frac{1}{4}(-2 - 4.5) = 4$$

$$o(0) = 1.5 + \frac{1}{2}(1 + 2) = 3$$

$$o(1) = 1.5 + \frac{1}{2}(2 + 3) = 4$$

$$o(2) = -4.5 + \frac{1}{2}(3 + 4) = -1$$

$$o(3) = -2 + \frac{1}{2}(4 + 4) = 2$$

The input $x(n)$ is obtained by interleaving $e(k)$ and $o(k)$.

The integer-to-integer DWT is obtained using the floor function defined as

$$\lfloor x \rfloor = k, \qquad k \leq x \quad \text{and} \quad k + 1 > x$$

where k is an integer. For example, $\lfloor 1.5 \rfloor = 1$ and $\lfloor -1.5 \rfloor = -2$.

$$X_\psi(2, k) = o(k) - \left\lfloor \frac{1}{2}\left(e\left(k\right) + e(k+1)\right) \right\rfloor, k = 0, 1, 2, 3$$

$$X_\phi(2, k) = e(k) + \left\lfloor \frac{1}{4}\left(X_\psi\left(2, k\right) + X_\psi(2, k-1)\right) + 0.5 \right\rfloor, k = 0, 1, 2, 3$$

The IDWT equations are

$$e(k) = X_\phi(2, k) - \left\lfloor \frac{1}{4}\left(X_\psi\left(2, k\right) + X_\psi(2, k-1)\right) + 0.5 \right\rfloor, k = 0, 1, 2, 3$$

$$o(k) = X_\psi(2, k) + \left\lfloor \frac{1}{2}\left(e\left(k\right) + e(k+1)\right) \right\rfloor, k = 0, 1, 2, 3$$

The steps of the DWT and IDWT algorithms can be verified.

$$X_\psi(2, k) = o(k) - \begin{cases} \frac{1}{2}\left(e\left(k\right) + e(k+1)\right) & \text{if} \quad (e(k) + e(k+1)) \quad \text{even} \\ \frac{1}{2}(e(k) + e(k+1)) - \frac{1}{2} & \text{if} \quad (e(k) + e(k+1)) \quad \text{odd} \end{cases}$$

$$X_\psi(2, k) = \begin{cases} o\left(k\right) - \frac{1}{2}(e(k) + e(k+1)) & \text{if} \quad (e(k) + e(k+1)) \quad \text{even} \\ o(k) - \frac{1}{2}(e(k) + e(k+1)) + \frac{1}{2} & \text{if} \quad (e(k) + e(k+1)) \quad \text{odd} \end{cases}$$

$$X_\psi(2, k) = \begin{cases} X_\psi^r\left(2, k\right) & \text{if} \quad (e(k) + e(k+1)) \quad \text{even} \\ X_\psi^r(2, k) + \frac{1}{2} & \text{if} \quad (e(k) + e(k+1)) \quad \text{odd} \end{cases}$$

where $X^r_\psi(2,k)$ is the real-valued transform value. If $(e(k)+e(k+1))$ is even

$$o(k) = X^r_\psi(2,k) + \frac{1}{2}(e(k)+e(k+1)) = o^r(k)$$

If $(e(k)+e(k+1))$ is odd

$$o(k) = X_\psi(2,k) + \frac{1}{2}(e(k)+e(k+1)) - \frac{1}{2}$$

$$o(k) = X^r_\psi(2,k) + \frac{1}{2} + \frac{1}{2}(e(k)+e(k+1)) - \frac{1}{2} = o^r(k)$$

The integer-to-integer mapping DWT is used to compute the transform for the same input values used in the previous example with whole-point symmetry extension

$$X_\psi(2,0) = 3 - \left\lfloor \frac{1}{2}(1+2) \right\rfloor = 2$$

$$X_\psi(2,1) = 4 - \left\lfloor \frac{1}{2}(2+3) \right\rfloor = 2$$

$$X_\psi(2,2) = -1 - \left\lfloor \frac{1}{2}(3+4) \right\rfloor = -4$$

$$X_\psi(2,3) = 2 - \left\lfloor \frac{1}{2}(4+4) \right\rfloor = -2$$

$$X_\phi(2,0) = 1 + \left\lfloor \frac{1}{4}(2+2) + 0.5 \right\rfloor = 2$$

$$X_\phi(2,1) = 2 + \left\lfloor \frac{1}{4}(2+2) + 0.5 \right\rfloor = 3$$

$$X_\phi(2,2) = 3 + \left\lfloor \frac{1}{4}(-4+2) + 0.5 \right\rfloor = 3$$

$$X_\phi(2,3) = 4 + \left\lfloor \frac{1}{4}(-2-4) + 0.5 \right\rfloor = 3$$

The IDWT is computed as

$$e(0) = 2 - \left\lfloor \frac{1}{4}(2+2) + 0.5 \right\rfloor = 1$$

$$e(1) = 3 - \left\lfloor \frac{1}{4}(2+2) + 0.5 \right\rfloor = 2$$

$$e(2) = 3 - \left\lfloor \frac{1}{4}(-4+2) + 0.5 \right\rfloor = 3$$

$$e(3) = 3 - \left\lfloor \frac{1}{4}(-2-4) + 0.5 \right\rfloor = 4$$

$$o(0) = 2 + \left\lfloor \frac{1}{2}(1 + 2) \right\rfloor = 3$$

$$o(1) = 2 + \left\lfloor \frac{1}{2}(2 + 3) \right\rfloor = 4$$

$$o(2) = -4 + \left\lfloor \frac{1}{2}(3 + 4) \right\rfloor = -1$$

$$o(3) = -2 + \left\lfloor \frac{1}{2}(4 + 4) \right\rfloor = 2$$

The input $x(n)$ is obtained by interleaving $e(k)$ and $o(k)$.

15.3 Recent Trends in Image Compression

The fact that images, in practical applications, have much redundancy coupled with efficient quantization, thresholding, and coding algorithms makes it possible to realize high compression ratio with acceptable image degradation. While the fundamentals of compression algorithms are presented in this chapter, we just mention some recent approaches. For details, the readers are advised to refer to the vast literature available. Coding the quantized transform coefficients requires some overhead such as code tables, location of the coefficients in the original image, and so on. The storage required by encoded data including the overhead should be smaller than that required by the original image.

In the tree structure shown in Figure 15.18, coefficients that belong to the spatial locality are all connected. The probability is very high that if a coefficient is less than a certain value, then all the related coefficients have less than that value. Using this fact, coding can be very efficient by using a bit to represent a large number of coefficients.

Sending the most significant bits of the largest coefficients quickly gives a coarse version of the image at a low bit rate. As more and more bits are sent in order of significance, the image

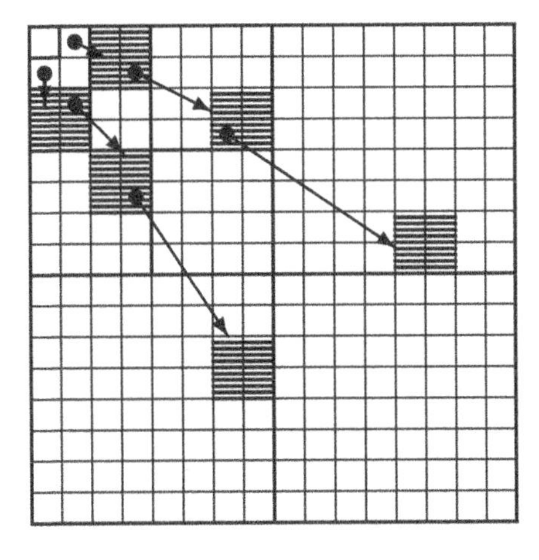

Figure 15.18 Coefficient trees spanning across different scales of a 2-D 16×16 image

gets progressively enhanced. Image transmission can be stopped as soon as the quality of the reconstructed image is adequate. This type of image coding is called embedded transform coding. Embedded transform coders can take advantage of prior knowledge of the location of large and small coefficients in the transform coefficients. The structure of the DWT coefficients provides this knowledge. Therefore, the DWT-based compression algorithms require less overhead and, hence, give a better compression ratio. This is the reason that one of the most often used applications of DWT is image compression. Furthermore, coefficients can be coded one at a time or groups of coefficients together. In general, coding groups of coefficients at a time provides better compression ratios. When a long sequence of coefficients has to be coded, the probability of occurrence of coefficients may change. In this situation, adaptive coding can be carried out by updating the probability from time to time. While variable-length coders provide efficiency, they are also more prone to noise. A single bit error changes the value of only one coefficient in fixed-length coding, whereas in variable- length coding, the values of a set of coefficients may be changed. Therefore, some error correction procedures should be used at the cost of some redundancy. While biorthogonal symmetric DWT filters are better for image compression, selection of appropriate number of levels of decomposition, quantization, thresholding, and coding methods is necessary for efficient compression.

15.3.1 The JPEG2000 Image Compression Standard

The JPEG2000 image compression standard is the current international standard for compressing digital still images. JPEG stands for Joint Photographic Experts Group. The JPEG2000 has many features. In this section, our main purpose is to point out that the two DWT filters, which we have presented in this book, are two of the main features of the standard. In addition, the integer-to-integer version of the DWT is presented. Some other salient features are also mentioned. The standard covers both lossless and lossy compression. For lossless compression, the standard recommends a version of the 5/3 spline filter. For lossy compression, the standard recommends the CDF 9/7 filter. Both of these filters were presented in Chapter 10, and the lossless version of the 5/3 filter has been presented in this chapter. As presented in Chapter 11, these filters provide an effective solution to the border problem. That is, the data expansion problem is avoided. Furthermore, they are symmetric filters with linear phase response. Due to the symmetry, only about $N/2$ multiplications are required for a filter with N coefficients in its implementation. The standard provides for progressive signal transmission. The binary representation of the pixel values is transmitted bit by bit starting from the most significant bit. With more and more bits transmitted, the resolution of the received image increases. The transmission can be stopped once the resolution is adequate. JPEG2000 allows tiling of images. An image can be split into nonoverlapping tiles, and each tile can be compressed independently. The maximum size of the image the standard deals with is $2^{32} - 1 \times 2^{32} - 1$. The standard supports up to 256 channels and also handles color images.

15.4 Summary

- Compression is essential for efficient transmission and storage of images.
- Compression consists of taking the transform, quantizing, thresholding, and coding.
- In lossless compression, quantizing and thresholding the coefficients are not carried out.

- By taking the DWT, a decorrelated representation of an image is obtained.
- In the transformed form, the approximation and detail parts are separated. Using the fact that the energy is mostly concentrated in the low-frequency part of the spectrum, the image can be coded efficiently.
- Biorthogonal symmetric filters are preferred for image compression.
- The current image compression standard for still images is JPEG2000

Exercises

15.1 Quantize the input $x(n)$ with the uniform quantization step size: (a) maximum of the magnitude of $2x(n)$ divided by 4 and (b) maximum of the magnitude of $2x(n)$ divided by 8. List the errors in the quantized representation $x_q(n)$.

15.1.1

$$x(n) = \{-93.8750, 119.4375, 134.4375, 140.6875\}$$

***15.1.2**

$$x(n) = \{136.6875, 160.6563, 36.7500, -53.1250\}$$

15.1.3

$$x(n) = \{142.3750, -33.7813, -81.9063, -50.4375\}$$

15.2 Apply hard and soft thresholding to $x(n)$ with the threshold level (a) 60 and (b) 100.

15.2.1

$$x(n) = \{-93.8750, 119.4375, 134.4375, 140.6875\}$$

15.2.2

$$x(n) = \{136.6875, 160.6563, 36.7500, -53.1250\}$$

***15.2.3**

$$x(n) = \{142.3750, -33.7813, -81.9063, -50.4375\}$$

15.3 Generate the Huffman code tree and the bit stream for the 4×4 image. What is the bpp? Decode the Huffman code and verify that the image is reconstructed. Assume top-left corner of the image as the origin.

15.3.1

$$x(n_1, n_2) = \begin{bmatrix} 1 & 4 & -1 & 2 \\ 2 & 3 & -1 & 1 \\ 2 & 1 & 1 & -3 \\ 1 & 2 & -1 & 3 \end{bmatrix}$$

***15.3.2**

$$x(n_1, n_2) = \begin{bmatrix} 2 & 1 & 3 & 4 \\ 4 & -3 & 1 & 2 \\ 3 & 1 & 3 & 4 \\ 4 & 1 & 2 & 3 \end{bmatrix}$$

15.3.3

$$x(n_1, n_2) = \begin{bmatrix} 2 & 2 & 1 & 1 \\ 4 & 1 & -2 & 2 \\ 1 & 3 & 1 & -1 \\ 4 & 4 & -1 & 3 \end{bmatrix}$$

15.4 Compute the entropy of $x(n)$.

***15.4.1** $x(n) = \{188, 190, 196, 197, 199, 193, 171, 124\}$
15.4.2 $x(n) = \{3, 3, 3, 3, 3, 3, 3, 3\}$
15.4.3 $x(n) = \{7, 7, 5, 4, 3, 3, 3, 0\}$
15.4.4 $x(n) = \{6, 7, 2, 2, 2, 2, 1, 3\}$
15.4.5 $x(n) = \{7, 7, 7, 7, 2, 2, 2, 2\}$

15.5 Compute the signal-to-noise ratio in decibels of the input $x(n)$ and its approximation $\widehat{x}(n)$.

15.5.1

$$x(n) = \{188, 190, 196, 197\} \quad \widehat{x}(n) = \{185.3773, 190.1587, 194.9401, 196.8527\}$$

15.5.2

$$x(n) = \{191, 193, 197, 199\} \quad \widehat{x}(n) = \{190.4775, 193.7448, 197.0121, 199.9606\}$$

15.5.3

$$x(n) = \{196, 199, 199, 189\} \quad \widehat{x}(n) = \{195.5777, 197.3308, 199.0840, 185.2179\}$$

***15.5.4**

$$x(n) = \{200, 200, 182, 130\} \quad \widehat{x}(n) = \{199.4028, 197.9684, 179.9583, 129.5938\}$$

15.5.5

$$x(n) = \{204, 178, 117, 85\} \quad \widehat{x}(n) = \{203.2280, 176.9300, 117.4808, 85.4452\}$$

15.6 Using the 1-level Haar DWT, uniform quantization with quantization levels from -63 to 63, and Huffman coding, find the compressed form of the 4×4 image. What are the bpp, compression ratio, and SNR? Assume top-left corner of the image as the origin.

***15.6.1** Table 15.15.

Table 15.15 Input image 15.6.1

146	151	153	151
153	150	150	150
164	158	154	148
165	164	154	157

15.6.2 Table 15.16.

Table 15.16 Input image 15.6.2

127	124	130	132
126	125	131	128
118	132	127	128
123	125	121	130

15.6.3 Table 15.17.

Table 15.17 Input image 15.6.3

207	183	141	143
203	167	144	147
186	153	146	151
165	154	148	150

15.7 Find the 1-level integer-to-integer DWT of $x(n)$ using the 5/3 spline filters. Compute the inverse DWT and verify that the input is reconstructed. Assume whole-point symmetry at the borders.

15.7.1

$$x(n) = \{7, 1, -5, 3, 4, 1, 2, 0\}$$

15.7.2

$$x(n) = \{2, 1, 3, 4, 5, 2, 3, 1\}$$

15.7.3

$$x(n) = \{3, 1, 2, 4, 2, 5, 1, 1\}$$

16

Denoising

Denoising a signal is estimating its true value from that of a noisy version. In all stages of processing a signal, the signal gets corrupted by some amount of noise. The model of the noisy signal is given by

$$y(n) = x(n) + s(n)$$

where $y(n)$ is the noisy signal, $x(n)$ is the true signal, and $s(n)$ is the noise. Assume that $s(n)$ is a Gaussian white noise. The samples are independent (uncorrelated) and normally distributed with mean zero and variance σ^2. In this chapter, we present an algorithm for denoising. Basically, denoising consists of taking the DWT of the noisy signal, estimating the noise level and, hence, a threshold, applying soft thresholding to the DWT detail coefficients, and taking the IDWT.

16.1 Denoising

The denoising algorithm estimates the true signal $x(n)$ from its noisy version $y(n) = x(n) + s(n)$, where $s(n)$ is the noise. As, in general, the noise component of a signal has relatively smaller magnitude and wider bandwidth, effective thresholding operation of the transformed signal results in reducing the noise. The steps of the denoising algorithm are as follows:

1. The k-scale DWT $Y(n)$ of the noisy signal $y(n)$ is computed, resulting in the approximation component $Y_a(n)$ and detail component $Y_d(n)$.
2. The detail component $Y_d(n)$ is subjected to a threshold operation to yield $Y_{dt}(n)$.
3. The modified DWT $Y_m(n)$ is constructed by concatenating $Y_a(n)$ and $Y_{dt}(n)$.
4. The k-scale IDWT of $Y_m(n)$ is computed, resulting in the denoised version $\hat{x}(n) \approx x(n)$ of $y(n)$.

The DWT using orthogonal filter banks is appropriate for denoising, as a Gaussian noise remains Gaussian after taking the DWT using an orthogonal filter bank. For denoising applications, soft thresholding is more suitable. The quality of the estimation of the true signal depends on the threshold level selection, which, in turn, depends on the characteristics of both

Discrete Wavelet Transform: A Signal Processing Approach, First Edition. D. Sundararajan.

Companion Website: www.wiley.com/go/sundararajan/wavelet

the signal and the noise. The estimation of a signal from its noisy version is with respect to the least squares error criterion. Let $\hat{x}(n)$ be an estimation of a true signal $x(n)$ with N samples. The square error between $x(n)$ and $\hat{x}(n)$ is defined as

$$E = \sum_{n=0}^{N-1} (x(n) - \hat{x}(n))^2$$

The lower the value of the error E, the closer $\hat{x}(n)$ is to $x(n)$.

As the DWT is a linear transform, the DWT $Y(n)$ of $y(n)$ is a linear combination of that of its components $x(n)$ and $s(n)$.

$$Y(n) = Wy(n) = W(x(n) + s(n)) = Wx(n) + Ws(n),$$

where W is the transform matrix. As the magnitude of the spectrum of signals tends to fall off with increasing frequencies, the energy of the approximation component of $X(n)$ (the DWT of $x(n)$), $X_a(n)$, is much greater than that of the detail component $X_d(n)$ of the DWT $X(n)$. On the other hand, the energy of the approximation component of the DWT of the Gaussian white noise $S(n)$, $S_a(n)$, is the same as that of the detail component $S_d(n)$. In taking the DWT of a signal, we decompose it in terms of components corresponding to parts of its spectrum. While the energy of most practical signals is concentrated in the low-frequency part of the spectrum, the average power of the Gaussian noise is uniformly distributed throughout the spectrum. Note that the autocorrelation of a Gaussian noise is an impulse. The Fourier transform of an impulse is a constant (Example 4.5). That is, the power spectral density is constant. The reduction in noise power increases with the number of levels of the DWT computed. For example, if we compute 1-level DWT, the detail component represents the second half of the spectrum. Therefore, by almost eliminating it by soft thresholding, the noise power is reduced approximately by a factor of 2. A 2-level DWT computation is expected to reduce the remaining noise by another factor of 2 and so on. Denoising is based on the assumptions that noise power is much smaller than that of the signal, the noise is of Gaussian type, and the signal energy is concentrated in the low-frequency components. Thresholding the detail components will also reduce the energy of the signal. It is almost impossible to reduce the noise without affecting the signal. Due to analytical convenience, the noise is assumed to be of Gaussian type. Furthermore, this assumption yields acceptable results in practice. The situation is similar to the case of assuming that the system is linear time invariant for analytical convenience, while all practical systems violate this assumption to some extent. Therefore, the detail component $Y_d(n)$ is essentially composed of noise. Consequently, a soft threshold operation on $Y_d(n)$ yields $Y_{dt}(n)$ with reduced noise component. The key factor in efficient denoising is, therefore, the proper selection of the threshold T.

16.1.1 Soft Thresholding

Figure 16.1(a) shows a sinusoidal signal, and the same signal soft thresholded with $T = 0.4$ is shown in Figure 16.1(b). All the values between 0.4 and -0.4 have been set to zero. Values greater than 0.4 and smaller than -0.4 have been changed, so they are closer to x-axis by 0.4.

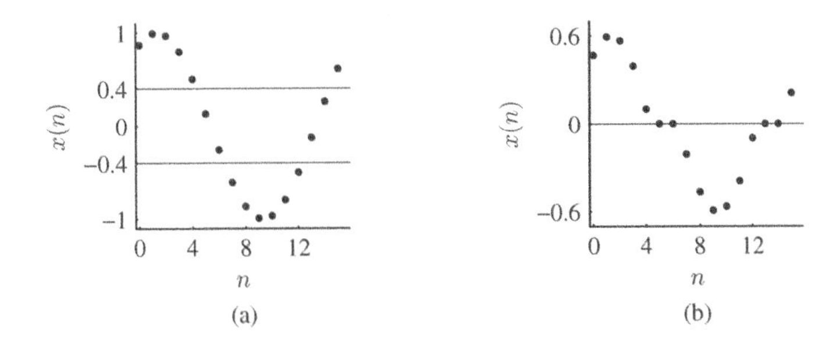

Figure 16.1 (a) A sinusoidal signal; (b) the signal in (a) soft thresholded with $T = 0.4$

16.1.2 Statistical Measures

Some measures used in statistical characterization of data are required in denoising algorithms. The mean of a list of N numbers

$$\{x(0), x(1), \ldots, x(N-1)\}$$

is defined as

$$\overline{x} = \frac{x(0) + x(1) + \cdots + x(N-1)}{N}$$

The median of a list of N numbers

$$\{x(0), x(1), \ldots, x(N-1)\}$$

is defined as the middle number $(N+1)/2$th of the sorted list of $x(n)$, if N is odd. If N is even, the median is defined as the mean of the two middle numbers $(N/2)$th and $(N/2+1)$th of the sorted list. The mean and median give an indication of the center of the data. The spread of the data is given by the variance and standard deviation. The variance σ^2 of $x(n)$ of N numbers is defined as

$$\sigma^2 = \frac{1}{N-1}\sum_{n=0}^{N-1}(x(n) - \overline{x})^2$$

Note that the denominator is $N-1$ rather than N. The standard deviation σ is the positive square root of the variance σ^2. The median of the magnitude of the deviation M of $x(n)$ is useful in estimating the noise level of signals and is defined as the median of the magnitude of the deviation of $x(n)$ from its median x_m.

$$M = \text{median of } \{|x_0 - x_m|, |x_1 - x_m|, \ldots, |x_{N-1} - x_m|\}$$

Example 16.1 Find the mean, median, variance, standard deviation, and median of the magnitude of the deviation for the signal

$$x(n) = \{2, 3, 4, 7, 6, -8\}$$

Solution

The mean is

$$\overline{x} = \frac{2+3+4+7+6-8}{6} = \frac{7}{3}$$

By sorting $x(n)$, we get

$$\{-8, 2, 3, 4, 6, 7\}$$

The median is the average of the middle two elements 3 and 4, which is 3.5. The variance is

$$\sigma^2 = \frac{\left(2-\frac{7}{3}\right)^2 + \left(3-\frac{7}{3}\right)^2 + \left(4-\frac{7}{3}\right)^2 + \left(7-\frac{7}{3}\right)^2 + \left(6-\frac{7}{3}\right)^2 + \left(-8-\frac{7}{3}\right)^2}{5} = 29.0667$$

The standard deviation $\sigma = \sqrt{29.0667} = 5.3914$. The magnitude of the deviation of $x(n)$ from its median is given by

$$\{1.5, 0.5, 0.5, 3.5, 2.5, 11.5\}$$

After sorting, we get

$$\{0.5, 0.5, 1.5, 2.5, 3.5, 11.5\}$$

The median of the magnitude of the deviation of $x(n)$, M, is the average of the middle two elements 1.5 and 2.5, which is 2. ■

16.2 VisuShrink Denoising Algorithm

Hard and soft thresholding were introduced in Chapter 15. In denoising applications, soft thresholding is found to be more effective. In VisuShrink algorithm, the threshold T is computed using the formula

$$T = \sigma\sqrt{2\log_e(N)}$$

where σ is the noise level and N is the number of elements in the signal. The noise level is estimated from the detail components of the 1-level DWT decomposition, $\psi^1(n)$. The estimated noise level $\overline{\sigma}$ is given by

$$\hat{\sigma} = \frac{M(\psi^1(n))}{0.6745}$$

where $M(\psi^1(n))$ is the median of the magnitude of the deviation of the detail component of the 1-level DWT decomposition from its median. A sinusoidal signal with 2048 samples, its noisy version, and the denoised signal are shown, respectively, in Figures 16.2(a)–(c). The amplitude of the sinusoid is 7. The mean of the Gaussian noise is zero, and its standard deviation σ is 1. The 1-level DWT, using D4 filter, of the sinusoid, the noise, and the noisy signal are shown, respectively, in Figures 16.3(a)–(c). The soft thresholded 1-level DWT of the signal, the 4-level DWT of the signal, and the soft thresholded 4-level DWT of the signal are shown, respectively, in Figures 16.4(a)–(c). It is evident that the noise power is reduced with more levels of DWT computation.

The disadvantage of this algorithm is that the threshold value is a function of N, the data size. The threshold value estimated tends to be higher than optimum for longer N. This results

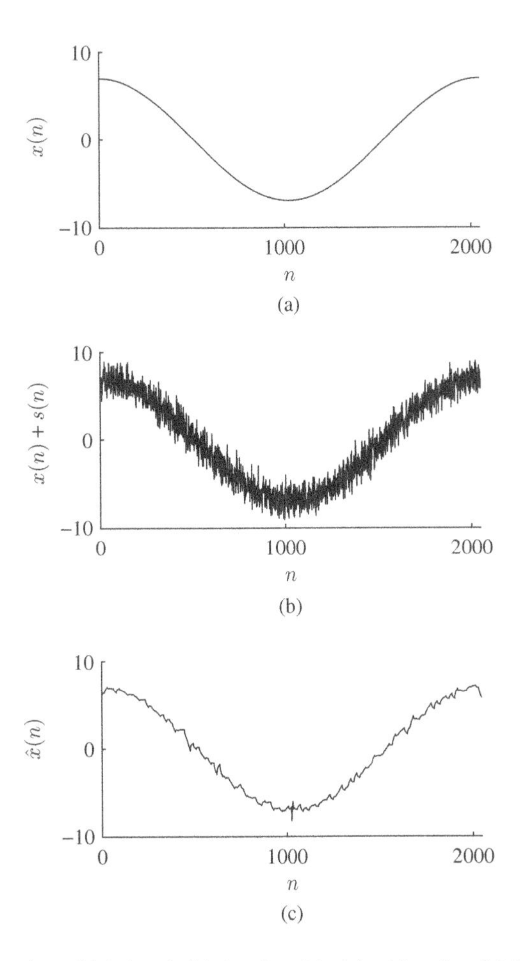

Figure 16.2 (a) A sinusoidal signal; (b) the signal in (a) with noise; (c) the denoised signal

in removing noise as well as affecting significant parts of the detail component. Due to this, we get a blurred version of the image. In such cases, other methods of determining the threshold have to be used.

Example 16.2 The samples of the signal $x(n)$ and the noise $s(n)$ are given. Add $x(n)$ and $s(n)$ to get the noisy signal $y(n)$. Compute the 1-level Haar DWT of $y(n)$. Determine the threshold T. Find the least square error in denoising $y(n)$ to obtain the estimated signal $\hat{x}(n)$ using 1-level and 2-level Haar DWT.

$$x(n) = \{7, 7, 7, 7, 7, 7, 7, 7, -7, -7, -7, -7, -7, -7, -7, -7\}$$

$$s(n) = \{-1.0667, 0.9337, 0.3503, -0.0290, 0.1825, -1.5651, -0.0845, 1.6039,$$
$$0.0983, 0.0414, -0.7342, -0.0308, 0.2323, 0.4264, -0.3728, -0.2365\}$$

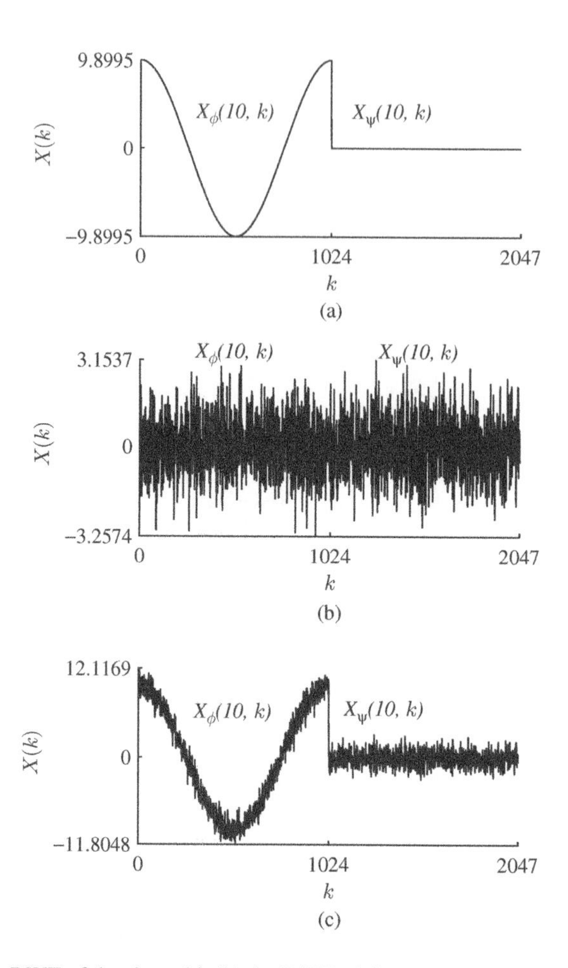

Figure 16.3 (a) The DWT of the sinusoid; (b) the DWT of the noise; (c) the DWT of the noisy signal

Solution

The noisy signal is

$$y(n) = x(n) + s(n) = \{5.9333, 7.9337, 7.3503, 6.9710, 7.1825, 5.4349,$$
$$6.9155, 8.6039, -6.9017, -6.9586, -7.7342,$$
$$-7.0308, -6.7677, -6.5736, -7.3728, -7.2365\}$$

The 1-level DWT of $y(n)$ is

$$X(k) = \{9.8055, 10.1267, 8.9218, 10.9739, -9.8007, -10.4404, -9.4337, -10.3303,$$
$$-1.4145, 0.2682, 1.2357, -1.1939, 0.0403, -0.4973, -0.1372, -0.0964\}$$

Sorting the eight detail coefficients, we get

$$\{-1.4145, -1.1939, -0.4973, -0.1372, -0.0964, 0.0403, 0.2682, 1.2357\}$$

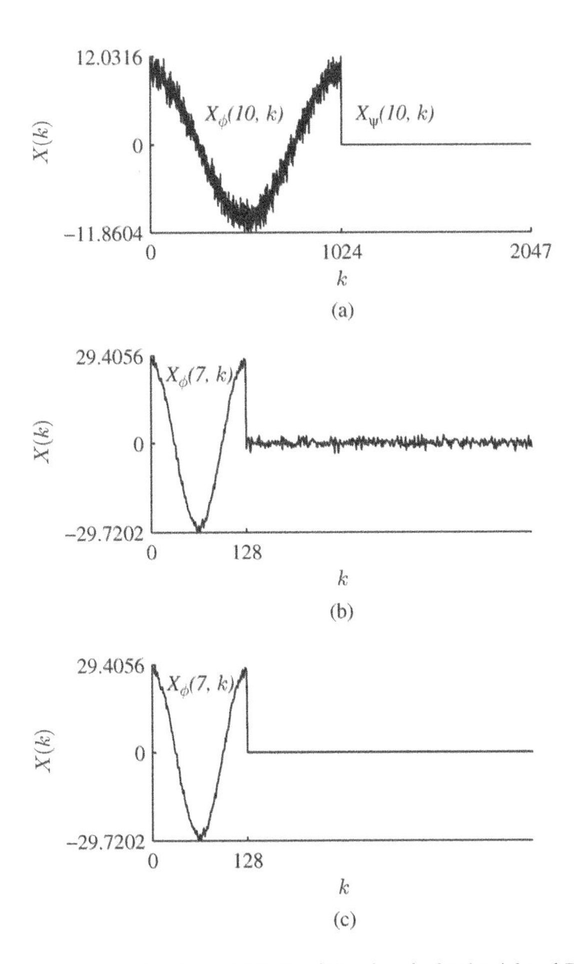

Figure 16.4 (a) The soft thresholded 1-level DWT of the signal; (b) the 4-level DWT of the signal; (c) the soft thresholded 4-level DWT of the signal

The median of the sorted coefficients is -0.1168. The magnitude of the deviation of the detail coefficients from the median is deviation

$$\{1.2977, 0.3850, 1.3525, 1.0771, 0.1571, 0.3805, 0.0204, 0.0204\}$$

and, after sorting, we get

$$\{0.0204, 0.0204, 0.1571, 0.3805, 0.3850, 1.0771, 1.2977, 1.3525\}$$

The median M is 0.3828. The estimated value of the noise level σ, $M/0.6575$, is 0.5675. The threshold, T, is 1.1573. The modified DWT coefficients, after soft thresholding the detail coefficients, are

$$\{9.8055, 10.1267, 8.9218, 10.9739, -9.8007, -10.4404, -9.4337, -10.3303,$$
$$-0.2572, 0, 0.0783, -0.0366, 0, 0, 0, 0\}$$

The 1-level Haar IDWT yields the estimated signal $\hat{x}(n)$ as

$$\{6.7517, 7.1154, 7.1607, 7.1607, 6.3641, 6.2533, 7.7338, 7.7856,$$
$$-6.9301, -6.9301, -7.3825, -7.3825, -6.6706, -6.6706, -7.3046, -7.3046\}$$

The square error, is 2.9491.

The 2-level Haar DWT coefficients of the noisy signal $y(n)$ are

$$\{14.0942, 14.0684, -14.3126, -13.9753, -0.2271, -1.4510, 0.4524, 0.6340,$$
$$-1.4145, 0.2682, 1.2357, -1.1939, 0.0403, -0.4973, -0.1372, -0.0964\}$$

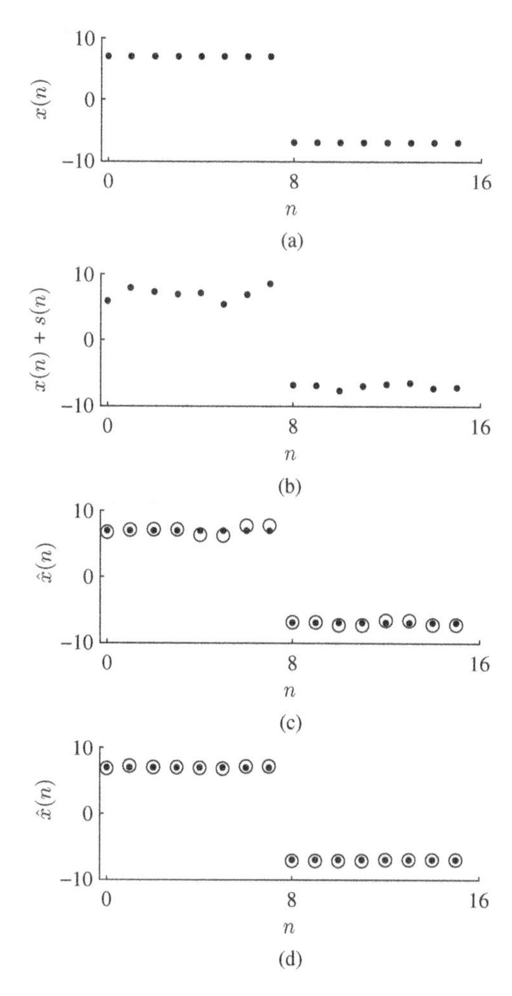

Figure 16.5 (a) A rectangular signal $x(n)$; (b) noisy version of $x(n)$; (c) $x(n)$ and its reconstructed version $\hat{x}(n)$ using 1-level Haar DWT; (d) $x(n)$ and its reconstructed version $\hat{x}(n)$ using 2-level Haar DWT

After soft thresholding, we get

$$\{14.0942, 14.0684, -14.3126, -13.9753, 0, -0.2937, 0, 0,$$
$$-0.2572, 0, 0.0783, -0.0366, 0, 0, 0, 0\}$$

The 2-level Haar IDWT yields the estimated signal $\hat{x}(n)$ as

$$\{6.8652, 7.2289, 7.0471, 7.0471, 6.9428, 6.8320, 7.1552, 7.2069,$$
$$-7.1563, -7.1563, -7.1563, -7.1563, -6.9876, -6.9876, -6.9876, -6.9876\}$$

The square error is 0.2717, which is far less than that of the 1-level DWT. Figures 16.5(a) and (b) shows, respectively, $x(n)$ and $y(n)$. The estimated signals using 1-level and 2-level Haar DWT are shown, respectively, in Figures 16.5(c) and (d). ∎

16.3 Summary

- The model of the noisy signal is given by

$$y(n) = x(n) + s(n)$$

  where $y(n)$ is the noisy signal, $x(n)$ is the true signal, and $s(n)$ is the noise.
- It is assumed that $s(n)$ is a Gaussian white noise. The samples are independent (uncorrelated) and normally distributed with mean zero and variance σ^2.
- The DWT of a noisy signal is taken. The detail components are soft thresholded. The IDWT of the approximation component and the modified detail components is the denoised signal.
- As most of the signal energy is in the approximation component and the energy of the noise is uniformly distributed, soft thresholding the detail component removes most of the noise.
- Selection of appropriate threshold level is important.
- In VisuShrink denoising algorithm, the noise level of the image is estimated using the median of the magnitude of the deviation value computed from the detail component of the 1-level DWT decomposition. Then, the threshold value is expressed as a function of the noise level and the data size.
- The disadvantage of this algorithm is that the denoised image is blurred, as the threshold tends to be much larger than the optimum for long data lengths.
- In any application, the characteristics of the class of signals have to be taken into account in determining the algorithm to be used.

Exercises

16.1 The samples of the signal $x(n)$ and the noise $s(n)$ are given. Add $x(n)$ and $s(n)$ to get the noisy signal $y(n)$. Compute the 1-level Haar DWT of $y(n)$. Determine the threshold T. Find the least square error in denoising $y(n)$ to obtain the estimated signal $\hat{x}(n)$ using 1-level and 2-level Haar DWT.

16.1.1

$$x(n) = \{8, 8, 8, 8, 8, 8, 8, 8, -8, -8, -8, -8, -8, -8, -8, -8\}$$

$$s(n) = \{0.5377, 1.8339, -2.2588, 0.8622, 0.3188, -1.3077, -0.4336, 0.3426, 3.5784, 2.7694, -1.3499, 3.0349, 0.7254, -0.0631, 0.7147, -0.2050\}$$

***16.1.2**

$$x(n) = \{6, 6, 6, 6, 6, 6, 6, 6, -6, -6, -6, -6, -6, -6, -6, -6\}$$

$$s(n) = \{-0.1241, 1.4897, 1.4090, 1.4172, 0.6715, -1.2075, 0.7172, 1.6302, 0.4889, 1.0347, 0.7269, -0.3034, 0.2939, -0.7873, 0.8884, -1.1471\}$$

16.1.3

$$x(n) = \{9, 9, 9, 9, 9, 9, 9, 9, -9, -9, -9, -9, -9, -9, -9, -9\}$$

$$s(n) = \{1.0933, 1.1093, -0.8637, 0.0774, -1.2141, -1.1135, -0.0068, 1.5326, -0.7697, 0.3714, -0.2256, 1.1174, -1.0891, 0.0326, 0.5525, 1.1006\}$$

Bibliography

Wavelet Transform

[1] Cohen, A, Daubechies, I, and Feauveau, J.C, *Biorthogonal bases of compactly supported wavelets*, Commun. Pure Appl. Math. **45** (1992), No. 5, 485–560.
[2] Daubechies, I, *Orthogonal bases of compactly supported wavelets*, Commun. Pure Appl. Math. **45** (1988), No. 41, 909–996.
[3] Jensen, A and Cour-Harbo, A. (2001) *Ripples in Mathematics: The Discrete Wavelet Transform*, Springer-Verlag, New York.
[4] Mallet, S. (2009) *A Wavelet Tour of Signal Processing*, Academic Press, San Diego, CA.
[5] Selesnick, W, Baraniuk, RG, and Kingsbury, NG, *The dual-tree complex wavelet transform*, IEEE Sig. Proc. Mag. **22** (2005), No. 6, 123–150.
[6] Strang, G and Nguyen, T. (1996) *Wavelets and Filter Banks*, Wellesley Cambridge Press, Wellesley, MA.
[7] The Mathworks, (2015) *Matlab Wavelet Tool Box User's Guide*, The Mathworks, Inc. USA.
[8] Van Fleet, PJ. (2008) *Discrete Wavelet Transformations*, John Wiley Sons, Inc., Hoboken, NJ.

Signal and Image Processing

[9] Gonzalez, RC and Woods, RE. (2008) *Digital Image Processing*, Prentice-Hall, Upper Saddle River, NJ.
[10] Mitra, SK. (2010) *Digital Signal Processing*, A Computer-Based Approach, McGraw Hill, USA.
[11] Proakis, JG and Manolakis, DJ. (2007) *Digital Signal Processing, Principals, Algorithms, and Applications*, Prentice-Hall, Hoboken, NJ.
[12] Sundararajan, D. (2008) *Signals and Systems – A Practical Approach*, John Wiley & Sons (Asia) Pte Ltd, Singapore.
[13] Sundararajan, D. (2001) *Discrete Fourier Transform, Theory, Algorithms, and Applications*, World Scientific, Singapore.
[14] Sundararajan, D. (2003) *Digital Signal Processing, Theory and Practice*, World Scientific, Singapore.
[15] The Mathworks, (2015) *Matlab Signal Processing Tool Box User's Guide*, The Mathworks, Inc. USA.
[16] The Mathworks, (2015) *Matlab Image Processing Tool Box User's Guide*, The Mathworks, Inc. USA.

Discrete Wavelet Transform: A Signal Processing Approach, First Edition. D. Sundararajan.

Companion Website: www.wiley.com/go/sundararajan/wavelet

Answers to Selected Exercises

Chapter 2

2.1.4

$$x_e(n) = \sqrt{3}\cos\left(2\frac{2\pi}{8}n\right) \qquad x_o(n) = -\sin\left(2\frac{2\pi}{8}n\right)$$

2.3.3

$$\{0.2215, 0.4333, 0.6262, 0.7842, 0.9259, 0.9890, 0.9999, 1\}$$

$$\{0.4140, 0.7817, 0.9786, 0.9865, 0.9931, 0.9975, 0.9995, 1\}$$

2.4.5

$$x(n) = 3\delta(n+3) + 2\delta(n-1) + 4\delta(n-2) + \delta(n-3)$$

2.5.1

$$\{1.7321, 1.0000, -1.7321, -1.0000, 1.7321, 1.0000, -1.7321, -1.0000\}$$

$$a = 2\cos(\pi/6) = 1.7321 \qquad b = 2\sin(\pi/6) = 1.0000$$

$$a\cos\left(2\frac{2\pi}{8}n\right) + b\sin\left(2\frac{2\pi}{8}n\right)$$

2.6.2

$$\{3.0000, 4.2426, 3.0000, 0.0000, -3.0000, -4.2426, -3.0000, -0.0000\}$$

$$x(n) = 4.2426\cos\left(\frac{2\pi}{8}n - 0.7854\right)$$

Discrete Wavelet Transform: A Signal Processing Approach, First Edition. D. Sundararajan.

Companion Website: www.wiley.com/go/sundararajan/wavelet

2.7.4

$$f = 1/8, \quad N = 8$$

2.8.3

$$2\left(e^{j\left(\frac{\pi}{4}n+\frac{\pi}{3}\right)} + e^{-j\left(\frac{\pi}{4}n+\frac{\pi}{3}\right)}\right)$$

2.9.3 Gets aliased. Impersonates component with frequency index one.

$$\{-3.0000, 2.1213, 0.0000, -2.1213, 3.0000, -2.1213, -0.0000, 2.1213\}$$

$$\{-3.0000, 0.0000, 3.0000, -0.0000\}$$

2.10.6

$$y(n) = \{y(-2) = 2, y(-1) = 2\}$$

Chapter 3

3.1.2

$$x^{-4} - 2x^{-2} + 6x^{-1} + 1 - 2x^{1} + 5x^{2} + x^{3}$$

3.2.3

$$y = \{2, -1, -3, 5, -8, 6, -1\}$$

3.3.2

$$y(n) = \{-3, -1, 6\}$$

3.4.3

$$y(n) = \{2, 4, -9, 10, -12, 14, -19, 12, -7, 6, -4, 2, 1\}$$

3.5.2

$$y(n) = \{-2, 6, 2, 10, 11, -1, 0, 8, 9 - 3\}$$

3.6.3

Periodic	**5**	**7**	**5**
Whole-point symmetric	**14**	**7**	−1
Half-point symmetric	**0**	**7**	**0**

3.7

$$y(n) = \{2, -3, 3, 1, -5, 6, -8, 10\}$$

3.13.7

$$\{0, 32, 0\}$$

3.13.10

$$\{0, 0, 0\}$$

Chapter 4

4.1.2 $a = 7/4$.

4.2.2 30.

4.3.4 The samples of the spectrum at $\omega = 0, \pi/2, \pi, 3\pi/2$ are $\{0, -j2, 0, j2\}$.

4.4.3

$$x(n/2) \leftrightarrow 2 + 3e^{-j2\omega}$$

4.7.2 10.

Chapter 5

5.1.3 $X(z) = 4z^2 + 3 + 3z^{-1}$.
$X(z) = 4z^4 + 3z^2 + 3z$.
$X(z) = 4z^{-1} + 3z^{-3} + 3z^{-4}$.

5.2.4 $\{x(-5) = 3, x(0) = 1, x(1) = 1, x(3) = -4\}$.

5.3.5 The nonzero values of $y(n)$ are $\{y(-6) = 9, y(-4) = 6, y(-3) = -12, y(-1) = -8\}$.

Chapter 6

6.1.4 Highpass filter.

$$\text{Magnitude} = \{0, 1.4142, 1.4142\}$$

$$\text{Angle} = \{180, 0, -180\} \text{ degrees}$$

6.1.6 Lowpass filter.

$$\text{Magnitude} = \{1.4142, 2, 0\}$$

$$\text{Angle} = \{0, -135, -270\} \text{ degrees}$$

Chapter 7

7.2.3

$$x_d(n) = 5\delta(n+1) + \delta(n) + 5\delta(n-1) \Longleftrightarrow 1 + 10\cos(\omega)$$

7.3.4

$$x_u(n) \leftrightarrow \frac{1}{1 - 0.8e^{-j2\omega}}$$

7.6.3

$$\{3, 10, 7, -3, -6, -8\}$$

7.7.2

$$3 \searrow 1 \nearrow 9 \searrow 6 \nearrow -3 \searrow 11 \nearrow -12 \searrow 5 \nearrow 3 \searrow -2$$

Chapter 8

8.1.1

$$\left\{-\frac{5}{\sqrt{2}}, \quad \frac{7}{\sqrt{2}}\right\}$$

8.2.2

$$\left\{\frac{4}{\sqrt{2}}, \quad \frac{2}{\sqrt{2}}, \quad \frac{2}{\sqrt{2}}, \quad \frac{-6}{\sqrt{2}}\right\}$$

$$\left\{3, \quad 1, \quad \frac{2}{\sqrt{2}}, \quad \frac{-6}{\sqrt{2}}\right\}$$

8.3.1

$$\left\{\frac{4}{\sqrt{2}}, \quad \frac{7}{\sqrt{2}}, \quad \frac{1}{\sqrt{2}}, \quad \frac{1}{\sqrt{2}}, \quad \frac{2}{\sqrt{2}}, \quad -\frac{3}{\sqrt{2}}, \quad \frac{5}{\sqrt{2}}, \quad \frac{-3}{\sqrt{2}}\right\}$$

$$\left\{\frac{11}{\sqrt{4}}, \quad \frac{2}{\sqrt{4}}, \quad -\frac{3}{\sqrt{4}}, \quad 0, \quad \frac{2}{\sqrt{2}}, \quad -\frac{3}{\sqrt{2}}, \quad \frac{5}{\sqrt{2}}, \quad \frac{-3}{\sqrt{2}}\right\}$$

$$\left\{\frac{13}{\sqrt{8}}, \quad \frac{9}{\sqrt{8}}, \quad -\frac{3}{\sqrt{4}}, \quad 0, \quad \frac{2}{\sqrt{2}}, \quad -\frac{3}{\sqrt{2}}, \quad \frac{5}{\sqrt{2}}, \quad \frac{-3}{\sqrt{2}}\right\}$$

8.6

8.11

$$\begin{bmatrix} 2 & 1 & 0 & 1 \\ 2 & 1 & 0 & 1 \\ 0 & 0 & 0 & 0 \\ 0 & 0 & 0 & 0 \end{bmatrix}$$

8.15 1-level 2-D DWT

$$\begin{bmatrix} 1.0 & 2.0 & 2.0 & -4.0 \\ -1.5 & 1.5 & 0.5 & 1.5 \\ -2.0 & 2.0 & 1.0 & 2.0 \\ -1.5 & 3.5 & 4.5 & -2.5 \end{bmatrix}$$

2-level 2-D DWT

$$\begin{bmatrix} 1.5 & -2.0 & 2.0 & -4.0 \\ 1.5 & 1.0 & 0.5 & 1.5 \\ -2.0 & 2.0 & 1.0 & 2.0 \\ -1.5 & 3.5 & 4.5 & -2.5 \end{bmatrix}$$

8.20

$$\{1, 1, 1, 1, -1, -1, -1, -1\}$$

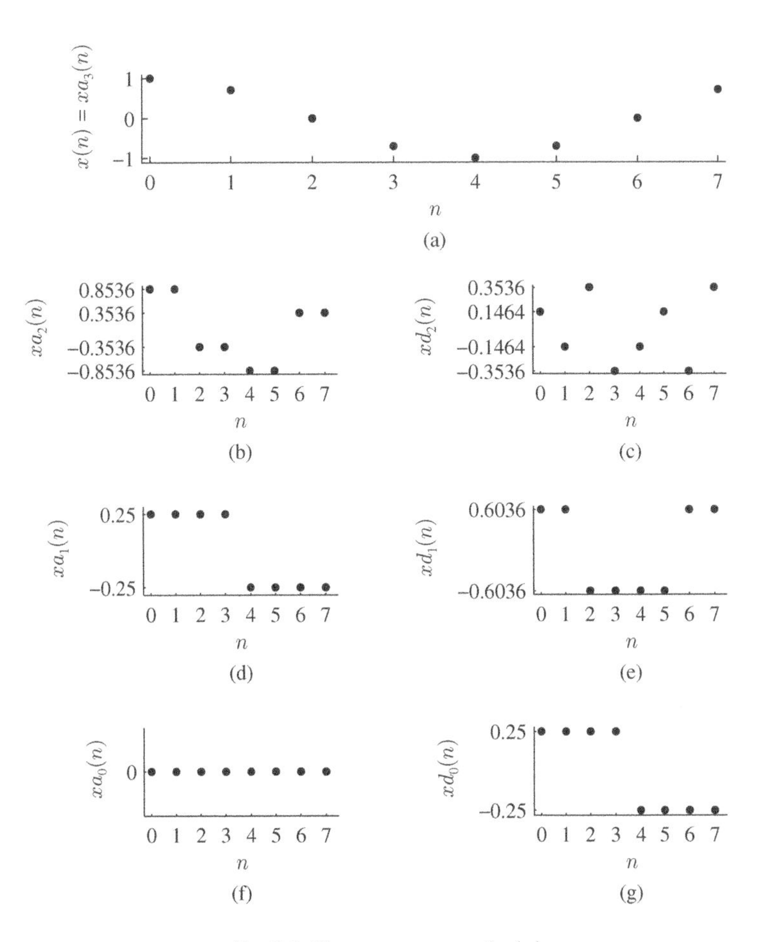

Ex. 8.6. The components of $x(n)$

Chapter 9

9.9.2

$$\{5.2779, -1.2247, 1.2247, 0.3789, 0.5176, 0.1895, -4.0532, 0.5176\}$$

9.10.1 Row transform

$$\begin{bmatrix} 2.8030 & -2.4495 & 2.2507 & 2.3455 & -0.0947 & 2.8284 & -0.7418 & 0.1294 \\ 3.9584 & 0.2588 & 0.0601 & 3.5009 & -2.9925 & -3.7944 & -0.6724 & 1.0953 \\ 1.8625 & 2.7337 & 3.0526 & 2.9578 & -0.2588 & -3.6996 & -0.0601 & 0.4830 \\ 2.6390 & 1.9572 & 3.1820 & 3.5355 & -2.6390 & 2.5442 & 0.8018 & 0.7071 \\ 3.9584 & 0.7071 & 1.0260 & 3.5009 & -2.9925 & -2.1213 & -0.9313 & 1.0953 \\ 1.0607 & -0.1641 & 2.6390 & 2.1213 & 3.4408 & 0.2842 & -4.0532 & 1.7424 \\ 4.1479 & 1.9665 & -2.8030 & 2.3455 & 0.1641 & 0.1294 & 1.5089 & -1.8024 \\ 0.7071 & 0.0947 & 2.0866 & 4.8897 & 3.5355 & 1.2501 & 0.5777 & 2.4148 \end{bmatrix}$$

2-D DWT

$$\begin{bmatrix} 4.7410 & -0.6071 & 1.4097 & 4.2667 & -2.2655 & -2.9665 & -1.0380 & 0.9955 \\ 3.8571 & 3.1373 & 4.0245 & 4.8962 & -3.4486 & -0.1708 & 0.9575 & 0.8448 \\ 3.6373 & 0.6328 & 1.8047 & 3.3583 & 1.0123 & -0.9196 & -3.5768 & 1.2700 \\ 2.7108 & 0.4465 & 0.8885 & 5.2958 & 3.4028 & 2.2333 & 1.1328 & 1.0368 \\ -0.9665 & 1.6005 & 0.7120 & -0.3215 & 1.7410 & -3.8391 & -0.1908 & -0.1998 \\ 1.9665 & -0.1217 & -1.5245 & 0.7288 & -3.5401 & -2.0033 & 1.0066 & -0.1462 \\ 2.3783 & 1.5446 & -4.0768 & -1.3280 & -1.9542 & -0.2847 & 2.0123 & -3.2063 \\ -0.2623 & -2.4498 & 1.7488 & -1.1283 & 0.5523 & 3.9016 & -0.6205 & -0.7288 \end{bmatrix}$$

The energy of the signal both in the time domain and in the frequency domain is 355.

9.12

$$\{-0.1250, -0.2165, -0.2745, -0.3415, 0.7075, 1.4575, -0.0915, -1.0245, -0.2165, 0.1250\}$$

9.13

$$\{0.2213, 0.5369, 0.8428, 1.2123, 0.9913, 0.5477, 0.1953, -0.2852, -0.2596, -0.0690, -0.0321, 0.0704, 0.0470, -0.0155, -0.0060, 0.0025\}$$

Chapter 10

10.9.2

$$\{32, 6, 8, -14, -8, 12, -12, -8\}(\sqrt{2}/8)$$

The energy of the signal is 48.
The energy of its DWT coefficients is 54.2500.

10.10.1 Row transform

$$\frac{\sqrt{2}}{8}\begin{bmatrix} 20 & -13 & 14 & 7 & -6 & 16 & 4 & -2 \\ 36 & 3 & -10 & 15 & -6 & -16 & -12 & -2 \\ 14 & 19 & 4 & 23 & 4 & -10 & -14 & 0 \\ 24 & -5 & 28 & 17 & -4 & -2 & 10 & 4 \\ 36 & 1 & 2 & 13 & -6 & -12 & -8 & -2 \\ 15 & 8 & 13 & -4 & 14 & 18 & -12 & -12 \\ 14 & 22 & -14 & 10 & -10 & -2 & 14 & -2 \\ 14 & 5 & 4 & 21 & 20 & 18 & 6 & 0 \end{bmatrix}$$

2-D DWT

$$X = \frac{1}{32}\begin{bmatrix} 192 & -103 & 82 & 81 & -2 & 112 & 12 & -14 \\ 148 & 122 & 44 & 182 & 16 & -100 & -84 & 8 \\ 266 & -29 & 104 & 71 & -10 & -28 & -52 & -26 \\ 86 & 170 & -66 & 74 & 20 & 56 & 76 & -32 \\ -76 & -0 & 76 & 0 & 20 & 76 & 28 & 4 \\ 4 & 60 & -100 & 4 & 12 & -36 & -84 & -20 \\ 40 & 14 & -76 & 62 & -88 & -100 & 60 & 40 \\ 12 & -2 & -16 & -50 & -112 & -44 & 12 & -8 \end{bmatrix}$$

The energy of the signal is 355.
The energy of its DWT coefficients is 394.5898.

10.11.3

$$\{64, 2, 48, -18, 10, -1, 8, 7\}(\sqrt{2}/8)$$

The energy of the signal is 126.
The energy of its DWT coefficients is 216.9375.

10.12.2 Row transform

$$\frac{\sqrt{2}}{8}\begin{bmatrix} 22 & 12 & -30 & 28 & 9 & 2 & -7 & -4 \\ 38 & -6 & 28 & 0 & 1 & 9 & 0 & 10 \\ 38 & 2 & 0 & 12 & 1 & 13 & 10 & 4 \\ 4 & 18 & -24 & 10 & -12 & 11 & 16 & 1 \\ -2 & -2 & 20 & 12 & -9 & -15 & 10 & 2 \\ -8 & 10 & 38 & -4 & -10 & -21 & 3 & 16 \\ 22 & -34 & 28 & 12 & 3 & -7 & -10 & 2 \\ -6 & 0 & 16 & 34 & -15 & -22 & -10 & -5 \end{bmatrix}$$

2-D DWT

$$X = \frac{1}{32}\begin{bmatrix} 296 & 32 & -44 & 76 & 88 & 84 & -42 & 38 \\ 180 & 136 & -240 & 108 & -50 & 156 & 136 & 6 \\ -112 & 80 & 340 & 4 & -96 & -224 & 66 & 102 \\ 68 & -248 & 248 & 228 & -70 & -136 & -112 & -42 \\ 4 & -56 & 190 & -62 & -40 & -14 & 1 & 33 \\ -62 & 44 & -64 & -18 & -29 & 18 & 8 & -1 \\ -36 & 88 & 2 & -50 & -18 & 0 & 5 & 41 \\ -114 & 100 & 32 & 34 & -73 & -68 & 10 & -1 \end{bmatrix}$$

The energy of the signal is 366.
The energy of its DWT coefficients is 793.1016.

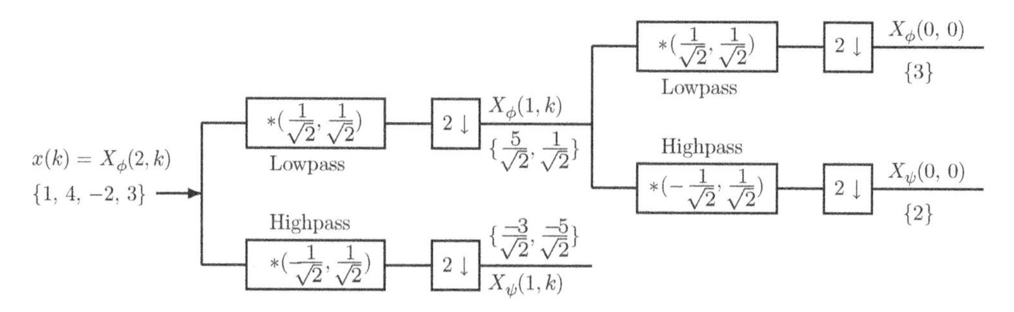

Ex. 11.1.2. Computation of a 2-level 4-point DWT using a two-channel Haar analysis filter bank

10.13.1

$$\{3.2220, 3.8583, 3.6548, 1.0988, 2.6933, 1.1510, 1.0822, -1.2039, -1.6317, -0.6019, -1.0158, 4.5471, -1.8786, 2.6573, -2.5422, -2.3626\}$$

The energy of the signal is 104.
The energy of its DWT coefficients is 98.3996.

10.15

$$\{0.1250,\ 0.2500,\ 0.3750,\ 0.5000,\ -0.5000,\ -1.5000,\ -0.5000,\ 0.5000,\ 0.3750,\ 0.2500,\ 0.1250\}$$

10.16

$$\{0.0625,\ 0.1875,\ 0.3750,\ 0.6250,\ 0.7500,\ 0.7500,\ 0.6250,\ 0.3750,\ 0.1875,\ 0.0625\}$$

Chapter 11

11.1.2

11.2.1

11.3.2

11.4.3

$$\{X_\phi(1,0), X_\phi(1,1), X_\phi(1,2)\} = \{4.5962, 4.7349, -3.8891\}$$
$$\{X_\psi(1,0), X_\psi(1,1), X_\psi(1,2)\} = \{-0.6124, -1.5089, -0.6124\}$$

11.5.1

$$\{X_\phi(2,0), X_\phi(2,1), X_\phi(2,2), X_\phi(2,3)\} = \{20, 9, 21, 30\}(\sqrt{2}/8)$$
$$\{X_\psi(2,0), X_\psi(2,1), X_\psi(2,2), X_\psi(2,3)\} = \{4, -6, -4, 8\}(\sqrt{2}/8)$$

11.6.2

$$\{X_\phi(3,0), X_\phi(3,1), X_\phi(3,2), X_\phi(3,3), X_\phi(3,4), X_\phi(3,5), X_\phi(3,6), X_\phi(3,7)\} = \{2.0991, 2.4790, 3.7378, 5.6660, 2.6088, 2.2249, 2.0835, 2.0708\}$$
$$\{X_\psi(3,0), X_\psi(3,1), X_\psi(3,2), X_\psi(3,3)X_\psi(3,4), X_\psi(3,5), X_\psi(3,6), X_\psi(3,7)\} = \{-0.5233, 1.2234, -0.2582, -0.3129, -1.4816, 2.7190, 1.5882, 5.4043\}$$

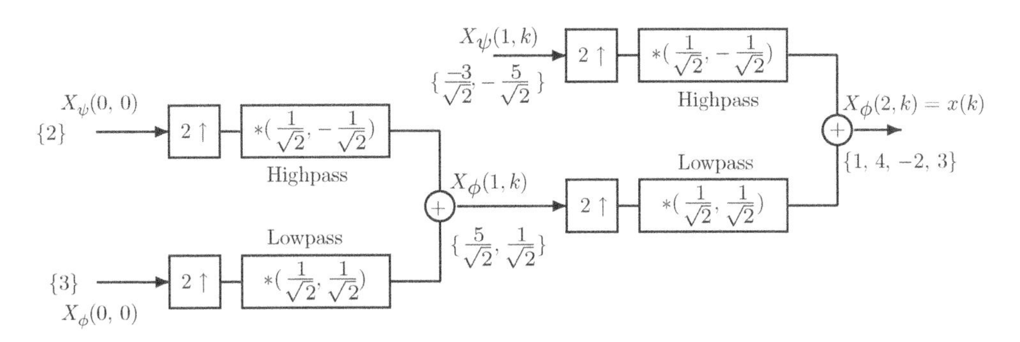

Ex. 11.1.2. Computation of a 2-level 4-point IDWT using a two-channel Haar synthesis filter bank

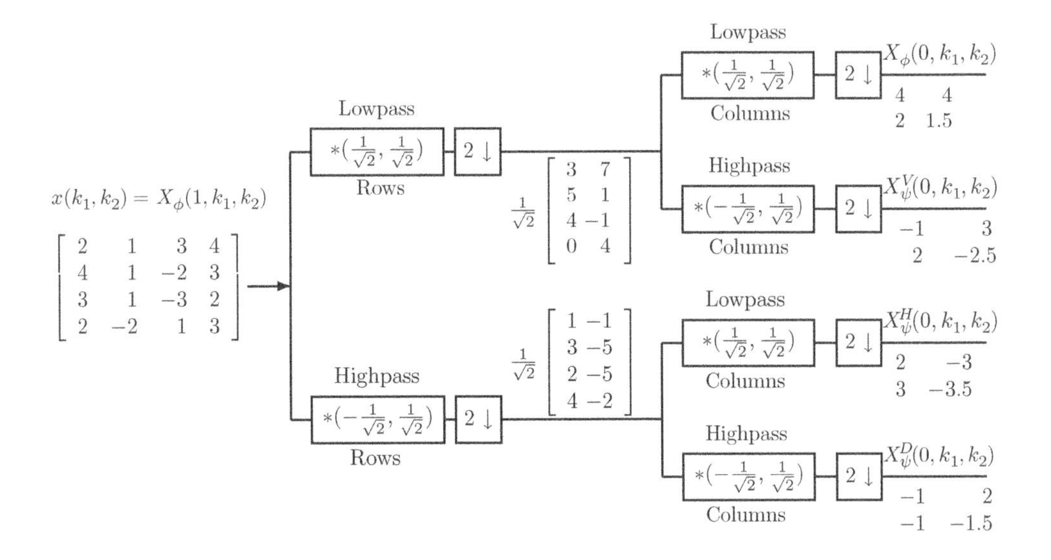

Ex. 11.2.1. Computation of a 1-level 4 × 4 DWT 2-D using a two-channel Haar filter bank

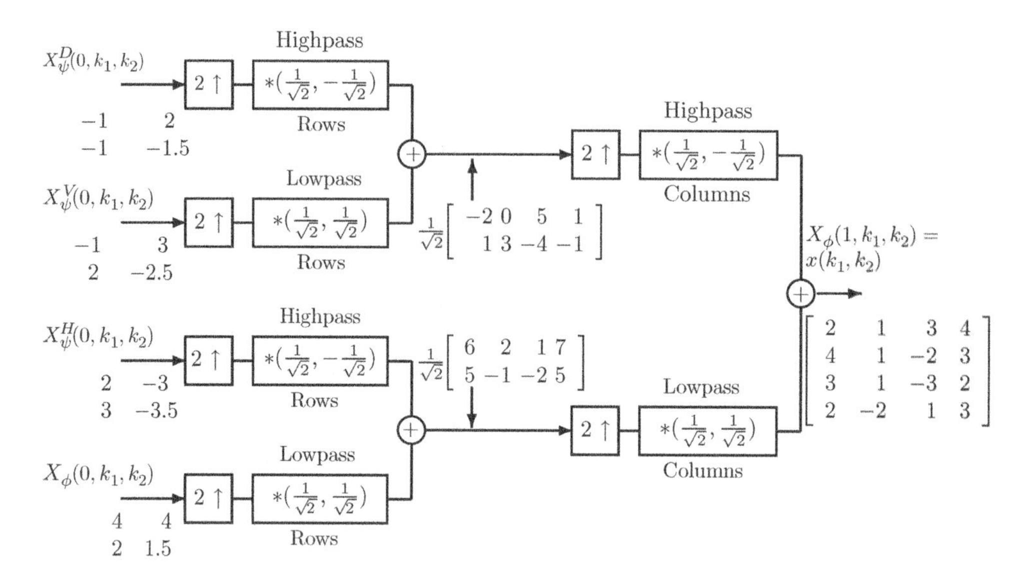

Ex. 11.2.1. Computation of a 1-level 4 × 4 2-D IDWT using a two-channel Haar filter bank

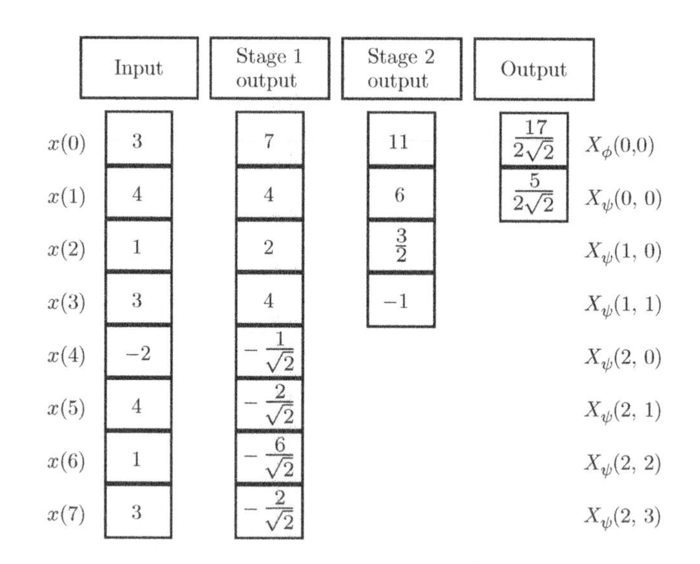

Ex. 11.3.2. The trace of the 1D Haar 3-level DWT algorithm, with $N = 8$

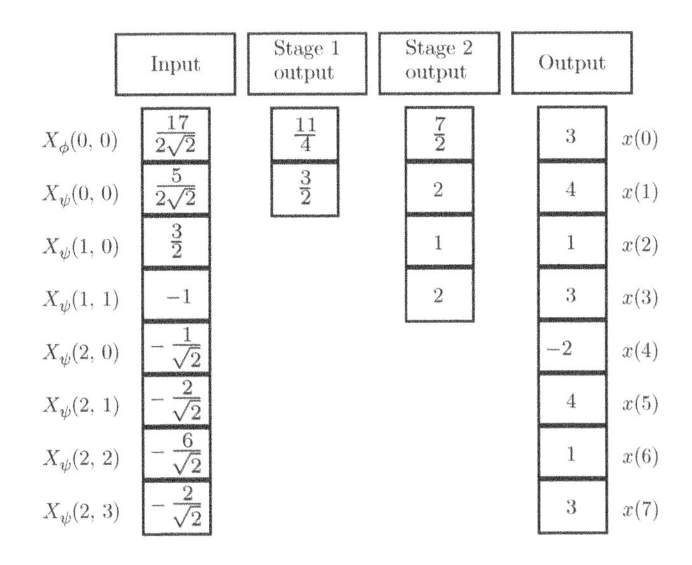

Ex. 11.3.2. The trace of the 1D Haar 3-level IDWT algorithm, with $N = 8$

Ex. 12.3.2. The 3-level Haar DWPT of the signal $\{1, 7, 2, 3, 1, 1, 4, 3\}$

1 7 2 3 1 1 4 3							
5.6569	3.5355	1.4142	4.9497	−4.2426	−0.7071	0	0.7071
6.5	4.5	1.5	−2.5	−3.5	0.5	−2.5	−0.5
7.7782	1.4142	−0.7071	2.8284	−2.1213	−2.8384	−2.1213	−1.4142

3							
2				**1**			
1		**1**		1		1	
1	**0**	0	1	1	1	1	0

11.7.3

$$\{X_\phi(2,0), X_\phi(2,1), X_\phi(2,2), X_\phi(2,3)\} = \{10, 0, 36, -14\}(\sqrt{2}/8)$$

$$\{X_\psi(2,0), X_\psi(2,1), X_\psi(2,2), X_\psi(2,3)\} = \{13, -10, 4, 11\}(\sqrt{2}/8)$$

11.8.2

$$x(n) = \{7, 3\}$$

$$\begin{array}{cccc} 5\sqrt{2} & 5 & 7 & 7 \\ -\frac{4}{\sqrt{2}} & -4 & -4 & 3 \end{array}$$

$$\begin{array}{cccc} 7 & 7 & 5 & 5\sqrt{2} \\ 3 & -4 & -4 & -\frac{4}{\sqrt{2}} \end{array}$$

Chapter 12

12.2.1

$$\begin{array}{cccccc} 2 & 2 & -3 & 4 & \rightarrow & 4 \\ 4/\sqrt{2} & 1/\sqrt{2} & 0/\sqrt{2} & -7/\sqrt{2} & \rightarrow & 2 \\ 2.5 & 1.5 & 0/\sqrt{2} & -7/\sqrt{2} & \rightarrow & 3 \\ 4/\sqrt{2} & 1/\sqrt{2} & -3.5 & 3.5 & \rightarrow & 3 \\ 2.5 & 1.5 & -3.5 & 3.5 & \rightarrow & 4 \end{array}$$

12.3.2

Chapter 13

13.2.2

$$xr(n) = \{1.25, 2.50, 0.75, -1.00, 0.25, 1.50, 0.75, 0\}$$

Chapter 15

15.1.2

$$\text{error}(n) = \{56.3594, 0, 36.7500, -53.1250\}.$$

$$\text{error}(n) = \{16.1953, 0, 36.7500, -12.9609\}.$$

15.2.3

$$x(n) = \{142.3750, 0, -81.9063, 0\}$$

$$x(n) = \{142.3750, 0, 0, 0\}$$

$$x(n) = \{82.3750, 0, -21.9063, 0\}$$

$$x(n) = \{42.3750, 0, 0, 0\}$$

15.3.2 bpp = 36/16=2.25

15.4.1 $E = 3$

15.5.4 SNR = 41.6867 dB

15.6.1 cr = 2.2857, bpp = 3.5, SNR = 47.0296 dB

Chapter 16

16.1.2 The 1-level square error is = 9.2305. The 2-level square error is 6.3023.

Index

Discrete Wavelet Transform: A Signal Processing Approach, First Edition. D. Sundararajan.

Companion Website: www.wiley.com/go/sundararajan/wavelet

Printed and bound by CPI Group (UK) Ltd, Croydon, CR0 4YY